FUNCTIONAL MATERIALS AND ELECTRONICS

FUNCTIONAL MATERIALS AND ELECTRONICS

Edited by
Jiabao Yi, PhD
Sean Li, PhD

Apple Academic Press Inc.
3333 Mistwell Crescent
Oakville, ON L6L 0A2 Canada

Apple Academic Press Inc.
9 Spinnaker Way
Waretown, NJ 08758 USA

Library and Archives Canada Cataloguing in Publication

Functional materials and electronics / edited by Jiabao Yi, PhD, Sean Li, PhD.
Includes bibliographical references and index.
Issued in print and electronic formats.
ISBN 978-1-77188-610-9 (hardcover).--ISBN 978-1-315-16736-7 (PDF)
1. Semiconductors--Materials. 2. Electronics--Materials. 3. Spintronics.
I. Yi, Jiabao, editor II. Li, Sean (Professor of materials science), editor
TK7871.85.F86 2017   621.3815'2   C2017-904220-3   C2017-904221-1

Library of Congress Cataloging-in-Publication Data

Names: Yi, Jiabao, editor. | Li, Sean (Professor of materials science), editor.
Title: Functional materials and electronics / editors, Jiabao Yi, PhD, Sean Li, PhD.
Description: Oakville, ON ; Waretown, NJ : Apple Academic Press, [2017] |
Includes bibliographical references and index.
Identifiers: LCCN 2017027576 (print) | LCCN 2017029475 (ebook) | ISBN 9781315167367 (ebook) | ISBN 9781771886109 (hardcover alk. paper) | ISBN 1771886102 (hardcover : alk. paper) | ISBN 1315167360 (eBook)
Subjects: LCSH: Semiconductors--Materials. | Electronics--Materials. | Spintronics.
Classification: LCC TK7871.85 (ebook) | LCC TK7871.85 .F84 2017 (print) | DDC 621.3815/20284--dc23
LC record available at https://lccn.loc.gov/2017027576

Apple Academic Press also publishes its books in a variety of electronic formats. Some content that appears in print may not be available in electronic format. For information about Apple Academic Press products, visit our website at **www.appleacademicpress.com** and the CRC Press website at **www.crcpress.com**

ABOUT THE EDITORS

Jiabao Yi, PhD

Dr. Jiabao Yi is a Senior Lecturer/Future Fellow in the School of Materials Science and Engineering, University of New South Wales (UNSW). Before he came to UNSW, he was a prestigious Lee Kuan Yew postdoctoral fellow in National University of Singapore. He joined UNSW in 2011 after he was awarded the Queen Elizabeth II fellowship. He obtained his PhD degree in 2008 from the National University of Singapore, Singapore. His research areas include oxide-based diluted magnetic semiconductors, two-dimensional materials, and soft and hard magnetic materials. He has published 120 papers which have been cited over 4500 times.

Sean Li, PhD

Professor Sean Li is currently leading a research group that consists of more than 40 researchers who work in the areas of advanced multifunctional materials at the University of New South Wales, Kensington, Australia. Their research activities have been funded by approximately $42 million from the Australian Research Council, the Australian Solar Institute, the Australian Nuclear Science and Technology Organization (ANSTO), the Commonwealth Science Industrial Research Organization (CSIRO), and various industries. Professor Sean Li's laboratory is equipped with a number of unique and world-class research facilities, which is specially designed and fully geared toward the development of advanced multifunctional and energy materials with a total value of AUD$12 million. It is one of the key research infrastructures at UNSW. Professor Li has published 3 textbooks, 1 edited book, 12 book chapters, and more than 325 scientific articles in international peer-reviewed journals.

CONTENTS

LIST OF CONTRIBUTORS

Anil Annadi
Department of Physics, National University of Singapore, Singapore 119260, Singapore

Ariando
Department of Physics, National University of Singapore, Singapore 119260, Singapore.
E-mail: ariando@nus.edu.sg

Dewei Chu
School of Materials Science and Engineering, University of New South Wales, Sydney 2052, NSW,
Australia. E-mail: d.Chu@unsw.edu.au

Xiang Ding
School of Materials Science and Engineering, UNSW, Kensington, NSW 2052, Australia

Y. Q. Fu
Faculty of Engineering and Environment, Northumbria University, Newcastle upon Tyne, NE1 8ST,
United Kingdom. E-mail: Richard.fu@northumbria.ac.uk

Nguyen Hoa Hong
Department of Physics and Astronomy, Seoul National University, Seoul, South Korea.
E-mail: nguyenhong@snu.ac.kr

Sean Li
School of Materials Science and Engineering, University of New South Wales, Sydney 2052, NSW,
Australia

Zhiqi Liu
School of Materials Science and Engineering, Beihang University, Beijing 100191, China.
E-mail: zhiqi@buaa.edu.cn

J. K. Luo
Centre for Material Research and Innovation, University of Bolton, Deane Road, Bolton BL3 5AB,
United Kingdom

Hua-Feng Pang
Department of Applied Physics, School of Science, Xi'an University of Science and Technology, Xi'an,
PR China; Faculty of Engineering and Environment, Northumbria University, Newcastle upon Tyne
NE1 8ST, United Kingdom

Junling Wang
Department of Materials Science and Engineering, Nanyang Technological University, Singapore
639798, Singapore. E-mail: jlwang@ntu.edu.sg

Yiren Wang
School of Materials Science and Engineering, UNSW, Kensington 2052, NSW, Australia

Jiabao Yi
School of Materials Science and Engineering, UNSW, Kensington 2052, NSW, Australia.
E-mail: jiabao.yi@unsw.edu.au

Lu You
Department of Materials Science and Engineering, Nanyang Technological University, Singapore 639798, Singapore

Adnan Younis
School of Materials Science and Engineering, University of New South Wales, Sydney 2052, NSW, Australia. E-mail: a.younis@unsw.edu.au

LIST OF ABBREVIATIONS

AFM	atomic force microscopy
AMR	anisotropic magnetoresistance
BFO	$BiFeO_3$
BKT	Berezinskii–Kosterlitz–Thouless
CBD	chemical bath deposition
CF	conductive filament
CMOS	complementary metal–oxide–semiconductor
CT	computed tomography
CTAB	cetyltrimethylammonium bromide
CT-AFM	conducting-tip atomic force microscopy
CTF	charge-transfer ferromagnetism
CVD	chemical vapor deposition
DFT	density functional theory
DLC	diamond-like carbon
DM	Dzyaloshinski–Moriya
DMSO	diluted magnetic semiconducting oxides
DMSs	diluted magnetic semiconductors
DOS	density of state
ECM	electrochemical metallization process
FBARs	film bulk acoustic resonators
FETs	field-effect transistors
FM	ferromagnetism
FPV	ferroelectric photovoltaic
FTJ	ferroelectric tunnel junction
GaN	gallium nitride
GGA	generalized gradient approximation
GMR	giant magnetoresistance
HER	hydrogen evolution reaction
HfO_2	hafnium oxide
HiTUS	high-target utilization sputtering
HMTA	hexamethylenetetramine
HRS	high-resistance state
IDTs	interdigital transducers
ITO	indium tin oxide

LAO	$LaAlO_3$
LDA	local density approximation
LIB	Li-ion battery
LOC	lab-on-a-chip
LRS	low-resistance state
LSDA	local spin density approximation
MBE	molecular beam epitaxy
ME	magnetoelectric
MEMS	micro-electromechanical systems
MFM	magnetic force microscopy
MIT	metal-to-insulator transition
ML	monolayer
MOCVD	metalorganic CVD
MOSFET	metal oxide semiconductor field-effect transistor
MPBs	morphotropic phase boundaries
MR	magnetoresistance
NCD	nanocrystalline diamond
NC-pp	norm-conserving pseudopotentials
NCs	nanocrystals
NPP	nanopyramid patterned
NRLs	nanorod layers
PAS	positron annihilation spectroscopy
PAW-pp	projector-augmented wave
PEG	polyethylene glycol
PEO	polyethylene oxide
PFM	piezoelectric force microscopy
PL	Photoluminescence
PLD	pulsed laser deposition
PS	polystyrene
PVD	physical vapor deposition
PW	plane wave function
PZT	$Pb(Zr,Ti)O_3$
QDs	quantum dots
RE	rare-earth
RF	radio-frequency
RHEED	reflection high-energy electron diffraction
RKKY	Ruderman–Kittel–Kasuya–Yosida
RRAMs	resistive random access memories
RS	resistive switching

RTFM	room temperature ferromagnetism
SAW	surface acoustic wave
SCFs	supercritical fluids
SCLC	space-charge-limited conduction
SiC	silicon carbide
SMRs	solidly mounted resonators
STO	SrTiO3
TFBAR	thin-film bulk acoustic resonator
TM	transition metal
TMDCs	transition metal dichalcogenides
TMs	transition metals
TR-FRET	time-resolved Förster resonance energy transfer
uc	unit cell
UNCD	ultra-nanocrystalline diamond
US-pp	ultrasoft pseudopotentials
VCM	valence change mechanism
VLS	vapor–liquid–solid
VS	vapor–solid
XAS	X-ray absorption spectra
XMCD	X-ray magnetic circular dichroism measurements
YSZ	yttrium-stabilized zirconia

PREFACE

Electronic devices have become a part of our daily modern life involving mobile phone, data storage, computers, and satellites. The basic component of the devices is transistors. The continual enhanced high performance and capacity of the devices require the transistors to be highly mobile, smaller size without much heat dissipation. Current semiconductor devices are all made by semiconductors. In order to achieve multifunctions, the semiconductors are often doped with impurities to induce electrons or holes. The collisions of the electrons or holes caused by their migration lower their mobility, and thus the performance of electronic devices. In the early 1980s, two-dimensional electron gas (2DEG) in the interface between AlGaAs/GaAs has been discovered, which exhibits very high mobility. Recently, the discovery of graphene demonstrates another kind of 2DEG, where the electrons with zero effective mass are confined in atomic thin layer. Due to the lack of bandgap, metal dichalcogenide such as molybdenum disulfide has been widely investigated. 2DEG has also been found in $LaAlO_3/SrTiO_3$ interface, which has shown multifunctionalities including ferromagnetism, ferroelectricity, superconductivity, and also strong electron and orbital coupling.

The relentless growth of microelectronics is popularly considered as following Moore's law, which predicts that the power of these semiconductor devices doubles in approximately 18 months. However, the shrink of chip size and limitation of the microprocessing techniques have pushed Moore's law to the end of traditional semiconductor roadmap. New physics or techniques must be developed to tackle this challenge. Spintronics devices that utilize the spin up and down of electronics as the logic "on" and "off" are believed as the promising devices to replace the current semiconductor devices due to its high performance, high capacity, low power loss, and no heat dissipation. In particular, the spin can be manipulated by a small electric field or magnetic field, making the devices having unique advantage, which the current semiconductor devices cannot achieve.

Diluted magnetic semiconductor, which possesses both semiconductor and spin behavior, has been attracted wide interest. Mn-doped GaAs is a typical example. However, due to its low Curie temperature, oxide-based diluted magnetic semiconductor has become the promised material for applications of spintronic devices, especially for the oxide semiconductors, such as ZnO, TiO_2, SnO_2, as well as In_2O_3. These materials also have a wide range of applications including electrochemical applications, sensors, and bioapplications. As the spintronic materials, spin manipulation is the essential property. Multiferroics materials are new materials, which can couple the spin and ferroelectricity manipulated by electric field. It is known that ferroelectric material is one of the materials for random access memory. The new property has provided one more freedom for the applications of both electronic and spintronic devices. Hence, it may be used for the fabrication of ferroelectric and spintronic-based memories. In addition, the simple resistive switching in oxide materials has attracted more attention in the research of oxide-based functional materials for electronics.

Therefore, this edited book focuses on the newly developed functional materials and their applications for the electronics and spintronics devices. Chapter 1 introduces the synthesis, crystalline structure, applications, and properties of the atomic thin materials—metal dichalcogenide. Chapter 2 describes the synthesis and properties of 2DEG in the interface of $LaAlO_3$/ $SrTiO_3$. Chapter 3 introduces the multiferroics and magnetoelectric applications in bismuth ferrite. Chapter 4 discusses the resistive switching for the applications of memories. Chapter 5 discusses the applications of ZnO for sensors. Chapter 6 focuses on the introduction of oxide-based diluted magnetic semiconductors and experimental results. Chapter 7 emphasizes on the theoretical study of oxide-based diluted magnetic semiconductors.

We would like to thank all the authors for their contributions and believe that this book can serve as a reference book for undergraduate, postgraduate, scientists, and engineers in the universities and industries, who need the information on the physical conception of functional materials.

—**Jiabao Yi**
—**Sean Li**
School of Materials Science and Engineering,
University of New South Wales

CHAPTER 1

SYNTHESIS AND APPLICATIONS OF 2D TRANSITION METAL DICHALCOGENIDES

XIANG DING and JIABAO YI*

School of Materials Science and Engineering, UNSW, Kensington, NSW 2052, Australia

*Corresponding author. E-mail: jiabao.yi@unsw.edu.au

CONTENTS

ABSTRACT

The discovery of graphene has attracted extensive research interest in 2D materials since 2D materials have shown many extraordinary properties different from bulk materials, which are promising for the applications in many areas including energy materials, sensors, and spintronics. Transition metal dichalcogenide (TMDC), one of the 2D materials, which has relatively large bandgap, strong spin-orbit coupling, and moderate mobility, is one of the major focuses in the research of 2D materials. In this chapter, we will introduce the basics of 2D and TMDC materials. We mainly give an introduction of the syntheses and characterizations of TMDCs with different approaches and techniques. In the last part of this chapter, we introduce the applications of TMDCs in the area of energy materials (hydrogen evolution, lithium batteries, and supercapacitors), bioapplications, nanodives, spintronics, and valleytronics.

1.1   INTRODUCTION

Two-dimensional (2D) materials are a group of materials which have one dimension in the nanometer scale. 2D materials with atomic thickness have attracted considerable attention since graphene, the monolayer counterpart of graphite, was discovered in 2004 [1]. Currently, graphene is still the most eminent 2D material due to its unique electronic structure, extraordinary strength, and electrical conductivity. However, graphene is far beyond perfect. For example, it has no bandgap, which restricts its application in logical circuits where the requirement of nonzero bandgap is almost mandatory. Although several possible solutions are proposed to manually introduce a bandgap into graphene, such as generating by antidote engineering or chemical doping, the prominent conductivity of graphene is lost after modification. Therefore, in this case, other 2D semiconductors with intrinsic bandgap may be a better choice.

Transition metal dichalcogenides (TMDCs or TMDs) are considered to be ideal substitutes for graphene. TMDCs are a group of materials with layered structures, which had a covalently bonded hexagonal network inside layers and piled by weak van der Waals forces between layers. 2D TMDCs have demonstrated excellent catalytic activity for hydrogen evolution reaction (HER), relatively high biosafety and biocompatibility, and ultrahigh on/off ration in field-effect transistors (FETs). These admirable performances have led to high expectations for 2D TMDCs, encouraging extensive research involving both theory and experiment.

The electronic and magnetic properties of 2D TMDCs can be significantly changed by defects and chemical doping. On the basis of theoretical research, transitional metal atom doping can effectively induce magnetism into MoS_2. For example, ferromagnetism is observed for Mn, Fe, Co, Cr, Zn, Cd, Hg, V, and Cu doping [2]. However, many results of magnetic behaviors of doping elements from different calculations are controversial and lack experimental evidences. Recently, Cu-doped MoS_2 nanosheets were fabricated by a hydrothermal method and magnetic measurement results confirm that doping of Cu ions can indeed induce ferromagnetism into MoS_2 nanosheets [3].

This chapter will first give a general introduction on the background of TMD materials, followed by the review of the fabrication and functionalization of 2D TMDCs. Finally, the last section will cover the applications of TMD materials in the area of energy and biomedical field, as well as state-of-art nanodevices.

1.2 TRANSITION METAL DICHALCOGENIDES

1.2.1 BULK TMDCS

TMDCs are a class of chemicals which share a formula MX_2, where M is a transition metal element, from group VIB (Mo, W) in most relevant research and may be also from group IVB or VB (Ti, Zr, V, Nb, and so on), meanwhile X is a chalcogen (S, Se, or Te). Depending on the combination of different transition metals and chalcogens, TMDCs possess more than 40 distinct categories. Most TMDCs have layered structures like a sandwich, stacking of hexagonally packed MX_2 layers and each MX_2 layer is composed of two chalcogen planes separated by a plane of M atoms [4]. The atoms within the same layer are connected to each other by strong ionic–covalent bonds, while only a weak van der Waals force exists between two neighbor layers. The minimum and maximum values of TMDCs' lattice constant **a** are around 3.1 Å (VS_2) and 3.7 Å ($TiTe_2$), while the interlayer spacing is about 6.5 Å [5].

Three phases with different stacking orders are commonly observed among TMDCs: 1T structure (tetragonal symmetry, one MX_2 layer per primitive cell, octahedral coordination), 2H structure (hexagonal symmetry, two MX_2 layers per primitive cell, trigonal prismatic coordination), and 3R structure (rhombohedral symmetry, three MX_2 layers per primitive cell, trigonal prismatic coordination). The unit cell structures of three phases are shown in Figure 1.1 [6]. Although TMDCs share similar crystal structures, they cover a wide range of electrical properties, from metals like VSe_2, semimetals like WTe_2, and semiconductors like MoS_2 to insulators like HfS_2. In this report, special emphasis is laid on the semiconductors MoS_2, the only TMDC occurring naturally in appreciable quantities. Bulk MoS_2 has applications in various fields such as dry lubrication and catalysis in hydrodesulfurization process [7].

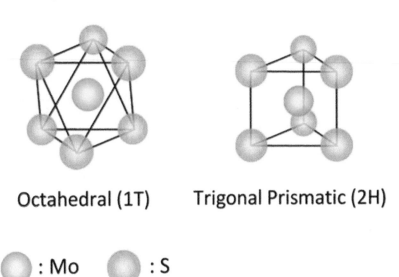

Octahedral (1T)   **Trigonal Prismatic (2H)**

⬤ : Mo   ⬤ : S

FIGURE 1.1   Octahedral (1T) and trigonal prismatic (2H and 3R) and unit cell structures. Reproduced with permission from Ref. [6]. Copyright 2011 American Chemical Society.

1.2.2 TWO-DIMENSIONAL TMDCS

One of the most intriguing properties of 2D TMDCs is the thickness-dependent band structure. Bulk MoS_2 has an indirect bandgap of 1.2 eV, and the bandgap increases with decreasing thickness. However, once the thickness is reduced to monolayer, a transition from indirect bandgap to direct bandgap will occur, as presented in Figure 1.2 [8]. Calculated by first principal methods, the direct bandgap of MoS_2 monolayer varies from 1.6 to 1.9 eV due to different approximations selected in different literature [9,10]. Meanwhile, the experimental data observed from photoluminescence are approximately 1.9 eV, which agree with the result based on *Perdew–Burke–Ernzerhof (PBE)* functional form of generalized gradient approximation [11]. This direct bandgap located in the visible frequency region grants great potentials in advanced electronic and photonic applications.

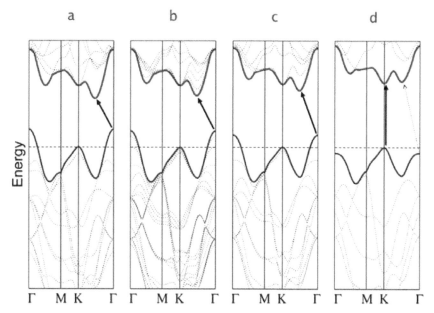

FIGURE 1.2   Transition of the band structure of MoS_2 from indirect to direct band gap. (a) bulk MoS_2, (b) four-layer MoS_2, (c) bilayer MoS_2, and (d) monolayer MoS_2. Reproduced with permission from Ref. [8]. Copyright 2010 American Chemical Society.

In addition to a direct bandgap, spin–orbit splitting is also revealed in monolayer TMDCs due to the presence of heavy metal atoms and broken inversion symmetry. The spin–orbit splitting increases with the size of the metal and the chalcogen atom, varying from 140–150 meV for MoS_2 mono-layers to nearly 500 meV for WTe_2 monolayers. Moreover, as demanded by time-reversal symmetry, the spin–orbit splitting in opposite valleys (valence band maximum and the conduction band minimum located at six corners of hexagonal Brillouin zone, K and K′) must be opposite, suggesting that the spin and valley of the valence bands are inherently coupled [12]. Therefore, the valley carriers can be also distinguished by their spin moments, which is the foundation of spintronics and valleytronics.

The electronic structure of 2D can be manipulated through external fields, such as strain and electric field. The influence of strain on the electronic properties of monolayer TMDCs was investigated by density functional calculations [13]. The results show that tensile and shear strain will reduce the bandgap. In addition, a tensile strain of 1–2% is strong enough to destroy the direct bandgap structure of monolayer MoS_2.

Most monolayer TMDs have two general polytypic structure: 2H (symmetry group $p6m2$) in most common cases and 1T (symmetry group $p3m2$) [14]. 2H and 1T MoS_2 are semiconducting and metallic because of the alteration in the crystal symmetry. Some treatments, such as Li intercalation, can cause the structure transition between two phases. The degree of transition can be controlled where the two phases can coexist to form monolayer MoS_2 heterostructures [15]. Another example is to induce S vacancies in 2H MoS_2 nanosheets via hydrothermal method, which prompts the transformation of the surrounding lattice to the 1T phase [16]. With 25% 1T structure transformation, the concentration of electron carrier can be increased by an order, creating an intrinsic ferromagnetic response of 0.25 μ_B per Mo atom.

Because TMDCs are a huge family with more than 40 distinct members, we can tune the band structures by delicately designing the composition, functionalizing, or applying various external fields. The synthesis approaches and potential applications will be discussed in the later paragraphs.

1.2.3 OTHER FORMS OF TMDCS

1.2.3.1 NANORIBBONS

TMDC nanoribbons usually describe the strips of TMDCs with width less than 100 nm, which can be considered as a kind of 2D TMDCs' derivatives. However, unlike 2D TMDCs, TMD nanoribbons are highly sensitive to their edge structures. Two most common edge types are armchair and zigzag, which can further display different geometries if the metal atoms and chalcogen atoms have different orientations. For example, armchair edges with both symmetric and asymmetric types can be constructed, which greatly expands the possible species of TMDC nanoribbons.

Edge structures are of considerable importance to TMDC nanoribbons because they directly determine their band structure and electronic properties. According to the DFT calculations, the MoS_2 zigzag nanoribbons show metallic and ferromagnetic behaviors, while armchair nanoribbons are semiconducting and nonmagnetic [17]. The calculation further indicates that the contribution to the magnetic moments is mainly from the d-electrons of edge Mo atoms and p-electrons of edge S atoms.

The unsaturated edge atoms of TMDC nanoribbons enable the attachment of other atoms, for example, hydrogen and carbon atoms. It was found that both electronic structure and magnetic properties of nanoribbons could be greatly modified. Passivation with two hydrogen atoms at one edge of a 3-Å wide zigzag-type nanoribbon can enhance the magnetic moment, but passivation with four hydrogen atoms reverses the trend. Furthermore, the zigzag nanoribbon suffers immense structural distortion with carbon passivation [18].

1.2.3.2 NANOCLUSTERS

TMDC nanoclusters are monolayer TMDCs with finite size. TMDC nanoclusters have extremely accurate stoichiometry, such as $Mo_{36}S_{104}$, which doesn't fit the TMDCs' general formula MX_2 due to unique edge conditions. TMDC nanoclusters are even more sensitive to their edge structures than TMDC nanoribbons with finite size usually less than 600 Å^2, so their edge dangling bonds must be removed either by excess chalcogen atoms or formation of fullerene-like nanostructures. It is proved that the saturation with excess chalcogen atoms is more likely to happen as a result of avoiding the immense expense of elastic bending energy [19].

Although all sorts of shapes are expected among TMDC nanoclusters, only triangular TMDC nanoclusters have ever been observed in reality [20]. Another amusing feature of TMDC nanoclusters is that only certain sizes are favored. For example, from experiment results, the number of Mo atoms along the edge of MoS_2 nanoclusters must be even. This can be explained by the preferred arrangement of S dimer pairs in a double period pattern, leading to edge reconstruction only when the number of S dimers is even. Therefore, the number of Mo atoms at the edge must be also even, equal to S dimers, otherwise it will cause edge frustration [21].

1.2.3.3 NANOTUBES

TMDC nanotubes can be occasionally found in some natural mineral deposits. In 1979, Chianelli et al. found TMDC nanosheets were often rolled into cylinders, which appeared as needles at low magnification [22]. Various TMDC nanotubes can be obtained by rolling up nanosheets along specific directions, which are usually described by two numeral indices (n, m) to indicate the tube chirality. TMDC nanotubes are typically chiral, with only two exceptions: the armchair type ($n = m$) and zigzag type ($n \neq 0$, $m = 0$).

The end of TMDC nanotubes can either be open or sealed by various caps. The closed caps can have positive or negative curvature according to topological defects appeared within the caps [23]. Furthermore, the cross-section shape of TMDC nanotubes can be either regular or irregular. Calculations claim that the geometry is highly dependent on the size of the nanotube. For instance, armchair MoS_2 nanotubes can only keep a circular shape when the indices are less than (50, 50) [24].

Dissimilar to carbon nanotubes, isolated single-wall TMDC nanotubes have not been experimentally observed yet. Theoretically, the wall numbers of TMDC nanotubes are commonly larger than four, as they are more thermodynamically stable [20].

1.2.3.4   FULLERENES

TMDC fullerenes are first synthesized by laser ablation of MoS_2 targets [25], the similar treatment to get fullerenes from graphite. But unlike C_{60}, which has a soccer shape, the shape of MoS_2 fullerenes is multiwalled octahedral with an average edge length (~50 Å) much larger than the diameter of C_{60}. The octahedral shape is considered as the result of self-assembling of TMDC triangular nanoclusters.

Only multiwalled octahedral TMDC fullerenes were observed when the size is small, but when the total number of atoms in TMDC fullerenes is larger than 100,000 (diameters ~440 Å), the sharp edges forged by two adjacent triangular nanoclusters will no longer be thermodynamically stable. Therefore, large TMDC fullerenes prefer a quasi-spherical shape due to bending edges. A series of hybrid nanostructures of MoS_2 fullerenes is also identified using ultrahigh-irradiance solar concentrator [26]. The hybrid nanostructures serve as transition phases between the multiwalled octahedral and quasi-spherical shells. Further observation show that the hybrid nanostructures are composed of both metallic and semiconducting phases, suggesting that metal–semiconductor junctions may exist inside individual MoS_2 fullerenes.

1.3   SYNTHESIS AND FUNCTIONALIZATION OF 2D TMDCS

To achieve potential applications of 2D TMDCs, developing controllable and scalable fabrication approaches is of top priority. Many strategies used in synthesizing graphene can be carried out with parallel effectiveness in the case of 2D TMDCs, such as mechanical or chemical exfoliation and chemical vapor deposition (CVD).

Prior to the review of these methods, several characterization techniques are briefly introduced on how to identify 2D TMDCs. Few-layer MoS_2 nanosheets can be simply observed via optical contrast in a microscope [27], and the thickness can be directly measured by atomic force microscopy (AFM) [28] (typical thickness of monolayer MoS_2 on a Si/SiO_2 substrate is ~0.7 nm). In addition, Raman spectroscopy is also very frequently used to determine the thickness of 2D MoS_2 [29]. As the layer number of MoS_2 decreases, the E^1_{2g} vibration mode near 382 cm^{-1} rises, whereas the A_{1g} vibration mode near 406 cm^{-1} falls. These peak position shifts as well as the distance between two peaks allow layer thicknesses to be identified via Raman spectroscopy, as illustrated in Figure 1.3.

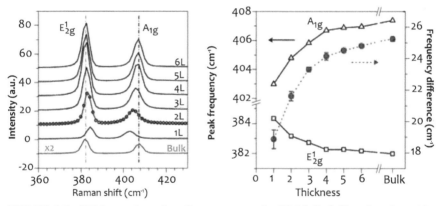

FIGURE 1.3 Thickness-dependent Raman spectra for 2D MoS_2 (left) and peak position shifts for the E^1_{2g} and A_{1g} modes as a function of MoS_2 layer thickness (right). Reproduced with permission from Ref. [29]. Copyright 2010 American Chemical Society.

1.3.1 TOP-DOWN METHODS

Since TMDC layers are held together by a weak van der Waals force, 2D TMDCs can be easily exfoliated from bulk crystals. The simplest method to achieve this goal is to use an adhesive tape to repeatedly cleave the material from bulk crystals until its thickness reduced to single or a few layers. 2D TMDCs produced by mechanical exfoliation are usually single-crystal nanosheets with high purity and cleanliness, which are favorable for fundamental characterizations and device manufacturing. Nevertheless, this method can hardly control the thickness and size of the nanosheets, and is not scalable. A thermal ablation method using a high-power laser can thin multilayer MoS_2 down to monolayer in a more controllable way; however, this technique is still of rather low in efficiency [30].

Liquid-phase exfoliation is considerably promising to obtain high-yield and large quantities of TMD nanosheets in solution. Exfoliation of bulk TMDCs to single- and few-layer nanosheets can be realized by simple one-step sonication in proper solvents [31]. Combined with the theoretical calculation, it is revealed that the key factor for effective exfoliation is the surface energies of TMDCs and solvent. Among all the studied solvents, N-methyl-pyrrolidinone is considered most suitable for MoS_2 exfoliation since its surface energy matches well with that of MoS_2. Generally, surface tensions of the solvents need to be close to 40 mJ/m^2 in order to exfoliate most 2D TMDCs. This makes water unsuitable for liquid-phase exfoliation since it has a large surface tension of 72 mJ/m^2. However, ethanol/water mixed solvent can be used to fabricate 2D TMDCs with thickness of 3–4 layers, although ethanol alone is also not a suitable solvent for sonication [32]. Polymer aqueous solutions are also confirmed to be available for liquid-phase exfoliation, but the problem is that some polymers may be left on TMDC nanosheets and are difficult to clean [33].

Due to the limits of sonication, as-prepared products regularly contain both nanosheets and microsheets, and the yield of few-layer TMDCs is especially low. Liquid-phase exfoliation process will be more effective with the assistance of ion intercalation. The Li ion intercalation method was first discovered in 1975 [34], by mixing bulk TMDCs and hexane solution with n-butyllithium for more than 24 h. During the intercalation, Li$^+$ can intercalate into the interlayer space of MoS_2 to form Li-intercalated compounds. The following violent reaction between Li-intercalated compounds and water generates hydrogen gas and separates the layers at the same time. It is worth noticing that Li intercalation exfoliation produces a much higher yield of monolayer TMDCs compared to mere sonication. In the case of MoS_2, the product has 1T structure, which can be restored to 2H structure by annealing at ~200°C with low vapor and oxygen levels.

Conventional Li intercalation is a rather time consuming process, and the degree of intercalation cannot be well controlled, sometimes resulting in either low yield of monolayer due to incomplete intercalation or decomposition into metal nanoparticles and Li_2S due to over insertion. Controllable lithiation can be realized by a Li-ion battery (LIB) design, using lithium foil and bulk TMDCs as the anode and cathode [35]. The electrochemical lithium interaction, which took only several hours, was conducted during the discharge process and could be modulated by monitoring the discharge voltage. Considering that Li is becoming an increasingly costly resource, it is reasonable to find alternative intercalants which are more abundant in earth. Recently, MoS_2 is successfully exfoliated by ion intercalation with Na and K, although the process is a bit complex [36].

1.3.2   BOTTOM-UP METHODS

1.3.2.1   CHEMICAL SOLUTION DEPOSITION

Bottom-up methods refer to the self-assembly of 2D TMDCs from atoms or molecules. Chemical solution deposition is an effective bottom-up approach to fabricate TMDC thin films. For example, chemical bath deposition has long been used to deposit polycrystalline films of metal sulfides or selenides [37]. For the deposition of MoS_2 thin film on glass substrate, a possible bath solution recipe is to mix ammonium molybdite, tartaric acid, hydrazine hydrate, and thiourea in ammonia solution [38]. Chemical bath deposition can be conducted at low temperature (<100°C). However, the method is not capable to deposit few-layer TMDC films because the thickness is usually in the range of several hundred nanometers. Compared to chemical bath deposition, a selective-area solution deposition method was reported to grow MoS_2 thin films with a minimum thickness of 11 nm [39]. The trick is the predeposition of Au electrodes, which act as catalyst as well as the source and drain contacts for the device. The obtained MoS_2 films have nonuniform thickness with the thickest places at the top and edge of the Au electrodes (~73 nm), and then the thickness gradually decreases to 11 nm at the center of two Au electrodes.

Hydrothermal synthesis is a kind of chemical solution methods conducting at high temperature and pressure, which is usually confined in a stainless steel autoclave to generate necessary high pressure above the boiling temperature of the solution. Ammonium molybdate and thiourea are the most common Mo and S precursors to synthesize MoS_2 nanosheets, while other chemicals like molybdic oxide and potassium thiocyanate are also available [40]. Hydrothermal method can produce large amount of few-layer MoS_2. However, the morphology of the nanosheets is not flat which often bends, wrinkles, or even agglomerates together to form more complicated structures, such as flower-like shape. Therefore, the products are usually used as catalysts for hydrogen generation or electrodes in LIBs where the shape of material has little effect on the performance [41].

1.3.2.2   CHEMICAL VAPOR DEPOSITION

CVD is claimed to be the most promising method to realize the growth of 2D TMDCs with large-scale, uniform thickness, and usually regular shapes. CVD can be divided into diverse categories depending on precursor type. Few-layer MoS_2 can be produced by simple sulfurization of pre-deposited Mo metal layer. The deposition rate of Mo layer can be controlled at a rate of ~0.1 Å/s by an e-beam evaporator [42]. The lateral size and thickness of the MoS_2 layer obtained by introducing sulfur vapor at 750°C merely depended on the pre-deposited Mo layer. In addition to Mo, MoO_2 layer reduced from MoO_3 thin film in Ar/H_2 atmosphere is another option for direct sulfurization [43]. In these reports, MoS_2 are polycrystalline with small crystal grain size (tens of nm), limiting their potential applications. To overcome this problem, Wang et al. employed an advanced method which began with thermal evaporation of MoO_3 powder and reduction by sulfur vapor to form MoO_2 rhomboidal microplates on SiO_2/Si substrate. Then, the MoO_2 microplates were further sulfurized at 850–950°C to generate a layer of MoS_2 thin film on top of MoO_2 microplates. The thickness of MoS_2 film depends on annealing time and its crystal domain size can reach 10 µm. Finally, the film could be peeled-off from MoO_2 microplates by poly(methyl methacrylate) and transferred onto other substrates. The fabrication procedure is presented in Figure 1.4 [44].

Epitaxial growth of few-layer MoS_2 is a process also utilized by the sulfurization of pre-deposited Mo. However, unlike those methods mentioned above in which sulfur vapor is generated by evaporating sulfur powder, in epitaxial growth, sulfur vapor is produced from the decomposition of MoS_2 powder at high temperature [45]. In this case, the sulfur pressure during the synthesis is kept at a very low level, which reduces the nuclei density of MoS_2 and hence, the nuclei can continue to grow into large crystalline domains.

Pyrolysis of a compound containing both metal and chalcogen elements, such as ammonium thiomolybdates, can be another straight way to produce TMDC thin films. For example, MoS_2 thin film can be produced by applying a two-step annealing on $(NH_4)_2MoS_4$ coating, which were performed at 500°C and 1000°C in Ar/H_2 and Ar/S atmosphere, respectively, as shown in Figure 1.5 [46].

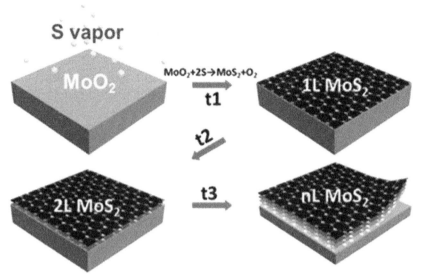

FIGURE 1.4 Schematics for the synthesis and cleavage of MoS_2. Reproduced with permission from Ref. [44]. Copyright 2013 American Chemical Society.

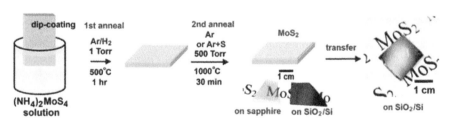

FIGURE 1.5 Schematic diagram of the two-step pyrolysis for the fabrication of MoS_2 thin film. Reproduced with permission from Ref. [46]. Copyright 2012 American Chemical Society.

Although the above methods can produce MoS_2 thin films with a large area, they cannot synthesize uniform monolayer MoS_2 because of the lattice mismatch problem between the pre-deposited materials and MoS_2. Monolayer MoS_2 can be realized by chemical reactions of gaseous Mo and sulfur precursors, in which MoS_2 can nucleate on a prepared substrate and then grow into monolayer flakes or film. This is more complicated than sulfurization of pre-deposited Mo film since both Mo and S sources are in vapor states during the reaction. However, on the other hand, the reaction parameters of different sources and environment can be altered individually. Therefore, MoS_2 monolayers are more likely to deposit via careful parameter adjustment.

MoS$_2$ monolayer flakes were first deposited by using MoO$_3$ and sulfur powder as Mo and S sources [47]. The powder was placed in two individual crucibles and the substrate was put facing down above MoO$_3$ powder. During the reaction, MoO$_3$ powder is first reduced to volatile MoO$_{3-x}$ which diffuses onto the substrate and further reacts with sulfur vapor to form star-shaped MoS$_2$ nanosheets. It is worth mentioning that a drop of hydrazine reduced graphene oxide, perylene-3,4,9,10-tetracarboxylic acid tetrapotassium salt (PTAS), or perylene-3,4,9,10-tetracarboxylic dianhydride solution, was dispersed on the substrate before the CVD growth. These materials, acting as some kinds of seeding promoters, are critical to grow monolayer MoS$_2$. Later, with optimized growth parameters, the same group obtained triangular-shaped MoS$_2$ nanosheets with much better effective field-effect mobility only using PTAS to pretreat the substrate [48]. Finally, a systematically analysis of the role of twelve seeding promoters demonstrates that only some aromatic chemicals could facilitate the growth of monolayer MoS$_2$. Among those, PTAS and F$_{16}$CuPc are the most favorable chemicals for hydrophilic and hydrophobic substrates [49].

Although seeding promoters can benefit the initial nucleation of MoS$_2$ nanosheets, they are not compulsory. For example, MoS$_2$ triangular nanosheets were preferentially nucleated and formed at the step edges of SiO$_2$/Si substrate when using highly crystalline MoO$_3$ nanoribbons as precursor, as a result of the smaller nucleation energy barrier compared to the flat surface [50]. It was also found that the coalescence of two nanosheets either led to the formation of five- or seven-fold rings along grain boundaries, or simply overlapping each other. Moreover, it is demonstrated that neither seeding prompters nor step edges are required if the substrates are ultraclean, like ultraclean SiO$_2$/Si substrate which can be obtained by cleaning sequentially in acetone, isopropanol, H$_2$SO$_4$/H$_2$O$_2$ (3:1), and O$_2$ plasma [51]. The side length of largest triangular MoS$_2$ nanosheet can reach 123 μm, 10 times longer than those in previous works.

Up to now, MoO$_3$ and sulfur powder are most commonly used in CVD process, other precursors such as MoCl$_5$ and H$_2$S are only occasionally reported [52, 53]. The morphology, crystallinity, and other properties like effective field-effect mobility vary tremendously between different reports, suggesting the difficulty in optimizing the growth parameters. There is evidence that MoS$_2$ is most sensitive to the concentration of precursors and the growth pressure. In many works, the MoS$_2$ monolayers have specific shapes like triangular or hexagon limited within a finite size (length <50 μm); thus, how to achieve uniform crystalline monolayer film is still an issue requiring much further research.

1.3.2.2 PHYSICAL VAPOR DEPOSITION

Physical vapor deposition (PVD) has been extensively used for 2D thin-film deposition. A simple physical vapor transport method using a MoS_2 powder source and Ar carrier gas is illustrated in Figure 1.6 [54]. The MoS_2 powder was evaporated at ~900°C in the center of furnace with pressure of ~20 Torr, diffused by flowing gas and deposited on the surface of various substrates at ~650°C. The fabricated MoS_2 was monolayer single crystals with size up to 25 µm and incredibly high optical quality.

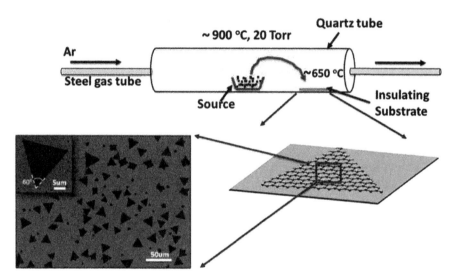

FIGURE 1.6 Schematic illustration of growth conditions and SEM image of the MoS_2 triangular flakes. Reproduced with permission from Ref. [55]. Copyright 2013 American Chemical Society.

Commercial PVD systems, such as pulsed laser deposition (PLD) and magnetron sputtering, are expected for precise manipulation of the MoS_2 film thickness and easy integration into various electronic device applications. Although these methods have long been an issue of great interest in single- and few-layer MoS_2 film deposition [56–58], MoS_2 thin film with high crystallinity over large area is just a very recent accomplishment by applying unique techniques. For example, instead of using pure MoS_2 target, extra sulfur was added by cold isostatic pressing of MoS_2 and S powder [59]. It is discovered that the best crystalline films with rocking curve full width half maxima of 0.01° were fabricated from the target with a Mo:S ratio of 1:4 by PLD. In addition, the monolayer MoS_2 film was highly smooth with a root mean square roughness of 0.27 nm. In another report, wafer-scale MoS_2 monolayers were fabricated by magnetron sputtering Mo target in a vaporized sulfur ambient [60]. Sulfur was vaporized in a container wrapping by heating tapes and the vapor leaking was controlled by a valve located between the container and the main chamber. These methods can be extended to the growth of other TMDCs and are considerably promising for the development of electronic and photonic devices.

1.3.3 FUNCTIONALIZATION

The performance of 2D TMDCs in various applications strongly depends on the electronic structures, which can be modulated by native defects or extra chemical doping. Existing theoretical research recognizes the critical role played by various defects in MoS_2, including vacancies, interstitials, antisites, and adatoms [61–65]. It is revealed that S vacancies are the most abundant defects, especially in the Mo-rich conditions. Under S-rich conditions, all formation energies are relatively high and good-quality crystal growth is expected [64]. However, calculation methods have great influence on the results. Instead of S vacancies, the adsorption of S adatoms is more favorable in monolayer MoS_2 in another report [62].

In contradiction to the theoretical calculations which suggest ideal 2D MoS_2 shall have nonmagnetic ground state, MoS_2 nanosheets prepared in laboratories display clear room-temperature ferromagnetism. For example, MoS_2 nanosheets fabricated by sonicating bulk MoS_2 in dimethylformamide for different times all exhibited room-temperature ferromagnetism [66]. Moreover, the saturation magnetization increased as the nanosheet size deceased, reaching 2.5×10^{-3} and 1.1×10^{-3} emu/g at 10 and 300 K for 10-h sonicated sample. CVD-prepared MoS_2 nanosheets by Yang et al. showed a larger saturation magnetization decreased from 1.08 emu/g at 10 K to 0.55 emu/g at 623 K [67]. In addition, the ferromagnetism had a very high Curie temperature, up to 865.5 K.

Although room-temperature ferromagnetism of MoS_2 is well demonstrated by a large body of literature, its origin has not yet been fully understood. To date, this phenomenon is widely explained by the presence of zigzag-edge states and MoS_2 triple vacancies [66,68,69]. However, this explanation is untrustworthy because simulation of MoS_2 zigzag nanoribbons suggests the magnetic moment per MoS_2 is only 0.037 μ_B with ribbon width of 6.43 nm, and it decreases with increasing width [70]. This value can be ignored in most situations because the grain size of MoS_2 nanosheets is commonly much larger than 6.43 nm. Another problem to comprehend ferromagnetism of MoS_2 is that compared to the considerable amount of theoretical studies, systematic characterizations of the defects are too scarce. Using atomic-resolution annular dark field imaging on an aberration-corrected scanning transmission electron microscope, six different types of point defects were commonly observed in CVD-grown monolayer MoS_2: monosulfur vacancy (V_S), disulfur vacancy (V_{S2}), vacancy complex of Mo and nearby three sulfur (V), vacancy complex of Mo nearby three disulfur pairs (V), and antisite defects where a Mo atom substituting a S2 column (Mo_{S2}) or a S2 column substituting a Mo atom ($S2_{Mo}$). MoS_2 triple vacancy (V_{MoS2}), which is expected as one of the ferromagnetism origins, has never been observed in the experiment because of the higher formation energy than V_{MoS3} [71].

Chemical doping for 2D TMDCs can be achieved by surface adsorption. Theoretically, study of adsorption of 16 individual adatoms (C, Co, Cr, Fe, Ge, Mn, Mo, Ni, O, Pt, S, Sc, Si, Ti, V, and W) on monolayer MoS_2 suggests Cr has the weakest binding energy ($E_b = 1.08$ eV) among all the adatoms, while elements with strong binding energy such as W ($E_b = 4.93$ eV) can create a local reconstruction on the S layer [69]. In experiment, 2D TMDCs can absorb both organic and inorganic dopants on the surface. For example, Mouri et al. used p-type dopants (F_4TCNQ and TCNQ) and n-type dopants (TCNQ) to decorate monolayer MoS_2 via solution-based chemical doping method [72]. Enhancement of photoluminescence (PL) intensity of mono-layer MoS_2 was observed by p-type decoration, but the trend was reversed by n-type doping. In another study, NO_2 and potassium were employed as p- and n-type dopants for few-layer WSe_2 [73]. Absorption of NO_2 was easily achieved by 10-min exposure to 0.05% NO_2 in N_2 gas, while the doping of potassium was realized by direct potassium dispensing with a high electron density of 2.5×10^{12} cm^{-2}.

Substitutional doping is considered as a more stable approach compared to surface adsorption because the dopants are secured and stabilized by covalent bonding. Theoretically, substituting S atoms in MoS_2 by Fe and V can induce magnetic moments [74]. Substitution of Mo atoms has also been studied by many researchers using theoretical methods [2,75]. A study of 27 different transition metals suggests that monolayer MoS_2 doped by Fe, Co, Mn, Zn, Cd, or Hg in Mo position is ferromagnetic [76]. Although some metal dopants have no contribution to magnetic moments, combining with vacancies may change the situation. For example, V-doped MoS_2 coupled with Mo and S vacancies indicates that $Mo_{14}VS_{32}$, $Mo_{15}VS_{31}$, and $Mo_{14}VS_{31}$ would gain a magnetic moment of 1, 1, and 0.95 μ_B, respectively. Among them, $Mo_{15}VS_{31}$ had a minimum energy state with a formation energy of 5.89 eV [77].

Experimentally, the atoms in 2D TMDCs can be easily substituted by the elements from the same groups, either metal or chalcogen atoms, forming alloys like $Mo_{1-x}W_xS_2$ and $MoS_{2x}Se_{2(1-x)}$. CVD method has been frequently used to prepare TMDC alloys. To synthesize $MoS_{2x}Se_{2(1-x)}$, a mixture of sublimated sulfur and selenium powder is used as chalcogen precursor. The optical bandgap of $MoS_{2x}Se_{2(1-x)}$ can be tuned between 1.85 and 1.60 eV by varying Se doping concentration [78].

Substitutional doping from the elements beyond the same groups has been proved to be much tougher [79,80]. Li et al. [81] synthesized hexagonal bilayer Co-doped MoS_2 nanosheets by CVD using MoO_3, Co_3O_4, and sulfur as precursors. Co atoms were mainly distributed at the edge of the nanosheets. However, when the temperature increased from 680 to 750°C, instead of further edge doping, a layer of CoS_2 film was observed on top of the MoS_2 nanosheets. In another report, Nb-doped MoS_2 nanosheets were prepared by a chemical vapor transport method [82]. The reaction was quite time consuming (>500 h) and the doping concentration could only reach 0.5%. The Nb-doped MoS_2 nanosheets displayed stable p-type conduction with a degenerate hole density of ~3 × 10^{19} cm^{-3}.

The effect of chemical doping can sometimes be strongly dependent on the choice of substrate. Mn-doped MoS_2 monolayers were grown on three substrates, graphene, sapphire, and SiO_2 via CVD using $Mn_2(CO)_{10}$ as Mn precursor [83]. Mn doping was only successful on graphene substrate, while the effective doping on sapphire and SiO_2 substrates could not be achieved no matter how to increase the amount of $Mn_2(CO)_{10}$ precursor or adjust other growth parameters such as temperature and carrier gas pressure. However, even on the graphene substrate, the maximum doping concentration was limited to 2%.

1.4   APPLICATION OF 2D TMDCS

In this part, a few crucial areas of applications of 2D TMDCs will be introduced and discussed, including those in energy conversion and storage, biomedical field, and nanodevices.

1.4.1 ENERGY-RELATED APPLICATIONS

1.4.1.1 HYDROGEN EVOLUTION

HER, $2H^+ + 2e^- \rightarrow H_2$, is a critical electrochemical process to generate hydrogen, which can be greatly prompted by electrochemical catalysts with excellent catalytic activity and electronic conductivity. It is suggested that MoS_2 may replace the most commonly used catalyst Pt because of the similar hydrogen-binding energies, abundancy, and low cost [84].

To date, many research works, both theoretically and experimentally, have confirmed the promising properties of 2D TMDCs for HER. From calculation, it indicates that HER catalysis mainly occurs at the edges of 2D TMDCs. The basal surface is almost inert but can be activated via introducing defects. For example, to create controllable defect-rich MoS_2, excess thiourea has been used during a hydrothermal reaction with $(NH_4)_6Mo_7O_{24}$, leading to the formation of certain degree of short-range disorder [85,86]. With the assistance of the defect sites on the surface as well as along the edges, as-formed MoS_2 nanosheets have shown outstanding HER activity with onset overpotential of 120 mV and Tafel slope of 50 mV per decade.

One drawback of 2D MoS_2 as the HER catalyst is its low conductivity, which causes slow electron transfer velocity and hinders the catalytic performance. Vanadium doping can be one solution [80]. Meanwhile, the metallic 1T TMDCs can also exhibit superb HER performance due to the advanced electrical conductivity and catalytic activity. Multilayered 2H MoS_2 nanosheets prepared by CVD method can be converted into 1T phase by lithium intercalation and enhance the performance of HER [87].

1.4.1.2 LITHIUM ION BATTERIES

Energy storage is an issue of great significance after energy conversion. LIBs are one of the most promising energy storage systems that we can rely on today, especially in portable electronics. Carbon-based materials have dominated the commercial LIB anodes for decades; however, their low capacities (theoretical capacity of graphite is 373 mA h/g) are far beyond satisfaction in the field of electric vehicles. MoS_2 has the theoretical capacity at least two times higher than that of graphite and suffers considerably less volume expansion (103%) during lithium-ion intercalation compared to many other anode materials, such as silicon [88]. However, MoS_2 electrodes also have several drawbacks which cannot be neglected. In addition to low electric conductivity which inevitably requires more conductive additives, the core disadvantage is their lithiation voltage, 1.1–2.0 V versus Li/Li$^+$, which is too low for practical applications. The commercial attempt of MoS$_2$ as anode was first patented in 1980 but it soon failed. Fortunately, recent advancements in nanocomposites have successfully revitalized the relative research.

Among all the MoS$_2$ nanocomposites, MoS$_2$/graphene composites have been intensively investigated. MoS$_2$/graphene composites can be synthesized by a hydrothermal method using graphene oxide sheets and Na$_2$MoO$_4$ as precursors and L-cysteine as reducing agent with the assistance of cetyltrimethylammonium bromide (CTAB) [89]. It was observed that single or few-layer MoS$_2$ nanosheets were grown in situ on the surface of graphene, and the number of layers could be controlled by the concentration of CTAB. The MoS$_2$/graphene composites exhibited a reversible capacity as high as 1020 mA h/g with outstanding cycle stability and high-rate performance.

It is noted that the disordered structure of 2D MoS$_2$, which can be stabilized by the integration of a polymer such as polyethylene oxide (PEO), is able to shorten the diffusion pathway for lithium ions and increase the performance. Various approaches have been employed to fabricate MoS$_2$/graphene/PEO composites. For example, Liu et al. designed a kind of hybrid paper which was composed of MoS$_2$ and graphene nanosheets cross-linked by PEO [90]. The hybrid paper was rather flexible and could be bent easily. Moreover, the hybrid paper could be directly affixed on the copper foil and acted as anode. With the optimum composition, the hybrid paper showed an initial discharge capacity of 1240 mA h/g with 71% retention after 100 cycles.

1.4.1.3   SUPERCAPACITORS

Supercapacitors, a class of energy storage devices with high energy densities and power, are expected to complement LIBs in many applications. Supercapacitors are generally divided into two categories, electrical double-layer capacitors and pseudocapacitors, depending on the source of the capacitance. For both types, electrode materials demand large surface area and superb electrical conductivity. 2D VS$_2$ with ultralarge surface area can be an ideal candidate. In one report, VS$_2$ nanosheets, fabricated by sonicating VS$_2$·3NH$_3$ icy solution, were filtrated over a cellulose membrane in vacuum to form a homogeneous thin film [91]. Electrochemical measurement revealed that the supercapacitor had a specific capacitance up to 4760 μF/cm^2 and an extraordinary cycle stability.

Although pristine MoS_2 nanosheets have low electric conductivity, their composites formed with a conducting substance, such as graphene, also present potential application in supercapacitors as well as LIBs. MoS_2/graphene composites can be prepared by direct microwave irradiation to a mixture of $MoCl_5$, butyl mercaptan, and a colloidal dispersion of graphene oxide in dimethyl fumarate [92]. The specific capacitance of composite was 265 F/g with a medium concentration of MoS_2 and only 8% of the capacitance was lost after 1000 cycles, indicating brilliant cycle stability.

Other than LIBs and supercapacitors, MoS_2/graphene composites also can be used in dye-sensitized solar cells, exhibiting relatively high electrocatalytic activity with power conversion efficiency as high as 6.07% [93].

1.4.2 BIOMEDICAL APPLICATIONS

The distinctive planar structure and diverse chemical compositions of 2D TMDCs make them promising for biomedical applications, including biosensing, drug delivery, photothermal/photodynamic therapy, and diagnostic imaging [94]. Moreover, they exhibit relatively high biosafety and biocompatibility.

MoS_2 behaves dissimilar affinity toward single- and double-stranded DNA. Based on this principle, MoS_2 nanosheets can be employed as electroactive labels to detect DNA hybridization, which is related to the diagnosis of Alzheimer's disease [95]. The range of detection varied from 0.03 to 300 nM with the optimum concentration of MoS_2 nanosheets in the biosensing assay of 0.018 mg/mL.

It is worth noticing that in addition to various biological species, MoS_2 nanosheets are highly sensitive to water and many gas molecules, entitling them good candidates for gas and pH sensors [96].

2D TMDCs have large surface-to-volume ratio, providing plentiful anchoring points available for drug loading. Functionalization of liquid-exfoliated MoS_2 nanosheets can be realized by the employment of lipoic-acid-conjugated polyethylene glycol (PEG) [97]. Outstanding loading capacities and physiological stabilities were verified by various therapeutic molecules on MoS_2–PEG nanosheets. Moreover, loaded with chemotherapy drugs doxorubicin, the combined photothermal and chemotherapy were demonstrated both in vitro and in vivo, demonstrating excellent synergistic anticancer effect in inhibiting tumor growth [98].

MoS_2 nanosheets can act as contrast agents for computed tomography (CT) imaging [99]. X-ray CT image of chitosan-decorated MoS_2 nanosheets shows apparent signal enhancement with increasing MoS_2 concentration. The slope of the CT value for MoS_2 nanosheets (~14.2) is slightly higher than that of the commercial contrast agent, Iopromide 300.

1.4.3 NANODEVICES

Generally speaking, any devices within nanoscale dimensions belong to the family of nanodevices, like the nanobiosensor mentioned above. But here only those applied in semiconductor industry are emphasized, including electronics, spintronics, and valleytronics.

1.4.3.1 ELECTRONICS

Transistor is considered as one of the most essential devices in digital electronics. In 1965, Gordon Moore published a paper in which he proposed that semiconductor industry would double the number of components on an integrated circuit proximately every 2 years. Past five decades have witnessed his prophecy with only slight modifications due to the continuing decreasing size of silicon-based metal oxide semiconductor field-effect transistor (MOSFET). Presently, state-of-the-art MOSFETs have a feature channel length of 14 nm, and it will soon meet the fundamental physical limits, motivating the exploration for new device designs and materials, such as 2D TMDCs channel.

Simulations of monolayer MoS_2-based FET have predicted outstanding electronic performances [100–102]. For example, a top-gate monolayer MoS_2-based FET with a gate length of 15 nm is expected to have large g_m (4.4 mS/μm), high on/off ratio ($>10^{10}$), and outstanding short-channel behavior due to low density of states, large bandgap, and advanced gate control. However, the main drawback of the transistor is its limited charge-carrier mobility. Calculated by DFT, the mobility of MoS_2 nanosheets, dominated by optical phonon scattering, is ~410 cm^2/V s at room temperature, which is comparable to that of silicon but much smaller than that of III–V semiconductors (few thousands of cm^2/V s) and graphene ($>10^4$ cm^2/V s) [103]. However, the low mobility of MoS_2 will not be a frustrating issue when the transport is almost ballistic which happens in diminished geometries, for example, extremely small-channel length. It has been demonstrated that monolayer MoS_2-based FET can comply with the current International Technology Roadmap down to a minimum channel length of 8 nm [104].

The first realization of TMDCs as the channel in FET was reported in 2004 by using 10–20-μm-thick WSe$_2$ film [105]. The mobility of the field-induced holes in FET reached 500 cm^2/V s at room temperature, which is comparable to that of electrons in the nonflexible Si MOSFETs. To date, a large body of studies have measured the mobility of 2D TMDCs using FETs or their analogues [106–107]. After the deposition of a 30-nm-thick layer of high-k HfO$_2$ insulator below the gate, the full potential of the monolayer MoS$_2$ was confirmed as the mobility prompted to at least 200 cm^2/V s and the current on/off ration exceeded 10^8.

Generally, mechanically exfoliated MoS$_2$ thin films exhibit the most excellent mobility among all fabrication methods, reaching several tens of cm^2/V s. However, the results between different groups may vary by two orders of magnitude, suggesting the mobility is highly dependent on the fabrication procedures. 2D MoS$_2$ with high crystallinity fabricated by other methods, like CVD and PVD, can also possess outstanding mobility. For example, Lee et al. employed two-probe measurement on liquid-exfoliated MoS$_2$ and obtained mobility of 162 and 195 cm^2/V s at negative and positive bias, respectively [108].

It is widely acknowledged that the electronic properties of 2D MoS$_2$ are thickness-dependent. However, the relationship between the number of layers and the mobility remains ambiguous [109–112]. In most research, monolayer MoS$_2$ has smaller mobility than few-layer FETs, which can be explained by the increased Coulomb scattering. However, Lembke opposed this idea by comparing the field-effect mobility of FETs after thorough in situ annealing in vacuum and found that the mobility of monolayer MoS$_2$ FET was significantly increased. It is suggested that gaseous adsorbates and water on the monolayer surface may degenerate its performance [113]. Since FETs in most applications are used in ambient environment, results argue that multilayer MoS$_2$ devices can outperform their single- or few-layer counterparts in certain aspects.

Recently, there has been profound interest in flexible electronics due to the increasing market of wearable devices. According to the indentation experiment conducted on monolayer MoS$_2$ using AFM cantilevers, monolayer MoS$_2$ has a Young's modulus of 270 ± 100 GPa, comparable to that of steel [114], entitling MoS$_2$ one of the strongest semiconductors and beneficial for flexible electronics. Electrical properties of MoS$_2$ during bending have also been widely explored. For example, Jiang et al. transferred MoS$_2$ thin films onto polyimide substrates and found that no obvious electrical degradation during bending at maximum curvature radius of 0.75 mm.

The variation of mobility (~3.01 cm^2/V s) as the function of bending was within 10% [115]. Similarly, a device containing MoS$_2$ channel and two-layer polyimide substrate can tolerate a bending radius of 1 mm, and, in the meantime, maintains the mobility of 30 cm^2/V s and on/off ratio larger than 10^7, as presented in Figure 1.7 [116].

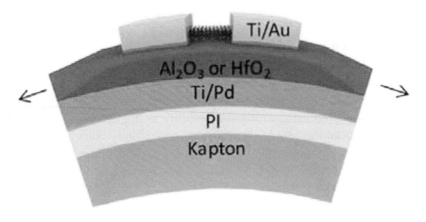

FIGURE 1.7   Illustration of a flexible MoS$_2$ device. Reproduced with permission from Ref. [116]. Copyright 2013 American Chemical Society.

Optoelectronic devices are a special group of electronics that can generate, detect, and interact with light, such as lasers, LEDs, and photo-detectors. Phototransistor-based MoS$_2$ has a similar structure of the FETs except that the incident illumination light source is used to tailor the photo-current, which is merely determined by the intensity of incident light at a fixed gate and drain voltage. Research shows that monolayer MoS$_2$-based phototransistor can have brilliant photo responsivity with switching time of photocurrent generation and annihilation less than 50 ms [117]. Photodetection of different light can be achieved by using MoS$_2$ thin films with various layer numbers due to the thickness-dependent bandgaps, for example, single-, double-, and triple-layer MoS$_2$ films have a bandgap of 1.8, 1.65, and 1.35 eV, respectively. As a result of the thickness-modulated optical gap of MoS$_2$ films, the phototransistor based on triple-layer MoS$_2$ is effective for red-light detection, while those based on single- and double-layer MoS$_2$ are more suitable for green-light detection [118].

1.4.3.2   SPINTRONICS AND VALLEYTRONICS

Spintronics is the study which mainly focuses on the intrinsic spin of electrons and associated magnetic moment. Spintronic devices utilize the spin degrees of freedom of electrons, rather than charge degrees of freedom. In theory, ideal spintronic devices can operate even without applying an electric current. However, it has never been realized in experiments.

The most successful application of spintronics is in the field of data storage followed discovery of giant magnetoresistance (GMR) via spin-dependent electron transport in metallic multilayers [119]. GMR is first observed in thin-film structures composed of alternating ferromagnetic and nonmagnetic conductive layers. The resistance of the material is determined by the alignment of spins in the ferromagnetic layers. A spin valve is a GMR-based device with a sandwich structure in which two ferromagnetic layers are separated by a nonmagnetic layer [120]. One of the ferromagnetic layers is pinned and the other one is free, in other words, only the magnetization of the free layer can be flipped by an applied magnetic field. As the orientations of magnetizations in two layers change from parallel to antiparallel, the resistance of the spin valve gives a slight rise from 5% to 10%. MoS_2-based spin valve can be constructed by using two Permalloy (Py, $Ni_{80}Fe_{20}$) electrodes as the ferromagnetic layers [121]. The MoS_2 films showed metallic behavior rather than semiconducting, due to the strong hybridization with Fe and Ni atoms in the electrodes. The highest magnetoresistance of the spin valve was ~0.73% at 10 K, which is far from satisfaction for practical applications. However, additional first principle calculation indicates that an ideal magnetoresistance of ~9% can be achieved for a perfect $Py/MoS_2/Py$ junction, suggesting promising candidate for spintronics applications.

Spin-based FETs is one of the most significant applications of spintronics devices, which were first envisioned by Datta in 1990 [122]. Unfortunately, unlike GMR which quickly finds its use in data storage after the discovery, realization of a functional spin-based FET at room temperature has not yet to be accomplished due to many unsolved problems. A basic structure of spin FET, very similar to conventional FET, contains a semiconductor channel, gate, and ferromagnetic source and drain contacts to inject and detect spin-polarized electrons. The total current through spin FET depends on the relative angle between magnetization direction at the drain contact and the end of the channel. Several alternative structures were also reported [123].

Bilayer TMDCs have been theoretically predicted as potential channel material of spin FET [124]. Although inversion symmetry forbids spin–orbit splitting of a bilayer TMD, this symmetry can be destroyed by applying external electric fields. The value of spin–orbit splitting of bilayer MoS_2 can reach 145 meV at an electric field less than 500 mV/Å. Thus, it is possible to switch the spin polarization on and off by tuning the gate voltage. Moreover, based on theoretical simulation, Li et al. proposed a normal/ferromagnetic/ normal monolayer MoS_2 junction and investigated the spin transport properties by tuning gate voltage by theory. It was argued that fully valley- and spin-polarized conductance can be achieved due to the spin–valley coupling of valence-band edges together with the exchange field, and both the amplitude and direction of the conductance can be modulated by the gate voltage [125]. Similarly, in a monolayer WSe_2-based spin FET, the gate field can introduce an additional spin–orbit interaction which rearranges the pattern of the spins into a Rashba-like texture [126].

Practically, a spin-based FET follows three operating principles: polarized spin injection into the channel, spin manipulation, and spin detection. So far, none of these operating principles has a perfect solution. For a long time, spin injection into semiconductors from a ferromagnetic metal electrode faces unconquerable obstacles due to the impedance mismatch problem. Studies of the contact behaviors in MoS_2-based FET confirm that the Schottky barrier height can be incredibly decreased by inserting a thin metal oxide layer between the contact and channel. For example, a layer of 2-nm-thick MgO could reduce the Schottky barrier between Co contact and monolayer MoS_2 from 35.9 to 9.7 meV. The Schottky barrier height was controlled by MgO thickness as well as back gate voltage and could even be completely eliminated at the optimum condition [127]. Similarly, a thin TiO_2 tunnel barrier (~13 nm) between the ferrimagnets and multilayer MoS_2 dramatically decreased the contact resistance, resulting in an enhancement of on-state current by two orders of magnitude and channel mobility by a factor of 6, respectively [128].

The transport of spins in monolayer TMDCs is often coupled with valleys due to broken inversion symmetry. The absorption and emission of light in monolayer MoS_2 follow a selection rule, in which optical field with right-handed circular polarization and frequency ω_u (ω_d) can generate spin-up (-down) electrons and spin-down (-up) holes in valley K, while the excitation in K' valley is the time-reversal of the above. Furthermore, researchers have also predicted the coexistence of valley Hall and spin Hall effects in n-doped and p-doped systems, and proposed photo-induced spin Hall and valley Hall effects for generating spin and valley accumulations on edges [129].

Successful valley polarization has been accomplished by several groups using optical methods by examining the circularly polarized luminescence spectra of mechanically exfoliated monolayer MoS_2 via circularly polarized excitation. It was found that the PL of monolayer MoS_2 has the identical helicity as the excitation laser [130]. The degree of circular polarization P, defined as $P = (I(\sigma+) - I(\sigma-)) / I(\sigma+) + I(\sigma-))$, is 32 ± 2% and −32 ± 2% with right- and left-handed circular excitation by HeNe laser (~1.96 eV) at 10 K, respectively. Furthermore, the PL polarization is independent on in-plane magnetic field, confirming that the polarized PL is indeed caused by valley polarization rather than spin polarization, because only valley polarization cannot be rotated by external magnetic field. Using similar method, Mak et al. reported nearly 100% valley polarization for photon energies in the range 1.90–1.95 eV at 14 K, and the polarization on resonance excitation with A exciton was retained longer than 1 ns [131].

Bilayer MoS_2 possesses no valley polarization due to restoration of inversion symmetry, but bilayer WS_2 shows anomalously robust valley polarization and coherence [132]. The degree of circular polarization of bilayer WS_2 (~95%) is much higher than that of monolayer WS_2 (~40%) and can even maintain at room temperature. Although the mechanism is still ambiguous, the short exciton lifetime, small exciton-binding energy, and extra spin-conserving channels are possible attributes to the robust circular polarization. On the other hand, inversion symmetry of bilayer MoS_2 can be destroyed by applying an electric field perpendicular to the plane. It was reported that the circularly polarized PL of bilayer MoS_2 can be tuned constantly from −15% to 15% as a function of gate voltage [133].

1.5 CONCLUSION

2D TMDCs offer vast opportunities for the investigation of fundamental principles and their practical applications. Their unique mechanical, catalytic, and electronic properties are promising for energy conversion and storage, biomedical field, and nanodevices. The loss of the inversion symmetry when 2D TMDCs are thinned to monolayer leads to more novel physical properties, like strong spin–orbital coupling with opposite spins in K and K′ valleys and a distinct optical excitation selection rule which couple the transport of spins and valleys. Furthermore, the electronic and magnetic properties of 2D TMDCs can be tuned by strain, external electronic field, and chemical doping. For example, ferromagnetism can be induced by Mn, Fe, Co, Cr, Zn, Cd, Hg, V, and Cu doping according to theoretical simulations. From experimental point of view, although several categories of methods, especially CVD, have been successfully employed to fabricate 2D TMDCs (mainly MoS_2), the controllable and scalable synthesis of highly qualified materials is still a challenge. Moreover, the influence of various defects on the magnetic properties remains ambiguous and how to increase the doping concentration also needs further research.

KEYWORDS

- two-dimensional (2D) materials
- graphene
- bandgap
- nanodevices
- transition metal dichalcogenides

REFERENCES

1. Novoselov, K. S.; Geim, A. K.; Morozov, S. V.; Jiang, D.; Zhang, Y.; Dubonos, S. V.; Grigorieva, I. V.; Firsov, A. A. *Science* **2004,** *306,* 666.
2. Fan, X.-L.; An, Y.-R.; Guo, W.-J. *Nanoscale Res. Lett.* **2016,** *11,* 154.
3. Xia, B.; Guo, Q.; Gao, D.; Shi, S. Tao, K. *J. Phys. D: Appl. Phys.* **2016,** *49,* 165003.
4. Wang, Q. H.; Kalantar-Zadeh, K.; Kis, A.; Coleman, J. N.; Strano, M. S. *Nat. Nanotechnol.* **2012,** *7,* 699.
5. Ataca, C.; Sahin, H.; Ciraci, S. *J. Phys. Chem. C* **2012,** *116,* 8983.
6. Eda G.; Yamaguchi H.; Voiry D.; Fujita T.; Chen M. W.; M. Chhowalla. *Nano Lett.* **2011,** *11,* 5111.
7. Lee, C.; Li, Q.; Kalb, W.; Liu, X.-Z.; Berger, H.; Carpick, R. W.; Hone, J. *Science* **2010,** *328,* 76.
8. Lee C.; Li Q.; Kalb W.; Liu X.-Z.; Berger H.; Carpick R. W.; Hone J. *Science* **2010,** *328,* 76.
9. Kadantsev, E. S.; Hawrylak, P. *Solid State Commun.* **2012,** *152,* 909.
10. Ping, Y.; Rocca, D.; Galli, G. *Chem. Soc. Rev.* **2013,** *42,* 2437.
11. Mak, K. F.; Lee, C.; Hone, J.; Shan, J.; Heinz, T. F. *Phys. Rev. Lett.* **2010,** *105,* 136805.
12. Feng, W.; Yao, Y.; Zhu, W.; Zhou, J.; Yao, W.; Xiao, D. *Phys. Rev. B* **2012,** *86,* 165108.
13. Johari, P.; Shenoy, V. B. *ACS Nano* **2012,** *6,* 5449.

14. Sorkin, V.; Pan, H.; Shi, H.; Quek, S. Y.; Zhang, Y. W. *Crit. Rev. Solid State Mater. Sci.* **2014**, *39*, 319.
15. Eda, G.; Fujita, T.; Yamaguchi, H.; Voiry, D.; Chen, M.; Chhowalla, M. *ACS Nano* **2012**, *6*, 7311.
16. Cai, L.; He, J.; Liu, Q.; Yao, T.; Chen, L.; Yan, W.; Hu, F.; Jiang, Y.; Zhao, Y.; Hu, T.; Sun, Z.; Wei, S. *J. Am. Chem. Soc.* **2015**, *137*, 2622.
17. Pan, H.; Zhang, Y.-W. *J. Mater. Chem.* **2012**, *22*, 7280.
18. Ataca, C.; Sahin, H.; Akturk, E.; Ciraci, S. *J. Phys. Chem. C* **2011**, *115*, 3934.
19. Sagynbaeva, M.; Panigrahi, P.; Li, Y.; Ramzan, M.; Ahuja, R. *Nanotechnology* **2014**, *25*, 165703.
20. Bertram, N.; Cordes, J.; Kim, Y. D.; Gantefor, G.; Gemming, S.; Seifert, G. *Chem. Phys. Lett.* **2006**, *418*, 36.
21. Lauritsen, J. V.; Kibsgaard, J.; Helveg, S.; Topsoe, H.; Clausen, B. S.; Laegsgaard, E.; Besenbacher, F. *Nat. Nanotechnol.* **2007**, *2*, 53.
22. Chianelli, R. R.; Prestridge, E. B.; Pecoraro, T. A.; Deneufville, J. P. *Science* **1979**, *203*, 1105.
23. Seifert, G.; Terrones, H.; Terrones, M.; Jungnickel, G.; Frauenheim, T. *Phys. Rev. Lett.* **2000**, *85*, 146.
24. Dallavalle, M.; Saendig, N.; Zerbetto, F. *Langmuir* **2012**, *28*, 7393.
25. Tenne, R.; Margulis, L.; Genut, M.; Hodes, G. *Nature* **1992**, *360*, 444.
26. Albu-Yaron, A.; Levy, M.; Tenne, R.; Popovitz-Biro, R.; Weidenbach, M.; Bar-Sadan, M.; Houben, L.; Enyashin, A. N.; Seifert, G.; Feuermann, D.; Katz, E. A.; Gordon, J. M. *Angew. Chem., Int. Ed.* **2011**, *50*, 1810.
27. Splendiani, A.; Sun, L.; Zhang, Y. B.; Li, T. S.; Kim, J.; Chim, C. Y.; Galli, G.; Wang, F. *Nano Lett.* **2010**, *10*, 1271.
28. Zeng, Z.; Yin, Z.; Huang, X.; Li, H.; He, Q.; Lu, G.; Boey, F.; Zhang, H. *Angew. Chem., Int. Ed.* **2011**, *50*, 11093.
29. Lee, C.; Yan, H.; Brus, L. E.; Heinz, T. F.; Hone, J.; Ryu, S. *ACS Nano* **2010**, *4*, 2695.
30. Castellanos-Gomez, A.; Barkelid, M.; Goossens, A. M.; Calado, V. E.; van der Zant, H. S. J.; Steele, G. A. *Nano Lett.* **2012**, *12*, 3187.
31. Coleman, J. N.; Lotya, M.; O'Neill, A.; Bergin, S. D.; King, P. J.; Khan, U.; Young, K.; Gaucher, A.; De, S.; Smith, R. J.; Shvets, I. V.; Arora, S. K.; Stanton, G.; Kim, H.-Y.; Lee, K.; Kim, G. T.; Duesberg, G. S.; Hallam, T.; Boland, J. J.; Wang, J. J.; Donegan, J. F.; Grunlan, J. C.; Moriarty, G.; Shmeliov, A.; Nicholls, R. J.; Perkins, J. M.; Grieveson, E. M.; Theuwissen, K.; McComb, D. W.; Nellist, P. D.; Nicolosi, V. *Science* **2011**, *331*, 568.
32. Zhou, K.-G.; Mao, N.-N.; Wang, H.-X.; Peng, Y.; Zhang, H.-L. *Angew. Chem. Int. Ed.* **2011**, *50*, 10839.
33. Smith, R. J.; King, P. J.; Lotya, M.; Wirtz, C.; Khan, U.; De, S.; O'Neill, A.; Duesberg, G. S.; Grunlan, J. C.; Moriarty, G.; Chen, J.; Wang, J.; Minett, A. I.; Nicolosi, V.; N. Coleman, J. *Adv. Mater.* **2011**, *23*, 3944.
34. Dines, M. B. *Mater. Res. Bull.* **1975**, *10*, 287.
35. Zeng, Z.; Sun, T.; Zhu, J.; Huang, X.; Yin, Z.; Lu, G.; Fan, Z.; Yan, Q.; Hng, H. H.; Zhang, H. *Angew. Chem. Int. Ed.* **2012**, *51*, 9052.
36. Zheng, J.; Zhang, H.; Dong, S.; Liu, Y.; Nai, C. T.; Shin, H. S.; Jeong, H. Y.; Liu, B.; Loh, K. P. *Nat. Commun.* **2014**, *5*, 2995.
37. Hodes, G. *PCCP* **2007**, *9*, 2181.

38. Garadkar, K. M.; Patil, A. A.; Hankare, P. P.; Chate, P. A.; Sathe, D. J.; Delekar, S. D. *J. Alloys Compd.* **2009,** *487,* 786.
39. Xi, Y.; Serna, M. I.; Cheng, L. X.; Gao, Y.; Baniasadi, M.; Rodriguez-Davila, R.; Kim, J.; Quevedo-Lopez, M. A.; Minary-Jolandan, M. *J. Mater. Chem. C* **2015,** *3,* 3842.
40. Matte, H. S. S. R.; Gomathi, A.; Manna, A. K.; Late, D. J.; Datta, R.; Pati, S. K.; Rao, C. N. R. *Angew. Chem. Int. Ed.* **2010,** *49,* 4059.
41. Wang, F.; Wang, Z.; Wang, Q.; Wang, F.; Yin, L.; Xu, K.; Huang, Y.; He, J. *Nanotechnology* **2015,** *26,* 292001.
42. Zhan, Y.; Liu, Z.; Najmaei, S.; Ajayan, P. M.; Lou, J. *Small* **2012,** *8,* 966.
43. Lin, Y.-C.; Zhang, W.; Huang, J.-K.; Liu, K.-K.; Lee, Y.-H.; Liang, C.-T.; Chu, C.-W.; Li, L.-J. *Nanoscale* **2012,** *4,* 6637.
44. Wang, X.; Feng, H.; Wu, Y.; Jiao, L. *J. Am. Chem. Soc.* **2013,** *135,* 5304.
45. Ma, L.; Nath, D. N.; Lee, II, E. W.; Lee, C. H.; Yu, M.; Arehart, A.; Rajan, S.; Wu, Y. *Appl. Phys. Lett.* **2014,** *105,* 072105.
46. Liu, K.-K.; Zhang, W.; Lee, Y.-H.; Lin, Y.-C.; Chang, M.-T.; Su, C.; Chang, C.-S.; Li, H.; Shi, Y.; Zhang, H.; Lai, C.-S.; Li, L.-J. *Nano Lett.* **2012,** *12,* 1538.
47. Lee, Y.-H.; Zhang, X.-Q.; Zhang, W.; Chang, M.-T.; Lin, C.-T.; Chang, K.-D.; Yu, Y.-C.; Wang, J. T.-W.; Chang, C.-S.; Li, L.-J.; Lin, T.-W. *Adv. Mater.* **2012,** *24,* 2320.
48. Lee, Y.-H.; Yu, L.; Wang, H.; Fang, W.; Ling, X.; Shi, Y.; Lin, C.-T.; Huang, J.-K.; Chang, M.-T.; Chang, C.-S.; Dresselhaus, M.; Palacios, T.; Li, L.-J.; Kong, J. *Nano Lett.* **2013,** *13,* 1852.
49. Ling, X.; Lee, Y.-H.; Lin, Y.; Fang, W.; Yu, L.; Dresselhaus, M. S.; Kong, J. *Nano Lett.* **2014,** *14,* 464.
50. Najmaei, S.; Liu, Z.; Zhou, W.; Zou, X.; Shi, G.; Lei, S.; Yakobson, B. I.; Idrobo, J.-C.; Ajayan, P. M.; Lou, J. *Nat. Mater.* **2013,** *12,* 754.
51. van der Zande, A. M.; Huang, P. Y.; Chenet, D. A.; Berkelbach, T. C.; You, Y.; Lee, G.-H.; Heinz, T. F.; Reichman, D. R.; Muller, D. A.; Hone, J. C. *Nat. Mater.* **2013,** *12,* 554.
52. Yu, Y.; Li, C.; Liu, Y.; Su, L.; Zhang, Y.; Cao, L. *Sci. Rep.* **2013,** *3,* 1866.
53. Dumcenco, D.; Ovchinnikov, D.; Sanchez, O. L.; Gillet, P.; Alexander, D. T. L.; Lazar, S.; Radenovic, A.; Kis, A. *2D Mater.* **2015,** *2,* 044005.
54. Wu, S.; Huang, C.; Aivazian, G.; Ross, J. S.; Cobden, D. H.; Xu, X. *ACS Nano* **2013,** *7,* 2768.
55. Shi, Y.; Li, H.; Li, L.-J. *Chem. Soc. Rev.* **2015,** *44,* 2744.
56. Muratore, C.; Voevodin, A. A. *Thin Solid Films* **2009,** *517,* 5605.
57. Muratore, C.; Varshney, V.; Gengler, J. J.; Hu, J. J.; Bultman, J. E.; Smith, T. M.; Shamberger, P. J.; Qiu, B.; Ruan, X.; Roy, A. K.; Voevodin, A. A. *Appl. Phys. Lett.* **2013,** *102,* 081604.
58. Late, D. J.; Shaikh, P. A.; Khare, R.; Kashid, R. V.; Chaudhary, M.; More, M. A.; Ogale, S. B. *ACS Appl. Mater. Interfaces* **2014,** *6,* 15881.
59. Serrao, C. R.; Diamond, A. M.; Hsu, S.-L.; You, L.; Gadgil, S.; Clarkson, J.; Carraro, C.; Maboudian, R.; Hu, C.; Salahuddin, S. *Appl. Phys. Lett.* **2015,** *106,* 052101.
60. Tao, J.; Chai, J.; Lu, X.; Wong, L. M.; Wong, T. I.; Pan, J.; Xiong, Q.; Chi, D.; Wang, S. *Nanoscale* **2015,** *7,* 2497.
61. Noh, J.-Y.; Kim, H.; Kim, Y.-S. *Phys. Rev. B* **2014,** *89,* 205417.
62. Santosh, K. C.; Longo, R. C.; Addou, R.; Wallace, R. M.; Cho, K. *Nanotechnology* **2014,** *25,* 375703.
63. Addou, R.; Colombo, L.; Wallace, R. M. *ACS Appl. Mater. Interfaces* **2015,** *7,* 11921.

64. Komsa, H.-P.; Krasheninnikov, A. V. *Phys. Rev. B* **2015,** *91,* 125304.

65. Najmaei, S.; Yuan, J.; Zhang, J.; Ajayan, P.; Lou, J. *Acc. Chem. Res.* **2015,** *48,* 31.

66. Gao, D.; Si, M.; Li, J.; Zhang, J.; Zhang, Z.; Yang, Z.; Xue, D. *Nanoscale Res. Lett.* **2013,** *8,* 129.

67. Yang, Z.; Gao, D.; Zhang, J.; Xu, Q.; Shi, S.; Tao, K.; Xue, D. *Nanoscale* **2015,** *7,* 650.

68. Sun, B.; Li, Q. L.; Chen, P. *Micro Nano Lett.* **2014,** *9,* 468.

69. Ataca, C.; Ciraci, S. *J. Phys. Chem. C* **2011,** *115,* 13303.

70. Li, Y.; Zhou, Z.; Zhang, S.; Chen, Z. *J. Am. Chem. Soc.* **2008,** *130,* 16739.

71. Zhou, W.; Zou, X.; Najmaei, S.; Liu, Z.; Shi, Y.; Kong, J.; Lou, J.; Ajayan, P. M.; Yakobson, B. I.; Idrobo, J.-C. *Nano Lett.* **2013,** *13,* 2615.

72. Mouri, S.; Miyauchi, Y.; Matsuda, K. *Nano Lett.* **2013,** *13,* 5944.

73. Fang, H.; Tosun, M.; Seol, G.; Chang, T. C.; Takei, K.; Guo, J.; Javey, A. *Nano Lett.* **2013,** *13,* 1991.

74. Fuhr, J. D.; Saul, A.; Sofo, J. O. *Phys. Rev. Lett.* **2004,** *92,* 026802.

75. Wang, Y.; Li, S.; Yi, J. *Sci. Rep.* **2016,** *6,* 24153.

76. Cheng, Y. C.; Zhu, Z. Y.; Mi, W. B.; Guo, Z. B.; Schwingenschloegl, U. *Phys. Rev. B* **2013,** *87,* 100401.

77. Xu, W.-B.; Li, P.; Li, S.-S.; Huang, B.-J.; Zhang, C.-W.; Wang, P.-J. *Phys. E: Low Dimens. Syst. Nanostruct.* **2015,** *73,* 83.

78. Gong, Y.; Liu, Z.; Lupini, A. R.; Shi, G.; Lin, J.; Najmaei, S.; Lin, Z.; Elias, A. L.; Berkdemir, A.; You, G.; Terrones, H.; Terrones, M.; Vajtai, R.; Pantelides, S. T.; Pennycook, S. J.; Lou, J.; Zhou, W.; Ajayan, P. M. *Nano Lett.* **2014,** *14,* 442.

79. Xiang, Z.; Zhang, Z.; Xu, X.; Zhang, Q.; Wang, Q.; Yuan, C. *PCCP* **2015,** *17,* 15822.

80. Sun, X.; Dai, J.; Guo, Y.; Wu, C.; Hu, F.; Zhao, J.; Zeng, X.; Xie, Y. *Nanoscale* **2014,** *6,* 8359.

81. Li, B.; Huang, L.; Zhong, M.; Huo, N.; Li, Y.; Yang, S.; Fan, C.; Yang, J.; Hu, W.; Wei, Z.; Li, J. *ACS Nano* **2015,** *9,* 1257.

82. Suh, J.; Park, T.-E.; Lin, D.-Y.; Fu, D.; Park, J.; Jung, H. J.; Chen, Y.; Ko, C.; Jang, C.; Sun, Y.; Sinclair, R.; Chang, J.; Tongay, S.; Wu, J. *Nano Lett.* **2014,** *14,* 6976.

83. Zhang, K.; Feng, S.; Wang, J.; Azcatl, A.; Lu, N.; Addou, R.; Wang, N.; Zhou, C.; Lerach, J.; Bojan, V.; Kim, M. J.; Chen, L.-Q.; Wallace, R. M.; Terrones, M.; Zhu, J.; Robinson, J. A. *Nano Lett.* **2015,** *15,* 6586.

84. Jaramillo, T. F.; Jorgensen, K. P.; Bonde, J.; Nielsen, J. H.; Horch, S.; Chorkendorff, I. *Science* **2007,** *317,* 100.

85. Xie, J.; Zhang, H.; Li, S.; Wang, R.; Sun, X.; Zhou, M.; Zhou, J.; Lou, X. W.; Xie, Y. *Adv. Mater.* **2013,** *25,* 5807.

86. Xie, J.; Zhang, J.; Li, S.; Grote, F.; Zhang, X.; Zhang, H.; Wang, R.; Lei, Y.; Pan, B.; Xie, Y. *J. Am. Chem. Soc.* **2013,** *135,* 17881.

87. Lukowski, M. A.; Daniel, A. S.; Meng, F.; Forticaux, A.; Li, L.; Jin, S. *J. Am. Chem. Soc.* **2013,** *135,* 10274.

88. Hwang, H.; Kim, H.; Cho, J. *Nano Lett.* **2011,** *11,* 4826.

89. Huang, G.; Chen, T.; Chen, W.; Wang, Z.; Chang, K.; Ma, L.; Huang, F.; Chen, D.; Lee, J. Y. *Small* **2013,** *9,* 3693.

90. Liu, Y.-T.; Zhu, X.-D.; Duan, Z.-Q.; Xie, X.-M. *Chem. Commun.* **2013,** *49,* 10305.

91. Feng, J.; Sun, X.; Wu, C.; Peng, L.; Lin, C.; Hu, S.; Yang, J.; Xie, Y. *J. Am. Chem. Soc.* **2011,** *133,* 17832.

92. da Silveira Firmiano, E. G.; Rabelo, A. C.; Dalmaschio, C. J.; Pinheiro, A. N.; Pereira, E. C.; Schreiner, W. H.; Leite, E. R. *Adv. Energy Mater.* **2014**, *4*, 1301380.

93. Lin, J.-Y.; Chan, C.-Y.; Chou, S.-W. *Chem. Commun.* **2013**, *49*, 1440.

94. Chen, Y.; Tan, C.; Zhang, H.; Wang, L. *Chem. Soc. Rev.* **2015**, *44*, 2681.

95. Loo, A. H.; Bonanni, A.; Ambrosi, A.; Pumera, M. *Nanoscale* **2014**, *6*, 11971.

96. Donarelli, M.; Prezioso, S.; Perrozzi, F.; Bisti, F.; Nardone, M.; Giancaterini, L.; Cantalini, C.; Ottaviano, L. *Sens. Actuators, B* **2015**, *207*, 602.

97. Liu, T.; Wang, C.; Gu, X.; Gong, H.; Cheng, L.; Shi, X.; Feng, L.; Sun, B.; Liu, Z. *Adv. Mater.* **2014**, *26*, 3433.

98. Liu, T.; Wang, C.; Cui, W.; Gong, H.; Liang, C.; Shi, X.; Li, Z.; Sun, B.; Liu, Z. *Nanoscale* **2014**, *6*, 11219.

99. Yin, W.; Yan, L.; Yu, J.; Tian, G.; Zhou, L.; Zheng, X.; Zhang, X.; Yong, Y.; Li, J.; Gu, Z.; Zhao, Y. *ACS Nano* **2014**, *8*, 6922.

100. Liu, L.; Kumar, S. B.; Ouyang, Y.; Guo, J. *IEEE Trans. Electron Devices* **2011**, *58*, 3042.

101. Yoon, Y.; Ganapathi, K.; Salahuddin, S. *Nano Lett.* **2011**, *11*, 3768.

102. Li, X.; Mullen, J. T.; Jin, Z.; Borysenko, K. M.; Nardelli, M. B.; Kim, K. W. *Phys. Rev. B* **2013**, *87*, 115418.

103. Kaasbjerg, K.; Thygesen, K. S.; Jacobsen, K. W. *Phys. Rev. B* **2012**, *85*, 115317.

104. Liu, L.; Lu, Y.; Guo, J. *IEEE Trans. Electron Devices* **2013**, *60*, 4133.

105. Podzorov, V.; Gershenson, M. E.; Kloc, C.; Zeis, R.; Bucher, E. *Appl. Phys. Lett.* **2004**, *84*, 3301.

106. Novoselov, K. S.; Jiang, D.; Schedin, F.; Booth, T. J.; Khotkevich, V. V.; Morozov, S. V.; Geim, A. K. *Proc. Natl. Acad. Sci. USA* **2005**, *102*, 10451.

107. Radisavljevic, B.; Radenovic, A.; Brivio, J.; Giacometti, V.; Kis, A. *Nat. Nanotechnol.* **2011**, *6*, 147.

108. Lee, K.; Kim, H.-Y.; Lotya, M.; Coleman, J. N.; Kim, G.-T.; Duesberg, G. S. *Adv. Mater.* **2011**, *23*, 4178.

109. Li, S. L.; Wakabayashi, K.; Xu, Y.; Nakaharai, S.; Komatsu, K.; Li, W. W.; Lin, Y. F.; Aparecido-Ferreira, A.; Tsukagoshi, K. *Nano Lett.* **2013**, *13*, 3546.

110. Min, S. W.; Lee, H. S.; Choi, H. J.; Park, M. K.; Nam, T.; Kim, H.; Ryu, S.; Im, S. *Nanoscale* **2013**, *5*, 548.

111. Yang, R.; Wang, Z. H.; Feng, P. X. L. *Nanoscale* **2014**, *6*, 12383.

112. Lin, M. W.; Kravchenko, II; Fowlkes, J.; Li, X. F.; Puretzky, A. A.; Rouleau, C. M.; Geohegan, D. B.; Xiao, K.. *Nanotechnology* **2016**, *27*, 165203.

113. Lembke, D.; Allain, A.; Kis, A. *Nanoscale* **2015**, *7*, 6255.

114. Bertolazzi, S.; Brivio, J.; Kis, A. *ACS Nano* **2011**, *5*, 9703.

115. Pu, J.; Yomogida, Y.; Liu, K.-K.; Li, L.-J.; Iwasa, Y.; Takenobu, T. *Nano Lett.* **2012**, *12*, 4013.

116. Chang, H.-Y.; Yang, S.; Lee, J.; Tao, L.; Hwang, W.-S.; Jena, D.; Lu, N.; Akinwande, D. *ACS Nano* **2013**, *7*, 5446.

117. Yin, Z.; Li, H.; Li, H.; Jiang, L.; Shi, Y.; Sun, Y.; Lu, G.; Zhang, Q.; Chen, X.; Zhang, H. *ACS Nano* **2012**, *6*, 74.

118. Lee, H. S.; Min, S.-W.; Chang, Y.-G.; Park, M. K.; Nam, T.; Kim, H.; Kim, J. H.; Ryu, S.; Im, S. *Nano Lett.* **2012**, *12*, 3695.

119. Baibich, M. N.; Broto, J. M.; Fert, A.; Vandau, F. N.; Petroff, F.; Eitenne, P.; Creuzet, G.; Friederich, A.; Chazelas, J. *Phys. Rev. Lett.* **1988**, *61*, 2472.

120. Dieny, B.; Speriosu, V. S.; Parkin, S. S. P.; Gurney, B. A.; Wilhoit, D. R.; Mauri, D. *Phys. Rev. B* **1991**, *43*, 1297.
121. Wang, W.; Narayan, A.; Tang, L.; Dolui, K.; Liu, Y.; Yuan, X.; Jin, Y.; Wu, Y.; Rungger, I.; Sanvito, S.; Xiu, F. *Nano Lett.* **2015**, *15*, 5261.
122. Datta, S.; Das, B. *Appl. Phys. Lett.* **1990**, *56*, 665.
123. Awschalom, D. D.; Flatte, M. E. *Nat. Phys.* **2007**, *3*, 153.
124. Zibouche, N.; Philipsen, P.; Kuc, A.; Heine, T. *Phys. Rev. B* **2014**, *90*, 125440.
125. Li, H.; Shao, J.; Yao, D.; Yang, G. *ACS Appl. Mater. Interfaces* **2014**, *6*, 1759.
126. Gong, K.; Zhang, L.; Liu, D.; Liu, L.; Zhu, Y.; Zhao, Y.; Guo, H. *Nanotechnology* **2014**, *25*, 435201.
127. Chen, J.-R.; Odenthal, P. M.; Swartz, A. G.; Floyd, G. C.; Wen, H.; Luo, K. Y.; Kawakami, R. K. *Nano Lett.* **2013**, *13*, 3106.
128. Dankert, A.; Langouche, L.; Kamalakar, M. V.; Dash, S. P. *ACS Nano* **2014**, *8*, 476.
129. Xiao, D.; Liu, G.-B.; Feng, W.; Xu, X.; Yao, W. *Phys. Rev. Lett.* **2012**, *108*, 196802.
130. Zeng, H.; Dai, J.; Yao, W.; Xiao, D.; Cui, X. *Nat. Nanotechnol.* **2012**, *7*, 490.
131. Mak, K. F.; He, K.; Shan, J.; Heinz, T. F. *Nat. Nanotechnol.* **2012**, *7*, 494.
132. Zhu, B.; Zeng, H.; Dai, J.; Gong, Z.; Cui, X. *Proc. Natl. Acad. Sci. USA* **2014**, *111*, 11606.
133. Wu, S.; Ross, J. S.; Liu, G.-B.; Aivazian, G.; Jones, A.; Fei, Z.; Zhu, W.; Xiao, D.; Yao, W.; Cobden, D.; Xu, X. *Nat. Phys.* **2013**, *9*, 149.

CHAPTER 2

TWO-DIMENSIONAL ELECTRON GAS AT LaAlO$_3$/SrTiO$_3$ INTERFACES

ZHIQI LIU[1*], ANIL ANNADI[2], and ARIANDO[2*]

[1]*School of Materials Science and Engineering, Beihang University, Beijing 100191, China*

[2]*Department of Physics, National University of Singapore, Singapore 119260, Singapore*

Corresponding authors. E-mail: zhiqi@buaa.edu.cn; ariando@nus. edu.sg

CONTENTS

ABSTRACT

In this chapter, we review the status and prospects of the two-dimensional electron gas (2DEG) at $LaAlO_3$/$SrTiO_3$ interfaces. We start by a brief introduction, and then turn to the fabrication of such heterostructures, emphasizing the significant role of fabrication conditions in its structural and physical properties. The main body of the review chapter is devoted to the understanding of abundant physical properties and origins of the oxide 2DEG. Subsequently, we will discuss the applications of the oxide 2DEG in oxide electronic devices. Finally, prospects and existing challenges in this field will be reviewed.

Strongly correlated complex oxides exhibit luxuriously abundant and exotic physical properties (such as high-temperature superconductivity [1,2], metal–insulator transition [MIT] [3,4], colossal magnetoresistance (MR) [5,6], electronic phase separation [7–9], ferroelectricity [10], and multiferroicity [11]) owing to their interacting and competing lattice, charge, spin, and orbital degrees of freedom. They have broad applications in both large scale (such as the Maglev train system, electricity grids, and high magnetic field generators) and various microscopic electronic devices (such as complementary metal–oxide–semiconductor, optoelectronic, and memory devices), as well as chemical catalysis, lithium batteries, and solar energy conversion.

Advances in the heteroepitaxy of complex oxides provide fantastic feasibility of fabricating epitaxial oxide heterostructures including strongly correlated oxides with atomic precision. In the last decade, interface electronics in epitaxial complex oxide heterostructures has been a research focus in oxide electronics. In fact, it had been significantly stirred by novel functionalities created at interfaces that do not exist in the bulk in nature [12]. For example, ferroelectricity emerges in $SrTiO_3$ thin films grown on $DyScO_3$ substrates owning to interface strain [13]; the bulk polarization of $BiFeO_3$ can be modulated by interface coupling in $BiFeO_3$/$La_{0.7}Sr_{0.3}MnO_3$ heterostructures [14]; and a magnetic proximity effect shows up in $YBa_2Cu_3O_7$/$La_{0.7}Ca_{0.3}MnO_3$ superlattices [15].

Particularly, the discovery of a high-mobility two-dimensional electron gas (2DEG) at the interface between two complex oxide insulators $SrTiO_3$ (STO) and $LaAlO_3$ (LAO) [16] fires the field of interface electronics in oxides. On one hand, high-mobility 2DEGs based on conventional semiconductors such as Si and GaAs have been utilized in high-mobility transistors that have built the entire semiconductor industry and enabled the discovery of the integer and fractional quantum Hall effects. On the other hand, the 2DEG at the LAO/STO interface exhibits a variety of functional properties such as superconductivity [17], the magnetic Kondo effect [18], and electronic phase separation [9] that are unachievable in conventional semiconductor 2DEGs. Hence, the oxide 2DEG is of significant interest in fundamental science and technology.

In this chapter, we will review the status and prospects of the 2DEG at LAO/STO interfaces. We start by a brief introduction, and then turn to the fabrication of such heterostructures, emphasizing the significant role of fabrication conditions in its structural and physical properties. The body of the review chapter is devoted to the understanding of abundant physical properties and origins of this oxide 2DEG. Subsequently, we will discuss the application of the oxide 2DEG in oxide electronic devices. Finally, prospects and existing challenges in this field will be reviewed.

2.1 INTRODUCTION

In 2004, Ohtomo and Hwang reported a high-mobility metallic 2DEG at the atomically flat interface between two oxide insulators LAO and STO [16]. STO is a band insulator with an indirect band gap of 3.25 eV, and LAO is a band insulator as well with a large band gap of 5.6 eV. It was therefore quite unexpected that a 2DEG emerged at the interface between these two insulators (Fig. 2.1). In addition, it was found that such a 2DEG is strictly sensitive to the STO surface termination when the LAO layer is deposited on top, that is, the 2DEG can only be realized in TiO_2-terminated but not for SrO-terminated STO substrates.

Two years later in 2006, Nakagawa et al. [19] proposed the polarization catastrophe model to explain the emergence of the 2DEG. As for this model, the (0 0 1) planes of ABO_3 perovskites can be divided into alternating AO and BO_2 planes. Correspondingly, STO can be thought as a perovskite containing alternating SrO and TiO_2 layers. Such sublayers are charge neutral and therefore STO is defined as a nonpolar insulator. On the contrary, LAO containing charged $(LaO)^{1+}$ and $(AlO_2)^{1-}$ sublayers is called as a polar insulator. What will happen at the interface if a polar oxide insulator faces a nonpolar insulator? Nakagawa et al. proposed that an interface electronic reconstruction will occur in this case to avoid the infinite potential build-up (Fig. 2.2a) in the polar LAO layers when a LAO film is deposited on a STO substrate. When a $(LaO)^{1+}$ layer is grown on a TiO_2-terminated STO substrate, half electron will be transferred from the $(LaO)^{1+}$ layer to the TiO_2 layer, resulting in the $(Ti^{3.5+}O_2^{4-})$ structure at the surface of STO. The interface STO layer becomes conducting as a result of free electrons occupying Ti 3d orbitals.

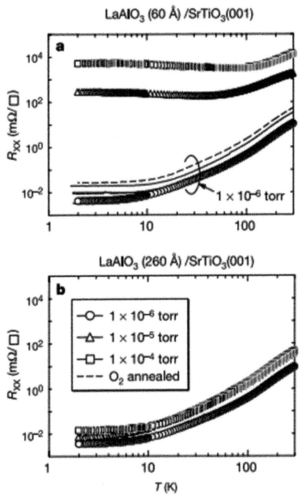

FIGURE 2.1 Temperature-dependent sheet resistance of (a) a 60-Å-thick and (b) a 260-Å-thick heterostructures fabricated at different oxygen partial pressures (reproduced from Ref. [16] with permission). Note: according to the corrigendum associated with Ref. [16], the y-axis of the above figures should be in units of ohm per square rather than milliohm per square. (Reprinted by permission from Macmillan Publishers Ltd: Ohtomo, A.; Hwang, H. Y. A high-mobility electron gas at the LaAlO: 3: /SrTiO: 3: heterointerface. *Nature* 2004, *427*, 423. http://www.nature.com/nature/journal/v427/n6973/abs/nature02308. html?foxtrotcallback=true.)

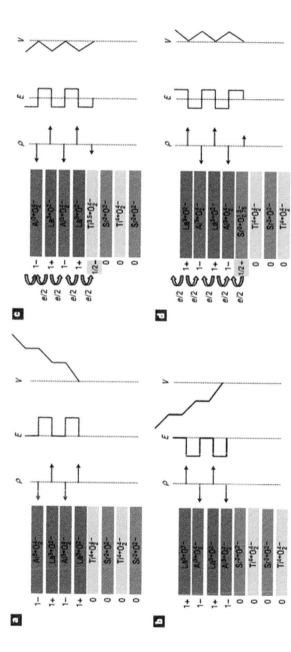

FIGURE 2.2 The polar catastrophe illustrated for atomically abrupt (0 0 1) interfaces between LAO and STO. (a) The unreconstructed interface has neutral (0 0 1) planes in STO, but the (0 0 1) planes in LAO have alternating net charges (ρ). This produces a nonnegative electric field (E), leading in turn to an electric potential (V) that diverges with thickness. (b) If the interface is instead placed at the AlO₂/SrO/TiO₂ plane, the potential diverges negatively. (c) The divergence catastrophe at the AlO₂/LaO/TiO₂ interface can be avoided if half an electron is added to the last Ti layer. This produces an interface dipole that causes the electric field to oscillate about 0 and the potential remains finite. The upper free surface is not shown, but in this simple model, the uppermost AlO₂ layer would be missing half an electron, which would bring the electric field and potential back to zero at the upper surface. (d) The divergence for the AlO₂/SrO/TiO₂ interface can also be avoided by removing half an electron from the SrO plane in the form of oxygen vacancies (reproduced from Ref. [19] with permission). (Reprinted by permission from Macmillan Publishers Ltd: Nakagawa, N.; Hwang, H. Y.; Müller, D. A. Why some interfaces cannot be sharp.*Nat. Mater.* 2006, 5, 204. http://www.nature.com/nmat/journal/v5/n3/full/nmat1569.html)

Later in 2006, Thiel et al. [20] found that a minimum thickness of 4 unit cell (uc) for the LAO layer is essential for the formation of a conducting interface. Density functional theory calculations based on the polarization catastrophe model by Pentcheva and Pickett [21] realized this critical thickness (Fig. 2.3). Therefore, these experimental and theoretical studies together strongly supported the polarization catastrophe model and the interface electronic reconstruction at the LAO/STO interface. Nevertheless, shortly upon the report of the 2DEG at this oxide interface system, some other defect-related mechanisms for the appearance of the interface conduction were also proposed. For example, oxygen vacancies in STO [22–25] can easily generate conductivity as they are shallow electron donors for STO [26]; another mechanism was proposed by Willmott et al. [27] based on structural analysis at the interface: the thermal intermixing of Ti/Al or Sr/La atoms at the interface as La-doped STO is conducting as well. As the origin of the 2DEG at the LAO/STO interface has been intensively explored in these years, we are able to clarify the controversy on this issue in this review, which will be elaborated in this chapter.

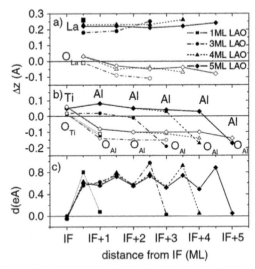

FIGURE 2.3 Vertical displacements of ions ΔZ in the (a) AO and (b) BO_2 layers with respect to the bulk positions in Å for 1–5 MLs of LAO and STO (0 0 1) obtained within GGA. Cation/oxygen relaxations are marked by full black/open red (gray) symbols. The corresponding layer-resolved dipole moments are displayed in (c). The x-axis shows the distance from the interface (I) TiO_2-layer (reproduced from Ref. [21] with permission). (Reprinted with permission from Pentcheva, R.; Pickett, W. E. Avoiding the polarization catastrophe in LaAlO3 overlayers on SrTiO3(001) through polar distortion. *Phys. Rev. Lett.* 2009, *102*, 107602.Copyright © 2009, American Physical Society.)

Regarding the ground state of the 2DEG at the LAO/STO interface, there were two interesting experimental perspectives unraveled in 2007. One was a resistance minimum (Fig. 2.4) observed by Brinkman et al. [18], which was attributed to the magnetic Kondo scattering. Furthermore, the hysteresis in the MR measurements suggested the magnetic order at the interface [18]. The other was a superconducting state found below 1 K by Reyren et al. [17], which was later explored to be electrically tunable by a back gate voltage [28] (Fig. 2.5). Naturally, ferromagnetism and superconductivity are antagonistic as spins require a parallel configuration for the magnetic order but antiparallel in the superconducting state for the sake of the cooper pair formation. Thus, whether the ground state of the 2DEG is ferromagnetic or superconducting had been a hotly debated issue in the field for a long time.

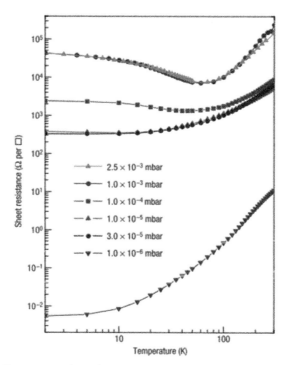

FIGURE 2.4 Temperature dependence of the sheet resistance, RS, for *n*-type LAO/STO conducting interfaces, grown at various oxygen partial pressures (reproduced from Ref. [18] with permission). (Reprinted by permission from Macmillan Publishers Ltd: Brinkman, A.; Huijben, M.; van Zalk, M.; Huijben, J.; Zeitler, U.; Maan, J. C.; van der Wiel, W. G.; Rijnders, G.; Blank, D. H. A.; Hilgenkamp, H. Magnetic effects at the interface between non-magnetic oxides. *Nat. Mater.* 2007, *6*, 493. http://www.nature.com/nmat/journal/v6/n7/full/nmat1931. html.)

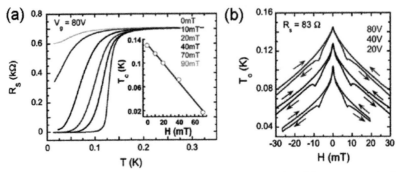

FIGURE 2.5 (a) Superconducting transition for $V_g = 80$ V, at a few different magnetic fields. Inset: T_c versus H obtained from these data. T_c is defined by the midpoint of the resistive transition. The line represents a linear fit to the data points with a slope of 1.6 mK mT^{-1}, giving a zero temperature coherence length of $\xi_0 = 64$ nm. (b) Phase diagrams at a bias point of $R_s = 83$ Ω, at the foot of the resistive transition, at three different gate voltages. Data for $V_g = 40$ V are shifted by 10 mK and data for $V_g = 80$ V by 20 mK for clarity. (Reprinted with permission from Dikin, D. A.; Mehta, M.; Bark, C. W.; Folkman, C. M.; Eom, C. B.; Chandrasekhar, V. Coexistence of Superconductivity and Ferromagnetism in Two Dimensions. *Phys. Rev. Lett.* 2011, *107*, 056802.© 2011 American Physical Society.)

The breakthrough in this issue came in 2011, when Ariando et al. [9] observed the coexistence of the ferromagnetic order and a superconducting-like state at the interface based on magnetometer measurements and proposed electronic phase separation to explain the observed phenomenon. Later in the same year, several experimental groups [29–31] reported the coexistence of ferromagnetism and superconductivity in this 2D system. Such experimental work excited the surge of magnetic studies on this interface system and theoretical studies on the coexistence of ferromagnetism and superconductivity. We will review these studies in this chapter as well.

Until 2012, almost all the studies on the LAO/STO interface had been focusing on the (0 0 1) orientation. That is because from the first beginning, the polarization catastrophe model was proposed to explain the emergence of the conductivity, where polar discontinuity is essential. A (0 0 1)-oriented STO single crystal consisting of alternating $(SrO)^0$ and $(TiO_2)^0$ layers is nonpolar. However, a (1 1 0)-oriented STO single crystal can be ionically divided into alternating $(O_2)^{4-}$ and $(SrTiO)^{4+}$ layers and thus it is polar. Similarly, a (1 1 1)-oriented STO single consisting of $(Ti)^{4+}$ and $(SrO_3)^{4-}$ is also polar. As schematized in Figure 2.6, (1 1 0)-oriented LAO films grown on (1 1 0)-oriented STO substrates similarly have $(O_2)^{4-}$ and $(LaAlO)^{4+}$ layers and therefore there is no polar discontinuity at the interface. Correspondingly, the interface electronic reconstruction would not occur and no interface conductivity could be expected.

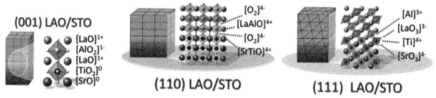

FIGURE 2.6 Schematics of LAO/STO interfaces with different orientations (reproduced from Ref. [32] with permission). (Reprinted from Herranz, G.; Sánchez, F.; Dix, N.; Scigaj, M.; Fontcuberta, J. High mobility conduction at (110) and (111) LaAlO3/SrTiO3 interfaces. *Sci. Rep.* 2012, *2*, 758. https://www.nature.com/articles/srep00758. https://creativecommons. org/licenses/by-nc-sa/3.0/. Ref. [32].)

However, in fact, the interface conductivity at the LAO/STO (1 1 0) interface has been observed by different groups. Herranz et al. [32] reported the high-mobility interface conduction at (1 1 0) and (1 1 1) LAO/STO interfaces. Annadi et al. [33] observed the anisotropic 2DEG at the LAO/STO (1 1 0) interface and suggested that the anisotropic interface conductivity still originates from polar discontinuity as a consequence of a buckled interface. In this review chapter, we will also include recent studies regarding such an emerging interface along these orientations.

2.2 FABRICATION

LAO/STO heterostructures have been predominantly fabricated by pulsed laser deposition (PLD) equipped with reflection high-energy electron diffraction (RHEED) due to its powerful capability of integrating distinct oxides with atomic precision. STO is a cubic perovskite with a lattice constant of 3.905 Å. LAO has a rhombohedrally distorted perovskite structure with a lattice constant of 5.357 Å and its pseudocubic cell has a lattice constant of 3.79 Å. The relative small lattice mismatch (~3%) between LAO and STO enables epitaxial growth of LAO films on STO substrates. Moreover, the similar crystal structure of LAO and STO allows a 2D layer-by-layer growth mode in certain optimized growth conditions, which provides the fantastic viability of fabricating atomically flat LAO/STO heterointerfaces. The layer-by-layer growth can be monitored by RHEED oscillations.

An interface between a LAO film and a TiO₂-terminated STO substrate is conducting, while an interface between a LAO film and a SrO-terminated STO substrate is insulating with much higher resistance [34]. The former is refereed as an *n*-type interface because the dominant carriers are electrons; the latter is called as a "*p*-type" interface since holes are expected in the

SrO layer to balance the polar discontinuity based on the polar catastrophe model (anyway, no reliable Hall data has been reported for this type of interface to prove that the conduction is intrinsically *p*-type, mainly due to the large resistance). The full TiO_2 termination on the STO surface can be realized by buffered HF treatment followed by thermal annealing (typically at 950°C in the oxygen atmosphere for 2 h). An atomic force image (AFM) of a TiO_2-terminated STO substrate is shown in Figure 2.7. The step height is 1-uc-high and the step width depends on the miscut angle. The larger the miscut angle, the smaller the step width is. The miscut angle α can be simply estimated by the geometric relationship $\tan(\alpha)$ = (step height)/(step width) = 3.905 Å/(step width).

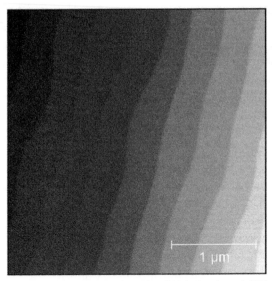

FIGURE 2.7   AFM image of a TiO_2-terminated and (0 0 1)-oriented STO substrate.

Nevertheless, up to now, any chemical or thermal treatment approach to achieve the full SrO termination on the STO surface is still lacking. Alternatively, the single SrO termination can be realized by depositing one monolayer (ML) of SrO on a TiO_2-terminated STO surface. In this section, we will focus on the fabrication of *n*-type interfaces that have been mostly studied.

Figure 2.8 shows the RHEED patterns of a TiO_2-terminated STO (0 0 1) substrate surface at different temperatures (a)–(c) and after the deposition of a 9-uc LAO layer (d). In the RHEED patterns at low temperatures, sharp diffraction points as well as Kikuchi lines are visible. As the temperature of the STO substrate increases to 750°C, the RHEED pattern becomes streaky. That is because the electron beam energy fluctuation increases as a result of the high temperature and thermal disturbance. In addition, streaky RHEED patterns indicate a flat sample surface and also a dominant 2D surface diffraction.

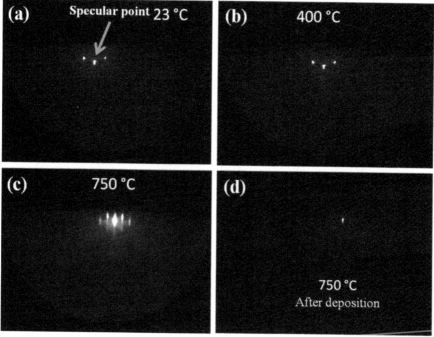

FIGURE 2.8   RHEED patterns of a TiO$_2$-terminated STO (0 0 1) surface at different temperatures (a)–(c) and after the deposition of a 9-uc LAO layer (d).

The RHEED intensity by integrating over a selected area on the specular spot during the deposition of the LAO layer is recorded in Figure 2.9. Each oscillation corresponds to 1-uc LAO. The periodic RHEED intensity oscillations indicate a layer-by-layer growth mode. After the deposition, the RHEED pattern shown in Figure 2.8d is obviously streaky, suggesting a flat surface of the LAO film. In what follows, we will summarize the role of several key deposition parameters (oxygen partial pressure, temperature, and laser fluence) and STO substrate itself in the fabrication of LAO/STO heterostructures.

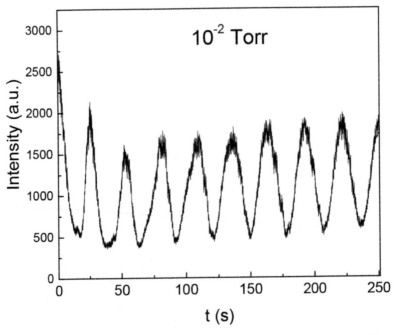

FIGURE 2.9   RHEED oscillations of a LAO film grown on a TiO_2-terminated STO (0 0 1) substrate at 750°C and 10^{-2} Torr oxygen partial pressure.

2.2.1   OXYGEN PARTIAL PRESSURE

Oxygen gas in the PLD chamber is essential for oxide thin film growth in PLD to maintain the correct oxygen composition in the deposited films. Oxygen partial pressure during the deposition directly affects the oxygen composition and hence the lattice structure of oxide films. Moreover, it also plays an important role in the defect formation in thin films. For example, oxygen vacancies could emerge at low oxygen partial pressure ($<10^{-4}$ Torr) while cationic defects tend to appear at high oxygen pressure ($>10^{-3}$ Torr). Any defect created in LAO films during deposition owing to the oxygen partial pressure cannot generate conductivity. That is because the energy levels of defects in LAO are at least 2 eV below its conduction band minimum [35,36], which are too low to serve as donors at room temperature.

However, the oxygen partial pressure during the LAO deposition at high temperatures (around 800°C) can largely affect the conductivity of STO substrates as oxygen vacancies are shallow electron donors for STO. STO single-crystal substrates become conducting even though a small amount of oxygen vacancies (less than 1%) is created. A low oxygen partial pressure (<10^{-4} Torr) during high-temperature deposition could induce oxygen vacancies in the STO bulk, leading to 3D metallicity in STO. In contrast, a STO single-crystal substrate annealed at oxygen partial pressures higher than 10^{-4} Torr and the typical deposition temperature is highly insulating (>GΩ). Hence, to avoid the introduction of 3D metallicity in the STO bulk, the LAO/STO heterostructures are generally fabricated at relatively high oxygen partial pressure (>10^{-4} Torr).

On the other hand, oxygen partial pressure affects the growth mode of thin films as well. For LAO films grown on STO substrates, 2D layer-by-layer growth can be achieved typically at 10^{-2} Torr oxygen pressure and below, while LAO films deposited at oxygen pressures above 10^{-2} Torr grow in a different fashion—3D island growth. For example, Maurice et al. [37] demonstrated that the surface of LAO/STO heterostructures fabricated at 40 Pa (3 × 10^{-1} Torr) exhibits 3D islands with large roughness and more importantly these high-pressure heterostructures are macroscopically insulating. Different types of conduction and growth mode for n-type LAO/STO interfaces are briefly schematized in Figure 2.10.

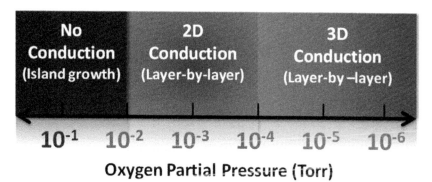

FIGURE 2.10 Categories of conduction and growth mode of a LAO film deposited on a TiO$_2$-terminated STO substrate depending on the deposition oxygen partial pressure.

2.2.2 GROWTH TEMPERATURE

A high temperature of a substrate surface is needed for oxide thin film growth so that atoms can obtain enough energy and mobility to crystallize. On the other hand, the substrate surface temperature has a large effect on the nucleation density that increases the smoothness of the deposited film. Generally, the nucleation density decreases with increasing temperature. Therefore, the growth temperature should be optimized to achieve high-crystallinity and high-quality films. PLD-deposited LAO films start to crystallize from 500°C. Most of LAO/STO heterostructures have been fabricated at the temperature range of 750–850°C. In addition, Caviglia et al. [38] reported that a reduced growth temperature of 650°C enhanced the crystalline quality of LAO films and the mobility of the 2DEG.

2.2.3 LASER FLUENCE

One of the most fascinating capabilities of PLD is that it can replicate the composition of bulk targets in thin films. To achieve this, two factors are crucial: laser fluence and the deposition angle (the angle between the target surface normal and the substrate–target line). Generally, a high laser fluence (e.g., for $YBa_2Cu_3O_7$, laser fluence should be higher than 1 J cm^{-2}) and a deposition angle smaller than 20° are needed to obtain stoichiometric films [39]. However, too high laser fluence could induce defects in substrates due to the bombardment of arriving high energy species. Especially for STO substrates, too high laser fluence could create oxygen vacancies, thus leading to bulk conductivity. Typical laser fluence adopted for LAO/STO heterostructure fabrication is 1–2 J cm^{-2}.

However in reality, exact replication of the composition of a target in thin films is nontrivial, which is affected by background gas pressure and substrate–target distance as well. The extension length and scope of laser plume strongly depend on background gas pressure and the substrate–target distance determines the cross section of laser plume that reaches on the substrate surface. From these aspects, laser fluence is not an independent parameter in PLD. Instead, it is closely associated with oxygen partial pressure and substrate–target distance. For example, Breckenfeld et al. [40] found that at 10^{-3} Torr and a substrate–target distance of 6.6 cm the cationic ration in LAO films grown on STO substrates can be tuned by laser fluence. Stoichiometric LAO films can only be achieved by a laser fluence of 1.6 J cm^{-2}, below which LAO films are Al-deficient while above which films are La-deficient.

2.2.4 SUBSTRATE MISCUT ANGLE

The TiO$_2$-terminated STO substrates used for n-type LAO/STO interface fabrication are atomically flat and any issue regarding the surface roughness is negligible. However, the substrate miscut angle could largely affect the growth and properties of LAO/STO interfaces. Straightforwardly, the miscut angle determines the step width of the TiO$_2$ termination. STO substrates with small vicinal angles (<0.2°, correspondingly the step width >100 nm) are commonly used for LAO/STO interface preparation. As reported by Fix et al. [41], a large vicinal angle of a STO substrate such as 8° can induce lattice distortions and vertical domain boundaries at the step edges in LAO films. On the other hand, highly miscut STO substrates result in mobility enhancement and carrier density suppression at low temperatures in the 2DEG at the LAO/STO interface. Basically, the resulted step edges due to the miscut of substrates plays a role in the scattering process at these step edges and thus affects the mobility of 2DEG. Overall, the STO miscut angle is another parameter that one can deal with to modulate the properties of the 2DEG.

2.3 ELECTRICAL PROPERTIES

2.3.1 CRITICAL THICKNESS

Not all the *n*-type LAO/STO interfaces are conducting. Instead, a minimum thickness of the LAO layer is needed for the appearance of conductivity. Thiel et al. [20] found that a minimum thickness of 4 unit cell (uc) is essential for the conductivity in oxygen-annealed LAO/STO heterostructures. Below 4 uc, the heterostructure is highly insulating and the carrier density is immeasurable, as discussed previously. Across the critical thickness, the MIT occurs with the conductance change over four orders of magnitude. Above the critical thickness, the room-temperature sheet carrier density is 1.5–2 × 10^{13} cm^{-2}, which is almost independent of the LAO layer thickness.

As the samples in the work of Thiel et al. were annealed at 600°C and 400 mbar oxygen for 1 h after deposition, the contribution of oxygen vacancies in STO substrates to the interface conductivity is negligible. The critical thickness of 4 uc can be well understood by the electrostatic charge transfer (polar catastrophe) model. As schematized in Figure 2.11, the charge transfer at the interface needed to avert the polar catastrophe is 0.5 e uc^{-1} or $\sigma = 0.5$ C m^{-2}, which results in an internal electric field in LAO of $E_{LAO} = \sigma/\varepsilon_0\varepsilon_{LAO} = 2.4$ V nm$^{-1} \approx 0.9$ V uc^{-1} ($\varepsilon_{LAO} = 24$). At a critical thickness, the valence O 2p band of LAO reaches the level of the STO 3d conduction band at the interface. Above the critical thickness, electrons will progressively transfer from the (LaO)$^+$ layer to the TiO$_2$ layer. The critical thickness is $\Delta E/(E_{LAO}e) = 3.3$ eV/(0.9 eV uc^{-1}) = 3.6 uc, where ΔE is the energy different between the LAO valence band and the STO conduction band [42]. This is in excellent agreement with the experiment. Meanwhile, a critical thickness was also realized in theoretical calculations [21] based on the polar catastrophe model, where the band gap in density of states closes with adding LAO layers (Fig. 2.12).

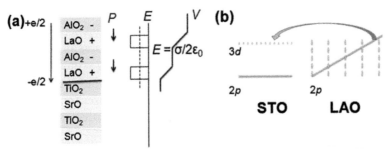

FIGURE 2.11 Schematics of the charge transfer model. (a) Half electron needs to be transferred from the $(LaO)^+$ layer to the TiO_2 layer to avoid the polar catastrophe in LAO. (b) The internal electrical field in LAO resulting from the charge transfer is ~0.9 eV uc^{-1}, which causes the band bending in LAO. Consequently, 4-uc LAO induces the charge transfer from the LAO valence band to the STO conduction band.

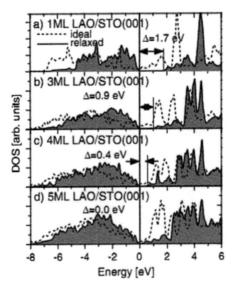

FIGURE 2.12 Density of states for the ideal (dashed line) and relaxed (solid line, gray filling) structure of 1–5 uc LAO on STO (0 0 1). Relaxation opens a band gap, but its size decreases with each added LAO layer (reproduced from Ref. [21] with permission). (Reprinted with permission from Pentcheva, R.; Pickett, W. E. Avoiding the Polarization Catastrophe in LaAlO3 Overlayers on SrTiO3(001) through Polar Distortion. *Phys. Rev. Lett.* 2009, *102*, 107602. © 2009 by the American Physical Society.)

The critical thickness for oxygen-annealed samples is found to be universal for other deposition oxygen pressures [43]. However, the oxygen vacancies in STO substrates could affect the critical thickness significantly in unannealed LAO/STO heterostructures. For example, we found that an unannealed 3-uc-thick LAO/STO heterostructure fabricated at 10^{-4} Torr and 750°C (laser fluence 1.3 J cm^{-2} and substrate–target distance 65 mm) is conducting with the room-temperature sheet carrier density of 8.8×10^{12} cm^{-2} (Fig. 2.13). Furthermore, if one performs deposition at 10^{-5} Torr oxygen pressure and below, oxygen vacancies will be introduced into the STO bulk and consequently the critical thickness could be even less.

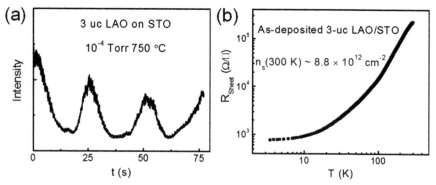

FIGURE 2.13 (a) RHEED oscillations of a 3-uc-thick LAO film grown on a TiO₂-terminated STO substrate in 10^{-4} Torr oxygen pressure at 750°C. (b) Temperature-dependent sheet resistance of an as-deposited 3-uc LAO/STO heterostructure.

2.3.2 CONFINEMENT DEPTH

2DEG means that the motion of electrons is confined in one of the three dimensions. For the LAO/STO interface, the electrons can freely move in the sample plane but are strongly restricted along the out-of-plane direction. A fundamental question related to the 2DEG is "how two-dimensional is it?," in another words, what is the confinement depth of the 2DEG? Regarding this question, both experimental and theoretical studies have been performed.

In 2008, Balestic et al. [44] mapped the spatial distribution of carriers in LAO/STO heterostructures by conducting-tip atomic force microscopy and found that the conducting depth of an oxygen-annealed interface is ~7 nm on the STO side (Fig. 2.14), corresponding to 18 uc of STO. Moreover, the 2DEG does not spread away from the interface but are still confined within

10 nm [45] at low temperatures. The Fermi wavelength estimated from the Hall measurements is ~16 nm for the 2DEG [45]. The conducting depth is smaller than the Fermi wavelength and thus it is meaningful to define such an electron system as "2D."

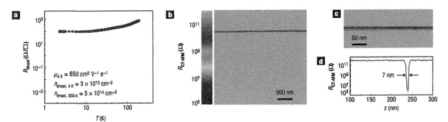

FIGURE 2.14 "In-situ-annealed" LAO/STO interface. (a) Temperature dependence of the sheet resistance. (b) Conducting-tip atomic force microscopy (CT-AFM) resistance mapping. (c) CT-AFM resistance image in high-resolution mode. (d) Resistance profile across the LAO/ STO interface extracted from (c) (reproduced from Ref. [44] with permission). (Reprinted by permission from Macmillan Publishers Ltd.: Basletic, M.; Maurice, J.-L.; Carrétéro, C.; Herranz, G.; Copie, O.; Bibes, M.; Jacquet, E.; Bouzehouane, K.; Fusil, S.; Barthélémy, A. Mapping the spatial distribution of charge carriers in LaAlO3/SrTiO3 Heterostructures. *Nat. Mater.* 2008, *7*, 621.)

Later, Sing et al. [46] profiled the 2DEG by hard X-ray photoelectron spectroscopy and inferred that the thickness of the 2DEG is smaller than 4 nm. Density functional calculations by Janicka et al. [47] predicted that the 2DEG is confined in STO within ~1 nm at the interface. By inserting Mn dopants in STO at different distances from the interface, Fix et al. [48] were able to locate the position of carriers within the STO surface layers and found that the majority of carriers in fully oxygenated LAO/STO samples are confined within 1 uc of the interface.

By infrared ellipsometry measurements, Dubroka et al. [49] found that the vertical carrier concentration profile has a strongly asymmetric shape with a rapid initial decay over the first 2 nm and a pronounced tail extending to ~11 nm, which suggests that the 2DEG is mainly confined in STO within 2 nm at the interface. First-principle calculations by Delugas et al. [50] found that electrons localize spontaneously in Ti $3d_{xy}$ levels within a thin (<2 nm) interface-adjacent STO region for $n_s < 10^{14}$ cm^{-2}. For $n_s > 10^{14}$ cm^{-2}, a portion of charge flows into Ti $3d_{xz}$–d_{yz} levels extending farther from the interface. This is in excellent agreement with experimental observation by Dubroka et al. [49]. In addition, the Seebeck effect measurements of the LAO/STO interface system indicate that the thickness of the conducting layer is ~7 nm [51]. Recent transport measurements by Huang et al. [52] of LAO/STO interfaces fabricated on NdGaO₃ (1 1 0) substrates with different STO thicknesses suggests a conducting depth of ~4 nm.

To summarize, all the above studies are in general agreement and the conducting depth of the 2DEG is 2–7 nm on the STO surface, which is smaller than the Fermi wavelength of free electrons at the interface. Thus, these studies justify the 2D nature of the LAO/STO interface electron system.

2.3.3 SHEET RESISTANCE, CARRIER DENSITY, AND MOBILITY

As shown in Figure 2.15, the sheet resistance of unannealed 10-uc LAO/STO heterostructures deposited at 10^{-2}–10^{-4} mbar is ~10 kΩ $\square^{-1}$ at room temperature. However, the sheet resistance of samples fabricated at 10^{-5} mbar is two orders of magnitude smaller, which is due to the introduction of 3D carriers by oxygen vacancies in STO. Correspondingly, the sheet carrier density of the unannealed 2DEG is ~10^{14} cm^{-2} at 300 K and ~2.5×10^{13} cm^{-2} at 2 K. The carrier density is almost constant above 150 K, while it has a significant decrease below 150 K. The mobility increases dramatically at low temperatures as a result of phonon scattering. The room-temperature mobility of the 2DEG is ~5–10 cm^2 (V·s)$^{-1}$, which is almost universal for most of groups. However, the low-temperature mobility varies over a large range of 500–10,000 cm^2 (V·s)$^{-1}$ [16,25,38].

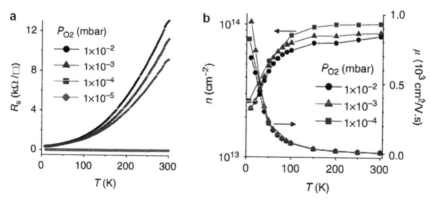

FIGURE 2.15 Sheet resistance (a), sheet carrier density and mobility (b) of 10-uc LAO/STO heterostructures deposited at different oxygen pressures (reproduced from Ref. [9] with permission). (Reprinted from Ariando; Wang, X.; Baskaran, G.; Liu, Z. Q.; Huijben, J.; Yi, J. B.; Annadi, A.; Barman, A. R.; Rusydi, A.; Dhar, S.; Feng, Y. P.; Ding, J.; Hilgenkamp, H.; Venkatesan, T. *Nat. Commun.* 2011, *2*, 188. With permission from Nature Publishing Group.)

Oxygen annealing has a significant effect on the room-temperature sheet resistance and carrier density. For example, Siemons et al. [22] found that the carrier density of as-deposited LAO/STO heterostructures is higher than 10^{14} cm^{-2} at 300 K, but it decreases to ~1.5×10^{13} cm^{-2} after oxygen annealing (Fig. 2.16). Nevertheless, the carrier density at low temperatures is relatively not sensitive to oxygen annealing. This indeed emphasizes the role of oxygen vacancies in as-deposited LAO/STO heterostructures. The sheet resistance of oxygen-annealed LAO/STO heterostructures is ~100 kΩ □$^{-1}$ at room temperature [20], which is one order of magnitude larger than that of as-deposited unannealed heterostructures. Anyway, the mobility is not affected too much by oxygen annealing both at room temperature and low temperatures [22,43].

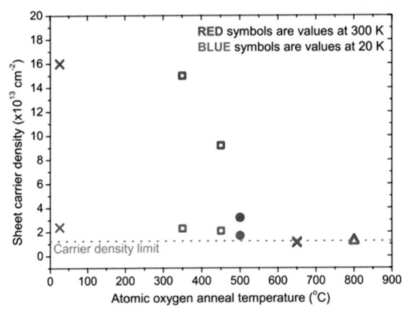

FIGURE 2.16 Sheet carrier densities at 20 (blue symbols) and 300 K (red symbols) as a function of annealing temperature in 600 W atomic oxygen, for samples made at 10^{-5} Torr of O$_2$. The values at 25°C indicate the as-deposited samples. The different symbol shapes indicate different samples made under similar conditions: two made at Stanford (circles and triangles), the other two made at the University of Twente (crosses and squares). Sample thickness ranges from 5 to 26 uc (reproduced from Ref. [22] with permission). (Reprinted with permission from Siemons, W.; Koster, G.; Yamamoto, H.; Harrison, W. A.; Lucovsky, G.; Geballe, T. H.; Blank, D. H. A.; Beasley, M. R. Origin of charge density at LaAlO3-on-SrTiO3 heterointerfaces; possibility of intrinsic doping. *Phys. Rev. Lett.* 2007, *98*, 196802. © 2007 by the American Physical Society.)

In addition, the mobility and sheet resistance of LAO/STO heterostructures could also be affected by scattering from step edges in substrates, leading to in-plane anisotropic transport properties [53].

2.3.4 SUPERCONDUCTIVITY

In 2007, Reyren et al. [17] reported superconductivity below 1 K in the oxygen-annealed LAO/STO heterostructures, as discussed previously. The bulk superconductivity in oxygen deficient and chemically doped STO is well known from 1964 [54]. However, the 2D superconductivity created by intrinsic interface doping is quite exciting. The extracted out-of-plane upper critical field at 0 K is H_{c2} (0 K) ≈ 65 mT and the corresponding in-plane coherence length ξ (0 K) ≈ 70 nm. Based on the $I–V$ and $R–T$ analyses, the superconducting transition was found to be a Berezinskii–Kosterlitz–Thouless (BKT) transition (T_{BKT} = 190 mK) expected for a 2D system, which demonstrated the 2D nature of the superconductivity at the LAO/STO interface. As a result, the superconductivity is highly anisotropic. The parallel upper critical field is ~20 times of the perpendicular upper critical field and the superconducting layer thickness is estimated to be ~10 nm [55].

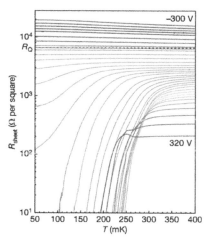

FIGURE 2.17 Field–effect modulation of the transport properties. Measured sheet resistance as a function of temperature, plotted on a semilogarithmic scale, for gate voltages varying in 10-V steps between −300 V and −260 V, 20-V steps between −260 V and 320 V, and for −190 V. The dashed line indicates the quantum of resistance R_Q = 6.45 kΩ □⁻¹ (reproduced from Ref. [28] with permission). (Reprinted by permission from Macmillan Publishers Ltd.: Caviglia, A. D.; Gariglio, S.; Reyren, N.; Jaccard, D.; Schneider, T.; Gabay, M.; Thiel, S.; Hammerl, G.; Mannhart, J.; Triscone, J.-M. Electric field control of the LaAlO: 3: /SrTiO: 3: interface ground state. *Nature* 2008, *456*, 624.)

Later, it was found that the superconducting state at the LAO/STO inter-face can be largely tuned by field effect [28]. The transition from the super-conducting phase to the insulating phase induced by the external electric field (Fig. 2.17) was found to be a 2D quantum phase transition with the transition exponent $z\bar{\nu} = 2/3$, indicative of a clean (or weakly disordered) 2D system in which quantum fluctuations dominate.

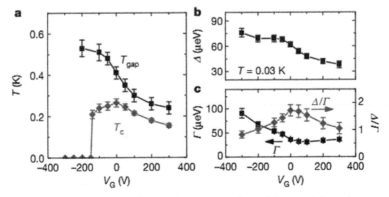

FIGURE 2.18 Dependences of T_c and T_{gap} on gate voltage. Measured dependence on gate voltage of the superconducting transition temperature, T_c, and the temperature at which the gap closes, T_{gap} (a); the superconducting gap, Δ (0 K) (b); and the coherence-peak-broadening parameter, Γ (0 K), and the ratio Δ (0 K)/Γ (0 K) (c). Error bars define the 90% confidence interval (reproduced from Ref. [56] with permission). (Reprinted by permission from Macmillan Publishers Ltd.: Richter, C.; Boschker, H.; Dietsche, W.; Fillis-Tsirakis, E.; Jany, R.; Loder, F.; Kourkoutis, L. F.; Muller, D. A.; Kirtley, J. R.; Schneider, C. W.; Mannhart, J. Interface superconductor with gap behaviour like a high-temperature superconductor. *Nature* 2013, 502, 528. © 2013.)

Recent tunnel spectroscopy [56] reveals a superconducting energy gap of 40 μeV in the density of states, whose shape is well described by the Bardeen–Cooper–Schrieffer superconducting gap function. In contrast to the dome-shaped dependence of the critical temperature, the gap increases with charge carrier depletion in both the underdoped region and the overdoped region (Fig. 2.18). These results are analogous to the pseudogap behavior of the high-transition-temperature copper oxide superconductors and imply that the smooth continuation of the superconducting gap into pseudogap-like behavior could be a general property of 2D superconductivity.

2.3.5 RESISTANCE MINIMUM

The resistance of a metallic system typically decreases with lowering temperature. However, a low-temperature resistance minimum was observed

(Fig. 2.4) at the LAO/STO interface by Brinkman et al. [18] when the LAO layer is as thick as 26 uc and deposited at relatively high oxygen pressures (10^{-4} mbar and above). Figure 2.19 shows the temperature-dependent sheet resistance down to 0.3 K on a semilogarithmic scale. The resistance minimum appears at ~70 K and the resistance upturn tends to saturate at low temperatures. The resistance minimum was attributed to the Kondo effect by Brinkman et al., which basically states the resistance upturn at low temperatures is due to the scattering of electrons from localized and noninteracting magnetic moments. At the LAO/STO interface, the localized moments are referring to localized Ti^{3+} $3d^1$ moments.

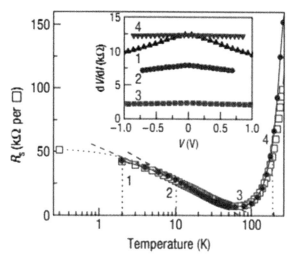

FIGURE 2.19 Temperature dependence of the sheet resistance, R_s, for two conducting interfaces, grown, respectively, at a partial oxygen pressure of 2.5×10^{-3} mbar (open squares) and 1.0×10^{-3} mbar (filled circles). The low-temperature logarithmic dependencies are indicated by dashed lines. Inset: Four-point differential resistance dV/dI as a function of applied voltage, at a constant temperature of 2.0 K (1), 10.0 K (2), 50.0 K (3), and 180.0 K (4) (reproduced from Ref. [18] with permission). (Reprinted by permission from Macmillan Publishers Ltd.: Brinkman, A.; Huijben, M.; van Zalk, M.; Huijben, J.; Zeitler, U.; Maan, J. C.; van der Wiel, W. G.; Rijnders, G.; Blank, D. H. A.; Hilgenkamp, H. Interface superconductor with gap behaviour like a high-temperature superconductor. *Nat. Mater.* 2007, *6*, 493.)

However, it is still puzzling that the Kondo temperature is so high and close to 100 K as the conventional Kondo effect found in metal doped with magnetic elements has a typical Kondo temperature of ~10 K. The reason could be that the Kondo effect in strongly correlated oxides is largely different from that in metals. On the other hand, the resistance minimum was argued to be due to the weak localization in a two-dimensional system [28].

Whether it is due to the Kondo effect or weak localization is still an open issue up to now as no systematical temperature- and field-dependent resistance data in the light of the Kondo model or the weak localization model have been reported to clarify this issue.

In addition, why the sheet resistance minimum only appears in LAO/STO heterostructures with thicker LAO layers is physically unclear so far. A perspective which may be useful for this issue is that microcracks could occur in LAO films during cooling down if the LAO layer is too thick due to the large lattice mismatch between LAO and STO. Experimentally, the sheet resistance minimum has been observed in LAO/STO heterostructures with the LAO layer larger than 15 uc [57]. In this sense, the sheet resistance minimum could be related to the disorder at the interface induced by structural degradation in thick LAO/STO heterostructures, which would be in agreement with the weak localization argument. Nevertheless, any direct investigation on the relationship between structural and electrical properties is still lacking.

2.3.6 LOCALIZATION AT THE INTERFACE

A fundamental issue related to the LAO/STO interface is that the measured carrier density for fully oxidized samples ($\sim1.5 \times 10^{13}$ cm^{-2}) is more than 20 times smaller than the expected carrier density ($\sim3.3 \times 10^{13}$ cm^{-2}) from the polar catastrophe arguments. DFT calculations by Popović et al. revealed that electrons occupy multiple Ti 3d t_{2g} subbands leading to a rich array of transport properties [58]. The xy bands are energetically lower than the yz and xz bands (Fig. 2.20) because of the larger hopping integrals along both directions in the plane. Above the bands of the first Ti layer, xy-, yz-, and xz-type bands of the second or the third Ti layer are located.

The lowest conduction subband-Ti$_1(xy)$ has a strong 2D character consisting mostly of Ti(d) states on the first layer. A small degree of disorder leads to Anderson localization in 2D and therefore these electrons are localized. States in the remaining bands spread several layers into the STO. Nevertheless, the xz/yz states are prone to localization as well due to their heavy masses along a planar direction. The intrinsic mobile carriers come from the Ti(xy) bands above the lowest conduction subband Ti$_1(xy)$, for example, Ti$_2(xy)$ and Ti$_3(xy)$, which spread over several Ti layers into the bulk. The electrons at these states have small masses along the plane and are consequently difficult to localize by disorder or electron–phonon coupling, thus contributing to transport properties. To briefly summarize, due to the occupation of multiple Ti 3d t$_{2g}$ bands near the interface, each with different effective mass, spatial extent, and susceptibility to localization by disorder, most of the charges are localized at the interface and only a small portion of electrons are delocalized to contribute measurable transport properties.

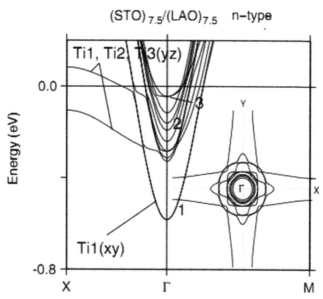

FIGURE 2.20 Conduction bands near the Fermi energy and their predominant orbital characters. Band dispersions are shown along the interface (*xy* plane) with $X = \pi/a(1,0,0)$ and $M = \pi/a(1,1,0)$, where "*a*" is the in-plane lattice constant. 1, 2, and 3 represent the Ti layer number from the interface (reproduced from Ref. [58] with permission). (Reprinted with permission from Popovic, Z. S.; Satpathy, S.; Martin, R. M. Origin of the Two-Dimensional Electron Gas Carrier Density at the LaAlO3 on SrTiO3 Interface. *Phys. Rev. Lett.* 2008, 101, 256801. © 2008 by the American Physical Society.)

Besides the effect of disorder on charge localization at the interface, disorder also results in the localization of the mobile carriers in conduction behavior. For example, Caviglia et al. found that the low-temperature negative MR at the LAO/STO interface is in agreement with the weak localization model [28]; Wong et al. found that the low-temperature resistance minimum and the in-plane magnetotransport data are related to weak localization induced by disorder at the LAO/STO interface [59]; disorder-induced strong localization (Fig. 2.21) of the 2DEG was observed by Hernandez et al. in the LAO/STO epitaxial thin-film heterostructures grown on (LaAlO$_3$)$_{0.3}$–(Sr$_2$AlTaO$_3$)$_{0.7}$ substrates [60]. Such disorder could arise from structural distortions at the interface due to the interface strain that arises from the lattice mismatch between LAO and STO.

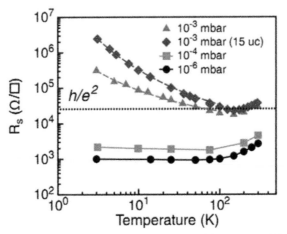

FIGURE 2.21 Temperature dependence of R_s. Two high-pressure (10^{-3} mbar) grown samples with 20- and 15-uc LAO thicknesses (triangles and diamonds, respectively), and two low-pressure samples with a LAO thickness of 20 uc grown at 10^{-4} and 10^{-6} mbar grown samples (squares and circles, respectively). The quantum of resistance $h/e^2 = 25.8$ kΩ is indicated as a dotted line. Dashed lines show fits for the Mott-type VRH model (reproduced from Ref. [60] with permission). (Reprinted with permission from Hernandez, T.; Bark, C. W.; Felker, D. A.; Eom, C. B.; Rzchowski, M. S. Localization of two-dimensional electron gas in LaAlO3/SrTiO3 heterostructures. *Phys. Rev. B* 2012, *85*, 161407. © 2012 by the American Physical Society.)

2.3.7 NONLINEAR HALL EFFECT

Depending on the carrier density, the occupation of Ti 3d subbands is different. At a certain carrier density range, multiband transport behavior could occur, leading to a nonlinear Hall effect. The multiple conducting carriers at the LAO/STO interface were first revealed by the comparison of the sheet carrier density and mobility obtained from optical transmission spectroscopy and dc transport measurements [62], where the low-density high-mobility electrons dominating electrical transport properties and the high-density low-mobility electron responsible for optical transmission were distinguished. The nonlinear Hall effect was later observed (Fig. 2.22) by Shalom et al. [61]. The fitted mobility of two types of carriers differs by a factor of 5.

The detailed study on the nonlinear Hall effect at the LAO/STO interface (Fig. 2.23) depending on the carrier concentration (tuned by back-gate voltage) by Joshua et al. [63] found that the critical carrier density for the multiple-carrier transport is 1.7×10^{13} cm^{-2}, above which multiple bands are involved in the transport properties due to the increase in the band filling level. Another interesting experiment is that the nonlinear Hall effect can be obtained by optical doping (Fig. 2.24) [64], which again demonstrates the changing occupation of Ti 3d subbands depending on the carrier density.

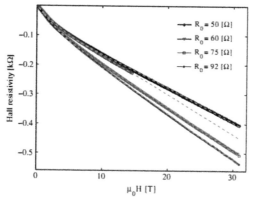

FIGURE 2.22 Hall resistivity as a function of magnetic field for various zero-field resistance values R_0 (controlled by gate voltage). The dashed lines are fittings based on a two-carrier model (reproduced from Ref. [61] with permission). (Reprinted with permission from Ben Shalom, M.; Ron, A.; Palevski, A.; Dagan, Y. Shubnikov-de Haas oscillations in SrTiO$_3$\LaAlO$_3$ interface. *Phys. Rev. Lett.* 2010, *105*, 206401. © 2010 by the American Physical Society.)

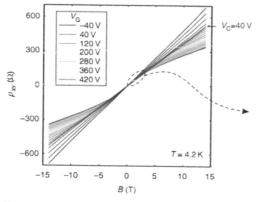

FIGURE 2.23 Hall resistance ρ_{xy} versus magnetic field B, for various back-gate voltages, V_G, in 20-V steps, at $T = 4.2$ K. At a critical value, $V_C = 40$ V, a transition is observed between two different types of B dependencies (reproduced from Ref. [63] with permission). (Reprinted by permission from Macmillan Publishers Ltd.: A. Joshua, S. Pecker, J. Ruhman, E. Altman; S. Ilani, A universal critical density underlying the physics of electrons at the LaAlO$_3$/SrTiO$_3$ interface. *Nat. Commun.* 2012, *3*, 1129. © 2012.)

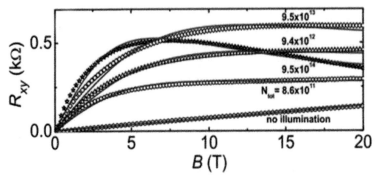

FIGURE 2.24 Hall resistance data as a function of the applied magnetic field, for illumination with different values of N_{tot} with energy of 3.65 eV at 4.2 K (open symbols). Solid lines: The two-band model fits to the experimental data (reproduced from Ref. [64] with permission). (Reprinted from Guduru, V. K.; Granados del Aguila, A.; Wenderich, S.; Kruize, M. K.; McCollam, A.; Christianen, P. C. M.; Zeitler, U.; Brinkman, A.; Rijnders, G.; Hilgenkamp, H.; Maan, J. C. Optically excited multi-band conduction in LaAlO₃/SrTiO₃ heterostructures. *Appl. Phys. Lett.* 2013, *102*, 051604.with the permission of AIP Publishing.)

2.3.8 SHUBNIKOV–de HASS EFFECT AND EFFECTIVE MASS

The Shubnikov–de Haas (SdH) effect is the oscillation in the resistance of a material at low temperatures under intense magnetic fields. It is a macroscopic demonstration of the inherent quantum mechanical nature of matter. The origin is the expansion of density states for each Landau level with increasing the magnetic field. As each Landau level passes through the Fermi energy with increasing the magnetic field, the electrons occupying the Landau level become free to flow as current. This induces the material's transport and thermodynamic properties to oscillate periodically. This quantum effect is typically used to determine the effective (cyclotron) mass of carriers.

To observe the SdH effect, both the mobility and magnetic field need to be high enough to fulfill quantum conductions [65]. The high-mobility 2DEG at the LAO/STO interface provides a fantastic platform to investigate the quantum transport. Caviglia et al. [38] observed the SdH effect in the LAO/STO heterostructures with low-temperature mobility larger than 5000 cm² (V·s)⁻¹, whose period depends only on the perpendicular component of the magnetic field (Fig. 2.23). Such an angular dependence of quantum oscillations directly demonstrates the 2D nature of the electron gas. In addition, the effective mass extracted from the temperature dependence of the SdH oscillations amplitude (Dingle plot) is $m^* = 1.45\ m_e$ (m_e—the electron rest mass). The SdH carrier density and mobility obtained by the Fourier transform of the oscillatory component of the MR is 1.05×10^{13} cm^{-2} and 2860 cm^2 (V·s)$^{-1}$, respectively, which are smaller than the values (Fig. 2.25) obtained in Hall measurements. This is due to the presence of nonoscillatory parallel transport channels resulting from multiple valley and spin degeneracy. The SdH oscillations observed in the LAO/STO heterostructures by Dagan et al. [61] revealed an effective mass of 2.1 ± 0.4 m_e and a SdH carrier density six times smaller than the measured Hall carrier density. The variation in the effective mass among different groups could be due to the difference in carrier concentration.

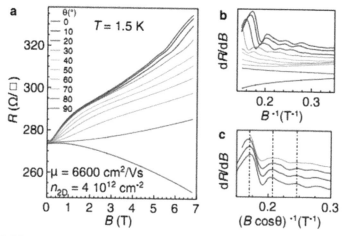

FIGURE 2.25 Angular dependence of the quantum oscillations. (a) Sheet resistance R as a function of magnetic field B recorded at different orientations (measured by the angle θ) with respect to the direction normal to the substrate. (b) Numerical derivative dR/dB as a function of the inverse of the magnetic field recorded at different orientations. (c) Numerical derivative dR/dB as a function of the inverse of the component of the magnetic field perpendicular to the plane of the interface. An offset has been introduced in each curve for clarity. The lines are a guide to the eye (reproduced from Ref. [38] with permission). (Reprinted with permission from Caviglia, A. D.; Gariglio, S.; Cancellieri, C.; Sacépé, B.; Fête, A.; Reyren, N.; Gabay, M.; Morpurgo, A. F.; Triscone, J.-M. *Phys. Rev. Lett.* Two-Dimensional Quantum Oscillations of the Conductance at LaAlO3\SrTiO3 Interfaces.2010, *105*, 236802. © 2010 by the American Physical Society.)

2.3.9 STRAIN EFFECT

In the above, we have summarized various electrical properties of the 2DEG at the *n*-type LAO/STO interface. Here, we would like to emphasize an

important factor in oxide heterostructures due to the lattice mismatch—
strain. It typically induces distortion in crystal lattices and therefore changes
the metal–oxygen bond angle and bond length, thus yielding the modulation
in band structures and various electrical properties. The effect of strain on
the electrical properties of the 2DEG at the n-type interface was initiated
by Bark et al. [66], where they fabricated LAO/STO interfaces on various
single-crystal substrates with different lattice constants and accordingly
controlled levels of biaxial epitaxial strain. It was found that tensile-strained
STO destroys the 2DEG, while compressively strained STO retains the
2DEG. The critical thickness for the appearance of conductivity increases
with compressive strain and the carrier density is reduced in strained LAO/
STO heterostructures compared with unstrained LAO/STO interfaces. It was
suggested that a strain-induced electric polarization in the STO film opposite
to that of the polar LAO layer reduces the polar discontinuity at the interface, leading to the increase in the critical thickness of the LAO layer and
decrease in the carrier density. Such experiment demonstrates that epitaxial
strain can be utilized to tailor the electronic properties of the 2DEG at the
LAO/STO interface.

2.3.10 INSULATING P-TYPE INTERFACE

All the above electrical properties are associated with n-type (TiO_2/LaO)
interfaces while the p-type (SrO/AlO_2) interface is insulating. Experimentally, the fabrication of p-type interface is nontrivial. Until now, any
chemical approach to achieve single SrO termination on the STO surface is
lacking, and the thermal treatment of as-received STO substrate typically
generates mixed terminations. Instead, a single-terminated SrO surface of
STO can only be obtained by depositing a SrO ML on a TiO_2-terminated
surface. In reality, a SrO target is susceptible to CO_2 and water absorption and consequently the relative density is low (60–70%). Moreover,
the epitaxial growth of SrO on STO can be achieved by molecular beam
epitaxy [67] and PLD [68] at relatively low temperatures 400–500°C, but
at normal LAO deposition temperatures 750–850°C, pulsed laser interval
deposition [69] as well as a high oxygen partial pressure of 0.13 mbar (M.
Huijben, *PhD Thesis, University of Twente, Enschede, The Netherlands,*
2006) are needed to produce atomically flat SrO termination with straight
steps.

In the polar catastrophe model [19], the interface electronic reconstruction at the p-type AlO_2/SrO interface requires the transfer of half hole from

the AlO$_2$ layer to the SrO layer. Meanwhile, oxygen vacancies appear in the SrO layer to compensate holes and consequently no mobile holes are existent at the interface, leading to the insulating behavior. Theoretical calculations by Pentcheva et al. [70] and Park et al. [71] support the charge compensation model. Later, first-principle calculations by Zhang et al. [72] suggested that due to the polarization in the LAO layer, the formation energy of oxygen vacancies at the p-type interface was much smaller than that at the LAO surface layer while it became larger at the n-type interface compared with that at the LAO surface. Hence, the transferred holes at the p-type interface were compensated by oxygen vacancies while the transferred electrons at the n-type interface were not affected. However, the recent second-harmonic optical spectroscopy study by Rubano et al. [73] found that the overall second harmonic generation in SrO-terminated interfaces had the same intensity as in bare STO surfaces, for which there is no charge injection. It was suggested that no charge transfer occurred at the p-type interface and the lack of conduction in SrO-terminated interfaces could not be ascribed to charge localization effect. This adds a question mark to the intrinsic insulating origin of the p-type LAO/STO interface and more experimental studies are needed to clarify this issue.

2.4 MAGNETIC PROPERTIES

2.4.1 PREDICTION OF THE FERROMAGNETIC ORDER

While both LAO and STO are nonmagnetic materials, theoretical calculations revealed that the ferromagnetic order could be realized in the 2DEG. In 2006, Pentcheva and Pickett [70] predicted that the spins occupying d_{xy} orbitals at the n-type interface are ferromagnetically aligned. Figure 2.26 shows the density of states of the n-type interface, where the t_{2g} degeneracy is lifted at the interface and the d_{xy} orbital is lower than other orbitals for Ti^{3+} ions. More importantly, the d_{xy} band is strongly spin-polarized, which indicates that the Ti^{3+} spin-half moments couple ferromagnetically. The magnetic moment of each Ti^{3+} ion was found to be 0.71 μ_B while a Ti^{4+} ion has very small magnetic moment of 0.05 μ_B.

In addition, Janicka et al. [74] performed spin-polarized calculations on LAO/STO superlattices and found that the n-type interface in a (LAO)$_3$/(STO)$_3$ superlattice is magnetic with a magnetic moment of each Ti^{3+} ion of 0.2 μ_B mainly due to the geometric confinement of the electron gas. However, fluctuation effects are more pronounced in 2D systems than in 3D and whether the ferromagnetic order can survive fluctuation effects experimentally was still an open issue.

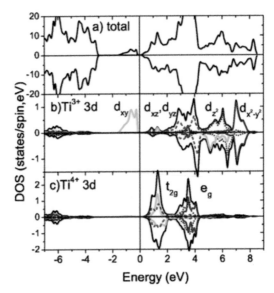

FIGURE 2.26 Density of states of the *n*-type interface: (a) total; (b) *d* states of magnetic Ti^{3+} showing the split-off majority d_{xy} band, with the corresponding minority states lying at +5 eV; the other 3*d* states are not strongly polarized; (c) the nonmagnetic Ti^{4+} ion, showing the conventional (although not perfect) t_{2g}–e_g crystal field splitting. d_{xy} orbitals are marked by a green (light gray) line, d_{xz} and d_{yz} states by a magenta dashed line, states with d_z^2 and $d_{x^2-y^2}$ character by a red dotted and blue (dark gray) solid lines, respectively (reproduced from Ref. [70] with permission). (Reprinted with permission from Pentcheva, R.; Pickett, W. Charge localization or itinerancy at $LaAlO_3/SrTiO_3$ interfaces: Hole polarons, oxygen vacancies, and mobile electrons. *Phys. Rev. B* 2006, *74*, 035112. © 2006 by the American Physical Society.)

2.4.2   IMPLICATIONS FROM MAGNETROTRANSPORT PROPERTIES

For a period, despite theoretical predictions of the ferromagnetic state at the LAO/STO interface, any direct probe on the magnetic properties of this system was lacking. One important reason could be that the superconductivity reported by Reyren et al. [17] can be well reproduced by many groups [28,75] and ferromagnetism and superconductivity were thought to be naturally antagonistic. Nevertheless, there were several studies on magnetotransport properties of the 2DEG, which shed light on the possible magnetic order. In 2007, Brinkman et al. [18] reported the resistance minimum at the LAO/

STO interface and attributed that to the Kondo effect. Moreover, negative and hysteretic magnetoresistance (MR) was observed at 0.3 K (Fig. 2.27), indicative of the ferromagnetic domain formation.

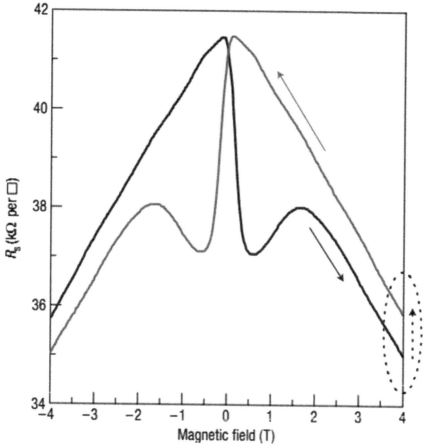

FIGURE 2.27 Magnetic field-dependent sheet resistance at 0.3 K of an *n*-type LAO/STO interface, grown at 10⁻³ mbar. The arrows indicate the direction of measurements (at a rate of 30 mT s⁻¹) (reproduced from Ref. [18] with permission). (Reprinted by permission from Macmillan Publishers Ltd: Brinkman, A.; Huijben, M.; van Zalk, M.; Huijben, J.; Zeitler, U.; Maan, J. C.; van der Wiel, W. G.; Rijnders, G.; Blank, D. H. A.; Hilgenkamp, H. Magnetic effects at the interface between non-magnetic oxides. *Nat. Mater.* 2007, *6*, 493. © 2007.)

Shalom et al. [75] found that the MR of the 2DEG is strongly anisotropic. The out-of-plane MR is positive, while the in-plane parallel MR is negative up to more than 60% (Fig. 2.28). The positive MR can be understood by the orbital effect. The relevant mechanisms that can produce negative MR are 2D weak localization, magnetic impurities, and the magnetic order itself. The weak localization effect is typically small, of the order of few percent. The magnetic impurity effect is usually isotropic, which is hard to coincide with the strongly anisotropic MR here. Finally, it was concluded that the large negative MR is due to the magnetic order formed at the interface. In addition, MR measurements by Seri and Klein [76] revealed that the out-of-plane MR has an antisymmetric term [namely R(H) ≠ R(−H)] which increases with decreasing temperature and increasing field. It was suggested that nonuniform field-induced magnetization exists at the interface. These pioneering studies largely stimulated magnetic investigations on the 2DEG at the LAO/STO interface later.

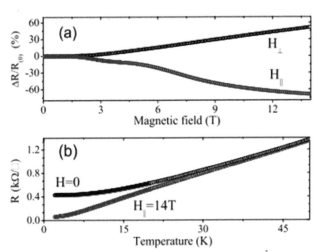

FIGURE 2.28   The n-type LAO/STO interface was grown at 5×10^{-5} Torr and 800°C. (a) Blue circles: MR as a function of magnetic field applied perpendicular to the interface. Red squares: MR data for field applied along the interface parallel to the current. Both are at 2 K. (b) The sheet resistance as a function of temperature at zero field (black circles) and at 14 T applied parallel to the current (red squares) (reproduced from Ref. [75] with permission). (Reprinted with permission from Ben Shalom, M.; Tai, C. W.; Lereah, Y.; Sachs, M.; Levy, E.; Rakhmilevitch, D.; Palevski, A.; Dagan, Y. Anisotropic magnetotransport at the SrTiO$_3$/LaAlO$_3$ interface. *Phys. Rev. B* 2009, *80*, 140403(R). © 2009 by the American Physical Society.)

2.4.3   DIRECT OBSERVATION OF FERROMAGNETISM

In 2011, Ariando et al. [9] fabricated LAO/STO heterostructures at a wide oxygen pressure range from 10^{-2} to 10^{-6} mbar and probed magnetic properties by SQUID–VSM (superconducting quantum interference device-vibrating sample magnetometer) measurements. It was found that magnetic moment of LAO/STO heterostructures strongly depends on the oxygen partial pressure used for LAO deposition (Fig. 2.29). The magnetic moment of LAO/STO heterostructures deposited at the high oxygen pressure of 10^{-2} mbar is one order of magnitude larger than the moment of those deposited at 10^{-6} mbar. This is consistent with the experimental fact that the Kondo-like resistance minimum was only observed at relatively high oxygen pressures.

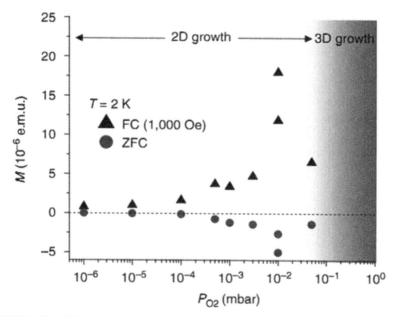

FIGURE 2.29   ZFC (zero-field-cooling) and FC (field-cooling) magnetic moment of 10-uc LAO/STO heterostructures at 2 K as a function of oxygen pressure. The data were taken while warming the samples from 2 to 300 K in a 0.1-kOe magnetic field (reproduced from Ref. [9] with permission). (Reprinted from Ariando; Wang, X.; Baskaran, G.; Liu, Z. Q.; Huijben, J.; Yi, J. B.; Annadi, A.; Barman, A. R.; Rusydi, A.; Dhar, S.; Feng, Y. P.; Ding, J.; Hilgenkamp, H.; Venkatesan, T. Electronic phase separation at the LaAlO₃/SrTiO₃ interface. *Nat. Commun.* 2011, *2*, 188. With permission. Nature Publishing Group.)

Detailed measurements revealed that LAO/STO heterostructures deposited at 10^{-2} mbar are clearly ferromagnetic with magnetic hysteresis loops showing up from 2 to 300 K (Fig. 2.30). Moreover, the large magnetization drop in the zero-field-cooling temperature dependence of magnetic moment at ~60 K is highly similar to the superconducting transition. It was suggested that the coexistence of ferromagnetism and the superconducting-like transition could be possible due to nanoscopic electronic phase separation, where some nanoscale regions are ferromagnetic, while some are superconducting owing to selective occupation of Ti 3d orbitals. Such an idea creatively shed light on the possible coexistence of ferromagnetism and superconductivity in this 2D system, which was quite exciting to promote one to further explore the ferromagnetic order in the presence of superconductivity. This study, in some sense, is a turning point from theoretical predictions to experimental investigations for magnetic studies of the 2DEG at the LAO/STO interface.

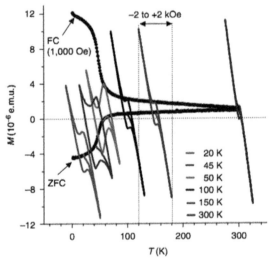

FIGURE 2.30   The 1-kOe FC and ZFC in-plane magnetic moment as a function of temperature and measured by a 0.1-kOe magnetic field applied while warming the sample from 2 to 300 K (solid black lines) for the 10-uc LAO/STO samples prepared at an oxygen partial pressure of 1 × 10⁻² mbar. In a separate measurement after ZFC, ferromagnetic hysteresis loops centered on the diamagnetic branch are observed when sweeping a ±2-kOe magnetic field applied at each temperature. Similar ferromagnetic loops are also observed on the paramagnetic branch when the hysteresis loops are collected after FC (not shown here for clarity) (reproduced from Ref. [9] with permission). (Reprinted from Ariando; Wang, X.; Baskaran, G.; Liu, Z. Q.; Huijben, J.; Yi, J. B.; Annadi, A.; Barman, A. R.; Rusydi, A.; Dhar, S.; Feng, Y. P.; Ding, J.; Hilgenkamp, H.; Venkatesan, T. Electronic phase separation at the LaAlO₃/SrTiO₃ interface. Nat. Commun. 2011, 2, 188. With permission. Nature Publishing Group.)

The study on the neutron spin-dependent reflectivity of LAO/STO superlattices obtained a magnetic signal several ten times smaller than that in SQUID magnetometer measurements [77]. It was therefore suggested that the large ferromagnetic signal obtained in magnetometer measurements are possibly due to experimental artifacts. Nevertheless, the *n*-type interfaces in LAO/STO superlattices could be different from the *n*-type interface between a LAO film and a STO single-crystal substrate. That is because STO films are significantly different from STO single crystals in terms of electronic and optical properties mainly due to the defective nature of thin films. For example, the insertion of homoepitaxial STO ultrathin films at LAO/STO interfaces degrades the 2DEG properties [78]. Therefore, magnetic proper-ties of an interface between a LAO film and a STO film could be largely different from that between a LAO film and a STO single-crystal substrate.

2.4.4 COEXISTENCE OF FERROMAGNETISM AND SUPERCONDUCTIVITY

Shortly upon the report of ferromagnetism and possible electronic phase sepa-ration at the LAO/STO interface, several groups found the coexistence of ferromagnetism and superconductivity below 300 mK. Dikin et al. [29] did magnetotransport studies and found that the field-dependent Curie temperature of the superconducting transition is hysteretic (Fig. 2.31), indicative of the ferro-magnetic order. Moreover, torque magnetometry measurements performed by Li et al. [31] revealed ferromagnetism in the presence of superconductivity.

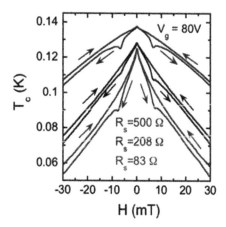

FIGURE 2.31 Phase diagram, T_c versus H, for a back gate voltage of $V_g = 80$ V, where the superconducting properties are maximal. The three curves represent different resistance bias points along the superconducting transition, with the normal state resistance per square being 704 Ω. The arrows mark the direction of the magnetic field ramp (reproduced from Ref. [29] with permission). (Reprinted with permission from Dikin, D. A.; Mehta, M.; Bark, C. W.; Folkman, C. M.; Eom, C. B.; Chandrasekhar, V. Coexistence of Superconductivity and Ferromagnetism in Two Dimensions. *Phys. Rev. Lett.* 2011, *107*, 056802. © 2011 by the American Physical Society.)

More straightforwardly, Bert et al. [30] imaged the coexistence of ferromagnetism and superconductivity by scanning SQUID that detects magnetic flux coming out of the sample. Figure 2.32a shows the image mapping of ferromagnetic dipoles in a LAO/STO heterostructure and Figure 2.32b shows the superfluid density image mapping of the same region. These studies together undoubtedly demonstrate the coexistence of ferromagnetism and superconductivity at the n-type LAO/STO interface, which excited the whole field quite a lot [79,80].

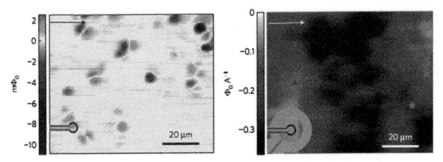

FIGURE 2.32 (a) LAO/STO magnetometry image mapping the ferromagnetic order. Inset: Scale image of the SQUID pick-up loop used to sense magnetic flux. (b) LAO/STO susceptometry image mapping the superfluid density at 40 mK. Inset: Scale image of the SQUID pick-up loop and field coil (reproduced from Ref. [30] with permission). (Reprinted by permission from Macmillan Publishers Ltd: Bert, J. A.; Kalisky, B.; Bell, C.; Kim, M.; Hikita, Y.; Hwang, H. Y.; Moler, K. A. Direct imaging of the coexistence of ferromagnetism and superconductivity at the LaAlO$_3$/SrTiO$_3$ interface. *Nat. Phys.* 2011, 7, 767. © 2011.)

Regarding physical understanding of the coexistence of ferromagnetism and superconductivity, Michaeli et al. [81] proposed that the transferred charge localizes and orders ferromagnetically via exchange with the conduction electrons; due to the strong spin–orbit coupling near the interface, the magnetism and superconductivity can coexist by forming a Fulde–Ferrell–Larkin–Ovchinnikov-type condensate of Cooper pairs at finite momentum, which is robust even in the presence of strong disorder. Consistently, Caprara et al. [82] suggested that the electronic phase separation is intrinsic for electron interfaces with strong Rashba spin–orbit coupling, which is proportional to the electric field perpendicular to the interface for LAO/STO.

The statistical study on the dipole distribution [30] indicated that most of ferromagnetic moments are in-plane. The dipole-moment density at the interface was inferred to be 0.4–1.1×10^{13} μ_B cm^{-2}, corresponding to 0.006–0.017 μ_B uc^{-1} provided that ferromagnetism is confined to 1 uc. In addition, the nuclear magnetic resonance study on LAO/STO superlattices [83] revealed a ferromagnetic moment density of 1.1×10^{12} μ_B cm^{-2}, which is much smaller than that in single LAO/STO interface samples. This again indicates that LAO/STO interfaces fabricated on STO thin films are largely different from those fabricated on STO single-crystal substrates.

2.4.5 CRITICAL THICKNESS FOR FERROMAGNETISM

Similar to the critical thickness for the appearance of conductivity, a critical thickness of 3–4 uc (Fig. 2.33) was also found for the ferromagnetism at the n-type LAO/STO interface [84] using scanning SQUID measurements. It was suggested that interfacial electronic reconstruction is needed for the appearance of ferromagnetism.

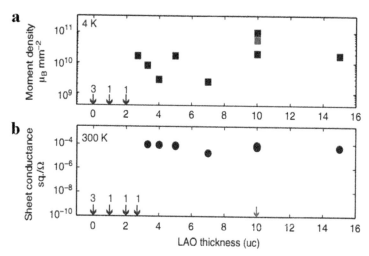

FIGURE 2.33 (a) Averaged moment density as a function of the thickness for TiO₂-terminated (red) and SrO-terminated (blue). Off-scale values are marked by an arrow, which includes the number of samples for that thickness. (b) Sheet conductance at room temperature measured by the Van der Pauw method (reproduced from Ref. [84] with permission). (Reprinted by permission from Macmillan Publishers Ltd: Kalisky, B.; Bert, J. A.; Klopfer, B. B.; Bell, C.; Sato, H. K.; Hosoda, M.; Hikita, Y.; Hwang, H. Y.; Moler, K. A. Critical thickness for ferromagnetism in LaAlO₃/SrTiO₃ Heterostructures. *Nat. Commun.* 2012, *3*, 922. © 2012.)

Besides, nuclear magnetic resonance experiment revealed that LAO/STO superlattices including LAO layers thicker than a certain threshold of 3 uc show ferromagnetic spin relaxation (Fig. 2.34). All these results imply that the ferromagnetism at the n-type interface is related to interface electronic reconstruction.

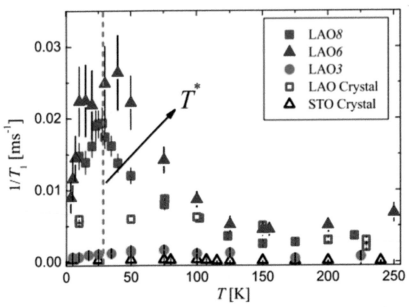

FIGURE 2.34 The spin-lattice relaxation rate $(1/T_1)$ as a function of temperature in 3 mT applied field. The solid red squares, blue triangles, and green circles are measurements in LAO8, LAO6, and LAO3, respectively. The open squares and triangles are reference measurements in LAO and STO bare crystals. LAOn refers to the $[(STO)_{10}(LAO)_n]_{10}$ superlattice structure (reproduced from Ref. [83] with permission). (Reprinted with permission from Salman, Z.; Ofer, O.; Radovic, M.; Hao,. H.; Ben Shalom, M.; Chow, K. H.; Dagan, Y.; Hossain, M. D.; Levy, C. D. P.; MacFarlane, W. A.; Morris, G. M.; Patthey, L.; Pearson, M. R.; Saadaoui, H.; Schmitt, T.; Wang, D.; Kiefl, R. F. Nature of Weak Magnetism in SrTiO₃/LaAlO₃ Multilayers. *Phys. Rev. Lett.* 2012, *109*, 257207. © 2012 by the American Physical Society.)

2.4.6 ORIGIN OF FERROMAGNETISM

Magnetic moments at the LAO/STO interface have been consistently attributed to localized electrons at the d_{xy} orbital [9,18,70,83,84]. Electric-field tuning of the carrier density at the interface does not affect ferromagnetic dipoles and ferromagnetism can also be observed in *p*-type heterostructures [84]. These experimental facts indicate that the ferromagnetism at the LAO/STO interface is not related to mobile carriers but localized electrons. Recent spectroscopic studies [85] of the ferromagnetism including X-ray magnetic circular dichroism and X-ray absorption spectroscopy observed direct evidence for in-plane ferromagnetic order at the interface, with Ti^{3+} character in the polarized d_{xy} orbital of the anisotropic t$_{2g}$ band. The energy of planar orbitals (d_{xy} and $d_{x^2-y^2}$) in both the t$_{2g}$ and e$_g$ bands was found to be lower than that of the out-of-plane orbitals (Fig. 2.35). Accordingly, the additional electron (3d^1) in the Ti^{3+} state occupies the polarized d_{xy} orbital. This basically proves that the ferromagnetism is from Ti d_{xy} orbitals.

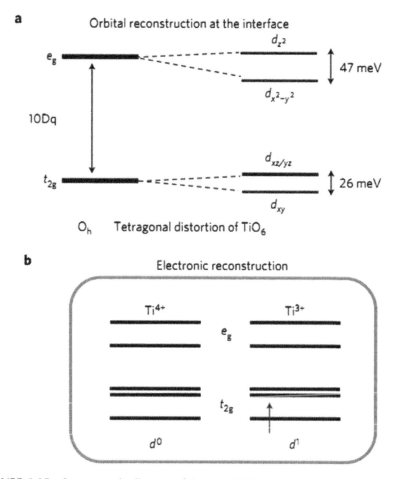

FIGURE 2.35 Spectroscopic diagram of the LAO/STO interface. (a) Schematic energy diagrams of the crystal field splitting and 3d orbital degeneracy, showing the orbital reconstruction at the interface and local bonding change. O_h denotes the octahedral environment. (b) The mixed valence Ti^{3+} and Ti^{4+} states at the interface (reproduced from Ref. [86] with permission). (Reprinted by permission from Macmillan Publishers Ltd: Lee, J.-S.; Xie, Y. W.; Sato, H. K.; Bell, C.; Hikita, Y.; Hwang, H. Y.; Kao, C.-C. Titanium dxy ferromagnetism at the LaAlO$_3$/SrTiO$_3$ interface. *Nat. Mater.* 2013, *12*, 703. © 2013.)

However, there is controversy on the origin of electrons in the localized d_{xy} orbital. Calculations by Pavlenko et al. [87] suggested that magnetism at the LAO/STO interface was not an intrinsic property of the interface electronic reconstruction but resulted from the orbital reconstruction induced by oxygen vacancies [88]. By using polarization-dependent X-ray absorption spectroscopy Salluzzo et al. [89] found that ferromagnetic moments in crystalline LAO/STO heterostructures were eliminated after oxygen annealing and therefore it was concluded that oxygen vacancies at the STO surface were crucial for the stabilization of interface ferromagnetism. Moreover, the oxygen-vacancy-induced orbital reconstruction at the LAO/STO interface has been experimentally observed by the recent resonant soft-X-ray scattering study [90], where the degeneracy of Ti t_{2g} orbitals is lifted due to oxygen vacancies, and the energy of d_{xy} orbital is lower than other two t_{2g} orbitals; oxygen-vacancy-induced ferromagnetism has been observed at the surface of Nb-doped STO single crystals [91]. These studies reveal that oxygen vacancies in STO are able to generate d_{xy} ferromagnetism. Therefore, whether the localized electrons in the Ti d_{xy} orbital are from oxygen vacancies or from electronic reconstruction is still an open issue. To clarify this issue, a thorough understanding of the origin of the 2DEG at LAO/STO interfaces would be helpful. In what follows, we will focus on the origin of the 2DEG at LAO/STO interfaces.

2.5   ORIGIN OF TWO-DIMENSIONAL ELECTRON GASES

2.5.1   THREE MECHANISMS

Since the first discovery of the 2DEG at the LAO/STO interface, a substantial body of experimental and theoretical work [19,21,22,24,25,42,43,58,92–108] has been devoted to the understanding of its origin. Generally, these studies provide three possible mechanisms. The first is interface electronic reconstruction to avoid the polarization catastrophe induced by the discontinuity at the interface between polar LAO and nonpolar STO [19]. The second is doping by thermal intermixing of Ti/Al or La/Sr atoms at the interface [93]. A third possible mechanism is creation of oxygen vacancies in STO substrates during the deposition process [22,25,92]. Oxygen vacancies are known to introduce a shallow intragap donor level close to the conduction band of STO [26], and their action may be specific to this one substrate. Although the interface mixing exists in oxide heterostructures, it was discounted in the recent work [109] regarding the origin of the interface conductivity, which studied the effect of a mixed interface layer and revealed a tunable critical thickness in accordance with the polar catastrophe model. It is also in conflict with the experimental results that p-type LAO/STO interfaces [16] and interfaces created by growing STO films on LAO are insulating [110].

2.5.2 BUILT-IN ELECTRIC POTENTIAL IN LAO

An essential prerequisite of the polar catastrophe mechanism is a built-in electric potential in the LAO layer according to its polar nature, which is expected to linearly increase with the LAO layer thickness. Experimental examination of such a potential build-up is therefore crucial to verify the validity of the polar catastrophe model. In 2009, Segal et al. [95] performed X-ray photoemission studies on the MIT in LAO/STO heterostructures. However, no band offset shift across the MIT was observed (Fig. 2.36).

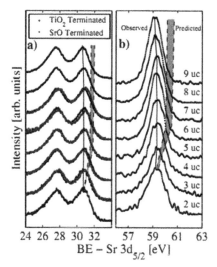

FIGURE 2.36   (a) La 4d and (b) Al 2p photoemissions in annealed LAO/STO samples. To cancel out charging and surface photovoltage effects, the x-axis is the binding energy relative to the Sr 3d 5/2 peak in the same sample. The thin solid line tracks the observed peak center, while gray bands follow the range of predicted shifts of the peak, according to the polar catastrophe picture for the TiO$_2$-terminated case. The predicted range is set by built-in fields of 0.6 and 0.9 V uc^{-1}. The upward shift expected to lead to the metal–insulator transition is not observed. Curves are shifted vertically for clarity (reproduced from Ref. [95] with permission). (Reprinted with permission from Segal, Y.; Ngai, J. H.; Reiner, J. W.; Walker, F. J.; Ahn, C. H. X-ray photoemission studies of the metal-insulator transition in LaAlO$_3$/SrTiO$_3$ structures grown by molecular beam epitaxy. *Phys. Rev. B* 2009, *80*, 241107. © 2009 by the American Physical Society.)

However, an obvious flaw of optical measurements is the photo-doping effect. STO is rather sensitive to photo radiation, leading to a large number of free carriers [111]. LAO/STO heterostructures exposed to high-energy X-ray are in the excited state and the large number of free carriers generated at the STO surface close to the interface could largely screen the polarization in the LAO layer. Thus, no thickness-dependent band offset shift could be detected.

In 2010, Singh-Bhalla et al. [100] carried out tunneling measurements between the 2DEG and metallic electrodes on LAO and a built-in electric field across LAO of 80.1 meV $Å^{-1}$ was observed. Later, using cross-sectional scanning tunneling microscopy and spectroscopy, Huang et al. [105] directly mapped out the electronic reconstruction and a built-in electric field of 30 ± 5 meV $Å^{-1}$ in the LAO layer. These studies were all conducted in the dark environment and the ground state of LAO/STO interfaces was explored, which clearly demonstrate the built-in electric potential in the LAO layer. The experimental values are much smaller than the theoretical values of 240 meV $Å^{-1}$ [42], which could be due to the partial screening of the electric potential by the 2DEG at the interface.

2.5.3 INTERFACE CONDUCTIVITY IN AMORPHOUS LAO/ STO

In 2007, Shibuya et al. [112] associated the metallic conductivity at interfaces between room temperature-deposited amorphous $CaHfO_3$ films and STO single crystal substrates with the bombardment of STO substrates by the plume during the PLD process. Nevertheless, a key experiment against the plume effect is that the growth of manganite films (such as $LaMnO_3$ [113] and $LaSrMnO_3$ [114]) on STO substrates always generates insulating interfaces. Later, Chen et al. [114] demonstrated metallic interfaces (Fig. 2.37) between STO substrates and various amorphous oxide overlayers including LAO, STO, and yittria-stabilized zirconia thin films fabricated by PLD. Instead, the origin of the 2DEG in such crystalline/amorphous heterostructures was attributed to formation of oxygen vacancies at the surface of the STO due to the strong chemical propensity of chemically active elements such as Al, Ti, and Zr [115] to oxygen atoms.

Moreover, metallic interfaces between Al-based amorphous oxides and STO substrates have also been realized by other less energetic deposition techniques such as atomic layer deposition [116] and electron beam evaporation [117]. The electronic properties of STO-based amorphous heterostructures

[114,116] are, to a large extent, similar to those of crystalline LAO/STO heterostructures [18,20], including the metallicity accompanied by the presence of Ti^{3+} ions and a sharp MIT as a function of overlayer thickness. These results call into question the polarization catastrophe model.

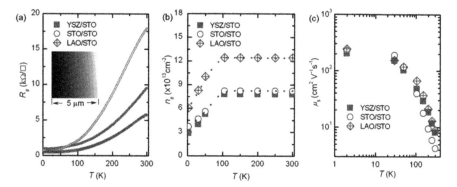

FIGURE 2.37 (a) Temperature-dependent sheet resistance, R_s, of the LAO/STO, STO/STO, and YSZ/STO heterostructures with about 8 nm amorphous capping films grown at $P_{O_2} \approx 1 \times 10^{-6}$ mbar. The inset shows a 5 × 5-µm atomic force microscopy image of the STO/STO sample with regular terraces of ~0.4 nm in height. (b,c) The sheet carrier density, n_s, and electron mobility, μ_s, versus temperature, respectively (reproduced from Ref. [114] with permission). (Reprinted with permission from Chen, Y.; Pryds, N.; Kleibeuker, J. E.; Koster, G.; Sun, J.; Stamate, E.; Shen, B.; Rijnders, G.; Linderoth, S. Metallic and Insulating Interfaces of Amorphous SrTiO3-Based Oxide Heterostructures. *Nano Lett.* 2011, *11*, 3774. © 2011 American Chemical Society.)

2.5.4 ROLE OF OXYGEN VACANCIES AND ELECTRONIC RECONSTRUCTION

To figure out the origin of 2DEGs at various LAO/STO interfaces, we performed a comprehensive comparison of (1 0 0)-oriented STO substrates with crystalline and amorphous overlayers of LAO of different thickness prepared under different oxygen pressures [43]. Photoluminescence (PL) in STO is a delicate way to detect oxygen vacancies inside [118]. First, we deposited 20-nm-thick amorphous LAO films on as-received STO substrates at room temperature and measured the PL properties of such heterostructures (aLAO/STO). It was found that the PL intensity in aLAO/STO heterostructures is increased by a factor of 5–9 relative to an as-received STO substrate (Fig. 2.38), depending on the oxygen partial pressure. This indicates the creation of oxygen vacancies in the STO substrates near their interface during the deposition process.

Electrical transport measurements revealed that such aLAO/STO hetero-structures have similar sheet resistance, carrier density, and mobility as crystalline LAO/STO heterostructures. In addition, there is also a critical thickness of the amorphous LAO layer for the appearance of interface conductivity (Fig. 2.39). However, the critical thickness in aLAO/STO heterostructures is strongly dependent on oxygen partial pressure. This is in sharp contrast to oxygen-annealed crystalline LAO/STO heterostructures, where the critical thickness being 4 uc is universal over a large oxygen pressure range.

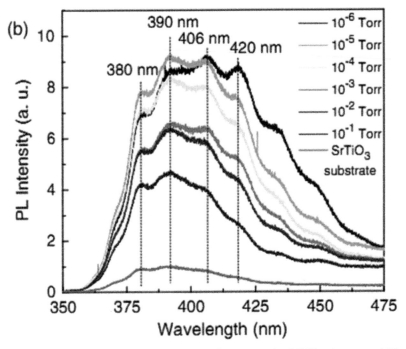

FIGURE 2.38 Room-temperature PL spectra of an as-received STO substrate and 20-nm amorphous LAO films deposited on untreated STO substrates at different oxygen partial pressure ranging from 10^{-1} to 10^{-6} Torr (reproduced from Ref. [43] with permission). (Reprinted from Liu, Z. Q.; Li, C. J.; Lü, W. M.; Huang, X. H.; Huang, Z.; Zeng, S. W.; Qiu, X. P.; Huang, L. S.; Annadi, A.; Chen, J. S.; Coey, J. M. D.; Venkatesan, T.; Ariando. Origin of the Two-Dimensional Electron Gas at LaAlO3/SrTiO3Interfaces: The Role of Oxygen Vacancies and Electronic Reconstruction. *Phys. Rev. X* 2013, *3*, 021010. https://creativecommons.org/licenses/by/3.0/)

Furthermore, the conductivity of all the aLAO/STO heterostructures vanishes after a 1-h post-anneal at 600°C in flowing oxygen (1 bar), as can be seen in Figure 2.40a. At the same time, the PL intensity of all oxygen-annealed amorphous LAO/STO heterostructures decreases significantly and approaches the intensity of the as-received substrate (Fig. 2.40b). This confirms that oxygen vacancies in STO create the conductivity. To compare the amorphous and crystalline LAO/STO heterostructures, 10-uc-thick crystalline LAO films were grown on TiO_2-terminated STO substrates under typical growth conditions, at 750°C and 10^{-3} Torr, and then post-annealed in oxygen as described above. They remain conductive, although there is a decrease in carrier concentration and the room-temperature sheet resistance increases by a factor of 7 (Fig. 2.40c).

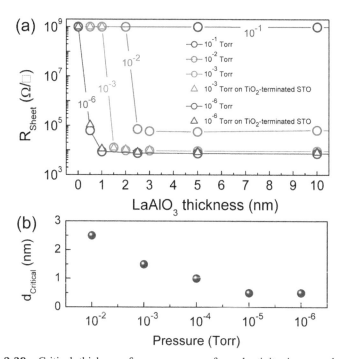

FIGURE 2.39 Critical thickness for appearance of conductivity in amorphous LAO/STO heterostructures. (a) Thickness dependence of room-temperature sheet resistance of amorphous LAO/STO heterostructures prepared at different oxygen pressures and on different STO substrates. Triangle symbols represent TiO_2-terminated STO substrates, while circles represent untreated STO substrates. (b) Critical thickness as a function of deposition oxygen pressure (reproduced from Ref. [43] with permission). (Reprinted from Liu, Z. Q.; Li, C. J.; Lü, W. M.; Huang, X. H.; Huang, Z.; Zeng, S. W.; Qiu, X. P.; Huang, L. S.; Annadi, A.; Chen, J. S.; Coey, J. M. D.; Venkatesan, T.; Ariando. Origin of the Two-Dimensional Electron Gas at LaAlO3/SrTiO3Interfaces: The Role of Oxygen Vacancies and Electronic Reconstruction. *Phys. Rev. X* 2013, *3*, 021010. https://creativecommons.org/licenses/by/3.0/)

Moreover, the n_s–T of the unannealed crystalline LAO/STO sample shows carrier freeze-out below about 100 K. In contrast, the carrier density of the post-annealed crystalline sample exhibits little temperature dependence. The carrier freeze-out effect in unannealed crystalline LAO/STO samples, which also exists in oxygen-deficient STO films [26], is characterized by an activation energy of 4.2 meV. In contrast, the activation energy of carriers in oxygen-annealed crystalline LAO/STO samples is even smaller, 0.5 meV. As shown in Fig. 2.40d, the PL intensity of the unannealed crystalline sample is greatly enhanced compared to that of its TiO_2-terminated STO substrate, which reveals the creation of a substantial amount of oxygen vacancies here too during deposition. After post-annealing, the PL intensity falls back to the substrate level, similar to the effect of post-annealing on the PL signal of amorphous samples.

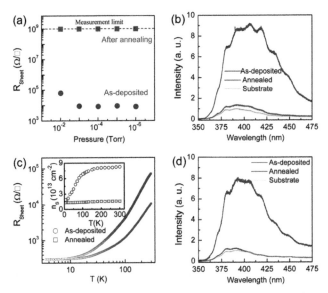

FIGURE 2.40 Oxygen-annealing effect. (a) Room-temperature sheet resistance of 20 nm amorphous LAO/STO heterostructures prepared at different oxygen pressures before and after oxygen-annealing in 1 bar of oxygen gas flow at 600°C for 1 h. (b) PL intensity of the 20 nm amorphous LAO/STO heterostructures fabricated at 10^{-6} Torr before and after oxygen-annealing. (c) R_s–T (inset) n_s–T, and (d) PL spectra of a 10-unit cell (uc) crystalline LAO/STO heterostructure prepared at 10^{-3} Torr and 750°C before and after oxygen-annealing in 1 bar of oxygen gas flow at 600°C for 1 h (reproduced from Ref. [43] with permission). (Reprinted from Liu, Z. Q.; Li, C. J.; Lü, W. M.; Huang, X. H.; Huang, Z.; Zeng, S. W.; Qiu, X. P.; Huang, L. S.; Annadi, A.; Chen, J. S.; Coey, J. M. D.; Venkatesan, T.; Ariando. Origin of the Two-Dimensional Electron Gas at LaAlO$_3$/SrTiO$_3$ Interfaces: The Role of Oxygen Vacancies and Electronic Reconstruction. *Phys. Rev. X* 2013, *3*, 021010. https://creativecommons.org/licenses/by/3.0/)

These results indicate that oxygen vacancies contribute significantly to the conductivity in *both* amorphous and unannealed crystalline LAO/STO heterostructures. Specifically, for amorphous LAO/STO samples, the existence of oxygen vacancies in STO substrates is the principal origin of the interface conductivity. For unannealed crystalline LAO/STO samples, oxygen vacancies are only partially responsible for a part of the interface conductivity, which can be eliminated by oxygen annealing.

Ar-milling experiments were performed to further explore the different mechanisms responsible for the interface conductivity in amorphous and oxygen-annealed crystalline LAO/STO heterostructures, fabricated at 10^{-3} Torr. It was found that after removing the top unit cell of LAO in a 4-uc oxygen-annealed LAO/STO heterostructure the conductivity disappears [red diamonds in Fig. 2.41(a)]. In contrast, Ar-milling the unannealed crystalline heterostructure from 10 uc down to 2 uc produces little change in conductance (blue circles), because the conduction is dominated by oxygen vacancies (Fig. 2.41a). Moreover, as the top amorphous LAO layer is removed, 1 nm at a time, from a 6-nm-thick aLAO/STO sample, the conductivity of the heterostructures is retained [green hollow diamonds in Fig. 2.41b].

These data clearly demonstrate that oxygen vacancies are the dominant source of mobile carriers when the LAO overlayer is amorphous, while both oxygen vacancies and polarization catastrophe contribute to the interface conductivity in unannealed crystalline LAO/STO heterostructures, and the polarization catastrophe alone accounts for the conductivity in oxygen-annealed crystalline LAO/STO heterostructures. In addition, it was found that the crystallinity of the LAO layer is crucial for the polarization catastrophe mechanism in the case of crystalline LAO overlayers [43].

2.6 2DEG AT THE LAO/STO (1 1 0) AND (1 1 1) INTERFACES

The research activities reviewed so far are focused to the (0 0 1)-oriented LAO/STO interfaces, and we now discuss new interfaces. Recent reports [32,33] have demonstrated that a similar 2DEG can be produced at the (1 1 0) and (1 1 1)-oriented LAO/STO interfaces. Herranz et al. [32] have prepared the LAO/STO (1 1 0) and (1 1 1) interfaces with different LAO layer thicknesses. Significantly these orientations are shown to exhibit the interface conductivity with a LAO critical thickness dependence similar to the (0 0 1) interfaces. However, the insulator to metal transition is shown to appears with a different critical thickness of LAO layers for (1 1 0) and (1 1

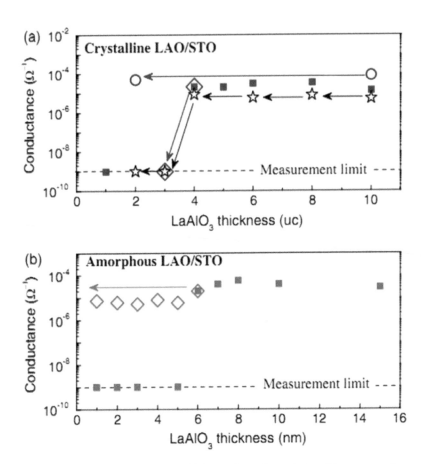

FIGURE 2.41 Ar-milling effect. (a) Thickness dependence (red solid squares) of room-temperature conductance of oxygen-annealed crystalline LAO/STO heterostructures fabricated at 10^{-3} Torr and 750°C, showing a critical thickness of 4 uc. The red hollow diamonds denote that the conductivity of the 4-uc sample disappears after the removal of the top 1 uc LAO by Ar-milling. Moreover, the blue hollow circles represent the conductance of an unannealed 10 uc crystalline LAO/STO heterostructure and after the removal of the top 8-uc LAO by Ar-milling. The black hollow stars represent the conductance of another oxygen-annealed 10-uc crystalline LAO/STO sample after step-by-step Ar milling. (b) Thickness dependence (green solid squares) of room-temperature conductance of amorphous LAO/STO heterostructures fabricated at 10^{-3} Torr, showing a critical thickness of 6 nm. The green hollow diamonds represent the conductivity of the 6-nm sample that remains after the removal of the top LAO layer 1 nm at a time by Ar-milling. All the arrows represent the Ar-milling process (reproduced from Ref. [43] with permission). (Reprinted from Liu, Z. Q.; Li, C. J.; Lü, W. M.; Huang, X. H.; Huang, Z.; Zeng, S. W.; Qiu, X. P.; Huang, L. S.; Annadi, A.; Chen, J. S.; Coey, J. M. D.; Venkatesan, T.; Ariando. Origin of the Two-Dimensional Electron Gas at LaAlO₃/SrTiO₃ Interfaces: The Role of Oxygen Vacancies and Electronic Reconstruction. *Phys. Rev. X* 2013, 3, 021010. https://creativecommons.org/licenses/by/3.0/)

1). The critical thickness of LAO are found to be seven MLs and nine MLs for (1 1 0) and (1 1 1) interfaces, respectively.

The LAO thickness dependent transport for the LAO/STO (1 1 0) and (1 1 1) interfaces is shown in Figure 2.42. Here, for both the orientations, thinner LAO samples show a typical metallic behavior down to low temperatures, whereas higher thickness samples shows an evolution of the localization, to insulator behavior at low temperatures. This thickness-dependent transport is also observed for the (0 0 1) case [57], however, with a different critical LAO layer numbers. In a subsequent report, the LAO/STO (1 1 0) interface is more extensively investigated by Annadi et al. [33] with oxygen growth pressure (P_{O_2}) in the range of 5×10^{-5}–10^{-3} Torr and with different LAO layer thicknesses. As shown in Figure 2.43a, the LAO/STO (1 1 0) interfaces supports the 2DEG, and the temperature-dependent sheet resistance (R_s) was found to be very sensitive to the P_{O_2}.

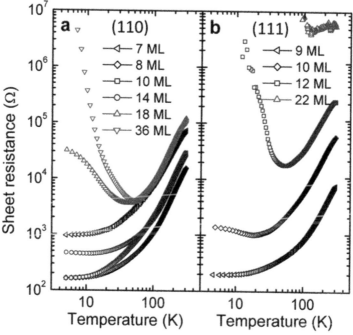

FIGURE 2.42 Figure parts (a) and (b) show sheet resistance variation with temperature for the LAO/STO (1 1 0) and (1 1 1) interfaces for various LAO thicknesses, respectively (reproduced from Ref. [32] with permission). (Reprinted from Herranz, G.; Sánchez, F.; Dix, N.; Scigaj, M.; Fontcuberta, J. High mobility conduction at (110) and (111) LaAlO₃/SrTiO₃ interfaces. *Sci. Rep.* 2012, 2, 758. https://www.nature.com/articles/srep00758.)

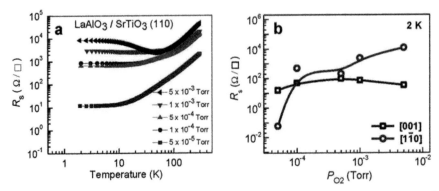

FIGURE 2.43 (a) Resistance (R_s) behavior with temperature for the LAO/STO (1 1 0) interfaces prepared at various oxygen growth pressure (P_{O_2}) and (b) sheet resistance measured with respect to the oxygen growth pressure (P_{O_2}) along crystallographic directions for the LAO/STO (1 1 0) interface (reproduced from Ref. [33] with permission). (Reprinted from Annadi, A.; Zhang, Q.; Renshaw Wang, X.; Tuzla, N.; Gopinadhan, K.; Lü, W. M.; Roy Barman, A.; Liu, Z. Q.; Srivastava, A.; Saha, S.; Zhao, Y. L.; Zeng, S. W.; Dhar, S.; Olsson, E.; Gu, B.; Yunoki, S.; Maekawa, S.; Hilgenkamp, H.; Venkatesan, T.; Ariando. Anisotropic two-dimensional electron gas at the LaAlO$_3$/SrTiO$_3$ (110) interface. *Nat. Commun.* 2013, *4*, 1838.Nature Publishing Group. https://www.nature.com/articles/ncomms2804)

The higher P_{O_2} (>1 × 10^{-4} Torr) samples show a localized behavior at low temperatures, whereas the low P_{O_2} sample show high conductivity with high mobility. This oxygen growth pressure dependent transport is also observed for conventional (1 0 0) interfaces, however, with a different critical thickness of LAO. An additional property of the (1 1 0) interface is the strong directional dependence of transport properties which is shown in Fig. 2.43b, and the R_s (at 2 K) measured along the [1–10] direction is found more sensitive to the P_{O_2} than the (0 0 1) direction. The anisotropic atomic arrangement of Ti–O bonding along these directions is proposed to influence the transport [33].

In the context of the origin of the conductivity at the interface, these results call a question on what has been learnt so far. According to electrostatics while the (1 1 1) interface exhibit a polarization discontinuity, it is not the scenario for the (1 1 0) interface where it should not have the polarization discontinuity, hence no conductivity is expected. In this scenario, it appears that the oxygen vacancy could be the origin for the experimentally observed conductivity at these interfaces. However, these crystalline (1 1 0) interfaces also show a post-annealing stability [32] brings again the same case to that of conventional (0 0 1) interfaces to make role of oxygen vacancies debatable.

Considering the instability of the polar stoichiometric STO termination in the ideal case of STO (1 1 0) surface, it is proposed that an alternative surface structure for STO (1 1 0) with TiO termination is energetically stable [33]. Significantly, in the DFT calculations, it is shown that the LAO/STO interface constructed with TiO-terminated STO (1 1 0) produces a polarization discontinuity and thereby a built in potential with the LAO layer number. Further, it induces an insulator to metal transition with a critical LAO thickness similar to the results of Petcheva et al. [21] for the (0 0 1) case. The possibility of this Ti-rich termination basis for the origin of conductivity at these polar surfaces needs further serious investigation. Recent experimental reports on STO (1 1 0) surfaces [119,120] hinted a high possibility to form various energetically stable Ti-rich reconstructed surfaces which further calls for an unprecedented look at these ideal polar surfaces.

Despite adding further argument to the origin of conductivity, these new findings at the LAO/STO (1 1 0) and (1 1 1) interfaces have lifted the constraint of the crystallography and stimulated the research activity along these crystallographic orientations. Recently, using DFT calculations, Doennig et al. [121] have reported a massive symmetry breaking along the LAO/STO (1 1 1) orientation with a Dirac point Fermi surface, charge-ordered flat band phases. Furthermore experimentally, Wang et al. [122] have reported anisotropic nature of band structure at the doped STO (1 1 0) surface, Herranz et al. [123] have reported superconductivity ($T_c \sim 200$ mK) at the LAO/STO (1 1 0) interface with avoiding the violation of the paramagnetic Pauli limit criterion. This is attributed to the extended nature of superconductivity in LAO/STO (1 1 0) interfaces when compared to (1 0 0) interfaces which obey the violation of the paramagnetic Pauli limit criterion. In another theoretical report [124], a different orbital hybridizations for the (1 1 0) and (1 1 1) interfaces is predicted and thus different magnetic ground states. All the above exciting new observations arise due to the complex orbital and band structure at these reduced-symmetric crystallographic orientations. It is further very stimulating to investigate the magnetic phase diagram and orbital nature through spectroscopy at these interfaces. This suggests a possibility of novel phenomena await at these new interfaces.

2.7 APPLICATIONS AND DEVICES

High mobility of the 2DEG at the LAO/STO interface can be utilized in field-effect transistor devices, which have been demonstrated by many groups. Nanoscale electronic devices based on the LAO/STO interface were pioneered by Levy's group, where they can control the MIT at room temperature on a nanoscale and build nanoscale field-effect transistors using an atomic force microscope lithography technique [125,126]. More excitingly, sketched single-electron transistors have been demonstrated by the same group for the LAO/STO interface [127]. It was found that such nanoscale transistors can be operated at frequencies in excess of 2 GHz [128]. In addition, based on the LAO/STO interface, Jany et al. [129] demonstrated diodes with breakdown voltages larger than 200 V and Förg et al. [130] illustrated field-effect devices (Fig. 2.44) which can be operated at temperatures up to 100°C (Fig. 2.45).

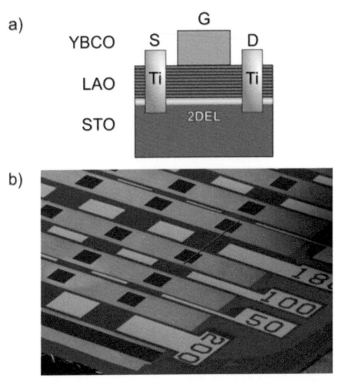

FIGURE 2.44 Sketch of a cross section of a device (a) and electron microscope image of a typical sample (b). The colors were added. The horizontal lines within the LAO-layer symbolize the 9 MLs of LAO, the standard thickness of gate dielectric in this study, grown on a STO substrate. The narrow, straight line (red) denotes the location of the cross section shown in (a); the numbers indicate the gate widths in microns. The two-dimensional electron liquid shown in (b) in pale gray is also present under the $YBa_2Cu_3O_7$ gates (dark gray) (reproduced from Ref. [130] with permission). (Reprinted from Förg, B.; Richter, C.; Mannhart, J. Field-effect devices utilizing $LaAlO_3$-$SrTiO_3$ interfaces. Appl. Phys. Lett. 2012, 100, 053506. © 2012 with the permission of AIP Publishing.)

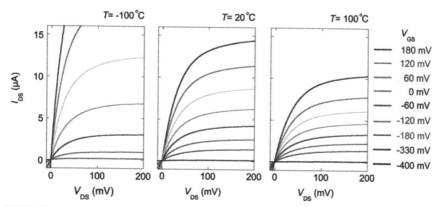

FIGURE 2.45 I_{DS} (V_{DS})-characteristics of a device measured in 4-point configuration at −100, 20, and 100°C. The measurement was done on a device with channel length of 40 μm and channel width of 1600 μm (reproduced from Ref. [130] with permission). (Reprinted from Förg, B.; Richter, C.; Mannhart, J. Field-effect devices utilizing LaAlO$_3$-SrTiO$_3$ interfaces. *Appl. Phys. Lett.* 2012, *100*, 053506. © 2012 with the permission of AIP Publishing.)

Especially, the successful integration of the 2DEG at the LAO/STO interface with Silicon by Park et al. [131] has created an exciting way of incorporating this multifunctional oxide interface system for large-scale oxide electronic device applications in Si-based platforms. Overall, oxide electronics on Si is rather promising but still lots of things need to be done in the future.

KEYWORDS

- two-dimensional electron gas
- LaAlO$_3$/SrTiO$_3$ interfaces
- electronic reconstruction
- ferromagnetism
- electronic devices

REFERENCES

1. Bednorz, J. A.; Müller, K. A. *Zeitschr. Phys. B: Condens. Matter* **1986**, *64*, 189.
2. Wu, M. K.; Ashburn, J. R.; Torng, C. J.; Hor, P. H.; Meng, R. L.; Gao, L.; Huang, Z. J.; Wang, Y. Q.; Chu, C. W. *Phys. Rev. Lett.* **1987**, *58*, 908.

3. Mott, N. F. *Rev. Mod. Phys.* **1968**, *40*, 677.
4. Imada, M.; Fujimori, A.; Tokura, Y. *Rev. Mod. Phys.* **1998**, *70*, 1039.
5. von Helmolt, R.; Wecker, J.; Holzapfel, B.; Schultz, L.; Samwer, K. *Phys. Rev. Lett.* **1993**, *71*, 2331.
6. Jin, S.; Tiefel, T. H.; Mccormack, M.; Fastnacht, R. A.; Ramesh, R.; Chen, L. H. *Science* **1994**, *264*, 413.
7. Yunoki, S.; Hu, J.; Malvezzi, A.; Moreo, A.; Furukawa, N.; Dagotto, E.; *Phys. Rev. Lett.* **1998**, *80*, 845.
8. Uehara, M.; Mori, S.; Chen, C. H.; Cheong, S.-W. *Nature* **1999**, *399*, 560.
9. Ariando; Wang, X.; Baskaran, G.; Liu, Z. Q.; Huijben, J.; Yi, J. B.; Annadi, A.; Barman, A. R.; Rusydi, A.; Dhar, S.; Feng, Y. P.; Ding, J.; Hilgenkamp, H.; Venkatesan, T. *Nat. Commun.* **2011**, *2*, 188.
10. Dawber, M.; Rabe, K. M.; Scott, J. F. *Rev. Mod. Phys.* **2005**, *77*, 1083.
11. Ramesh, R.; Spaldin, N. A. *Nat. Mater.* **2007**, *6*, 21.
12. Hwang, H. Y.; Iwasa, Y.; Kawasaki, M.; Keimer, B.; Nagaosa, N.; Tokura, Y. *Nat. Mater.* **2012**, *11*, 103.
13. Haeni, J. H.; Irvin, P.; Chang, W.; Uecker, R.; Reiche, P.; Li, Y. L.; Choudhury, S.; Tian, W.; Hawley, M. E.; Craigo, B.; Tagantsen, A. K.; Pan, X. Q.; Streiffer, S. K.; Chen, L. Q.; Kirchoefer, S. W.; Levy, J.; Schlom, D. G. *Nature* **2004**, *430*, 758.
14. Yu, P.; Luo, W.; Yi, D.; Zhang, J. X.; Rossell, M. D.; Yang, C.; You, L.; Singh-bhalla, G.; Yang, S. Y.; He, Q.; Ramasse, Q. M.; Erni, R.; Martin, L. W.; Chu, Y. H.; Pantelides, S. T.; Pennycook, S. J.; Ramesh, R. *Proc. Natl. Acad. Sci. USA* **2012**, *109*, 9710.
15. Satapathy, D.; Uribe-Laverde, M.; Marozau, I.; Malik, V.; Das, S.; Wagner, T.; Marcelot, C.; Stahn, J.; Brück, S.; Rühm, A.; Macke, S.; Tietze, T.; Goering, E.; Fraño, A.; Kim, J.; Wu, M.; Benckiser, E.; Keimer, B.; Devishvili, A.; Toperverg, B.; Merz, M.; Nagel, P.; Schuppler, S.; Bernhard, C. *Phys. Rev. Lett.* **2012**, *108*, 197201.
16. Ohtomo, A.; Hwang, H. Y. *Nature* **2004**, *427*, 423.
17. Reyren, N.; Thiel, S.; Caviglia, A. D.; Kourkoutis, L. F.; Hammerl, G.; Richter, C.; Schneider, C. W.; Kopp, T.; Ruetschi, A.-S.; Jaccard, D.; Gabay, M.; Muller, D. A.; Triscone, J.-M.; Mannhart, J. *Science* **2007**, *317*, 1196.
18. Brinkman, A.; Huijben, M.; van Zalk, M.; Huijben, J.; Zeitler, U.; Maan, J. C.; van der Wiel, W. G.; Rijnders, G.; Blank, D. H. A.; Hilgenkamp, H. *Nat. Mater.* **2007**, *6*, 493.
19. Nakagawa, N.; Hwang, H. Y.; Müller, D. A. *Nat. Mater.* **2006**, *5*, 204.
20. Thiel, S.; Hammerl, G.; Schmehl, A.; Schneider, C. W.; Mannhart, J. *Science* **2006**, *313*, 1942.
21. Pentcheva, R.; Pickett, W. E. *Phys. Rev. Lett.* **2009**, *102*, 107602.
22. Siemons, W.; Koster, G.; Yamamoto, H.; Harrison, W. A.; Lucovsky, G.; Geballe, T. H.; Blank, D. H. A.; Beasley, M. R. *Phys. Rev. Lett.* **2007**, *98*, 196802.
23. Herranz, G.; Basletic, M.; Bibes, M.; Carre, C.; Tafra, E.; Jacquet, E.; Bouzehouane, K.; Deranlot, C.; Barthe, A.; Fert, A.; Hamzic, A.; Broto, J.-M.; Barthelemy, A. *Phys. Rev. Lett.* **2007**, *98*, 216803.
24. Eckstein, J. N. *Nat. Mater.* **2007**, *6*, 473.
25. Kalabukhov, A.; Gunnarsson, R.; Olsson, E.; Claeson, T.; Winkler, D. *Phys. Rev. B* **2007**, *75*, 121404(R).
26. Liu, Z. Q.; Leusink, D. P.; Wang, X.; Lü, W. M.; Gopinadhan, K.; Annadi, A.; Zhao, Y. L.; Huang, X. H.; Zeng, S. W.; Huang, Z.; Srivastava, A.; Dhar, S.; Venkatesan, T.; Ariando. *Phys. Rev. Lett.* **2011**, *107*, 146802.

27. Willmott, P. R.; Pauli, S. A.; Herger, R.; Schleputz, C. M.; Martoccia, D.; Patterson, B. D.; Delley, B.; Clarke, R.; Kumah, D.; Cionca, C.; Yacoby, Y. *Phys. Rev. Lett.* **2007**, *99*, 155502.

28. Caviglia, A. D.; Gariglio, S.; Reyren, N.; Jaccard, D.; Schneider, T.; Gabay, M.; Thiel, S.; Hammerl, G.; Mannhart, J.; Triscone, J.-M. *Nature* **2008**, *456*, 624.

29. Dikin, D. A.; Mehta, M.; Bark, C. W.; Folkman, C. M.; Eom, C. B.; Chandrasekhar, V. *Phys. Rev. Lett.* **2011**, *107*, 056802.

30. Bert, J. A.; Kalisky, B.; Bell, C.; Kim, M.; Hikita, Y.; Hwang, H. Y.; Moler, K. A. *Nat. Phys.* **2011**, *7*, 767.

31. Li, L.; Richter, C.; Mannhart, J.; Ashoori, R. C. *Nat. Phys.* **2011**, *7*, 762.

32. Herranz, G.; Sánchez, F.; Dix, N.; Scigaj, M.; Fontcuberta, J. *Sci. Rep.* **2012**, *2*, 758.

33. Annadi, A.; Zhang, Q.; Renshaw Wang, X.; Tuzla, N.; Gopinadhan, K.; Lü, W. M.; Roy Barman, A.; Liu, Z. Q.; Srivastava, A.; Saha, S.; Zhao, Y. L.; Zeng, S. W.; Dhar, S.; Olsson, E.; Gu, B.; Yunoki, S.; Maekawa, S.; Hilgenkamp, H.; Venkatesan, T.; Ariando *Nat. Commun.* **2013**, *4*, 1838.

34. Huijben, M.; Brinkman, A.; Koster, G.; Rijnders, G.; Hilgenkamp, H.; Blank, D. H. A. *Adv. Mater.* **2009**, *21*, 1665.

35. Luo, X.; Wang, B.; Zheng, Y. *Phys. Rev. B* **2009**, *80*, 104115.

36. Liu, Z. Q.; Leusink, D.; Lü, W. M.; Wang, X.; Yang, X.; Gopinadhan, K.; Lin, Y.; Annadi, A.; Zhao, Y.; Barman, A.; Dhar, S.; Feng, Y.; Su, H.; Xiong, G.; Venkatesan, T. *Phys. Re. B* **2011**, *84*, 165106.

37. Maurice, J.-L.; Herranz, G.; Colliex, C.; Devos, I.; Carrétéro, C.; Barthélémy, A.; Bouzehouane, K.; Fusil, S.; Imhoff, D.; Jacquet, É.; Jomard, F.; Ballutaud, D.; Basletic, M. *Europhys. Lett.* **2008**, *82*, 17003.

38. Caviglia, A. D.; Gariglio, S.; Cancellieri, C.; Sacépé, B.; Fête, A.; Reyren, N.; Gabay, M.; Morpurgo, A. F.; Triscone, J.-M. *Phys. Rev. Lett.* **2010**, *105*, 236802.

39. Venkatesan, T.; Wu, X. D.; Inam, A.; Wachtman, J. B. *Appl. Phys. Lett.* **1988**, *52*, 1193.

40. Breckenfeld, E.; Bronn, N.; Karthik, J.; Damodaran, A. R.; Lee, S.; Mason, N.; Martin, L. W. *Phys. Rev. Lett.* **2013**, *110*, 196804.

41. Fix, T.; Schoofs, F.; Bi, Z.; Chen, A.; Wang, H.; MacManus-Driscoll, J. L.; Blamire, M. G. *Appl. Phys. Lett.* **2011**, *99*, 022103.

42. Lee, J.; Demkov, A. *Phys. Rev. B* **2008**, *78*, 193104.

43. Liu, Z. Q.; Li, C. J.; Lü, W. M.; Huang, X. H.; Huang, Z.; Zeng, S. W.; Qiu, X. P.; Huang, L. S.; Annadi, A.; Chen, J. S.; Coey, J. M. D.; Venkatesan, T.; Ariando. *Phys. Rev. X* **2013**, *3*, 021010.

44. Basletic, M.; Maurice, J.-L.; Carrétéro, C.; Herranz, G.; Copie, O.; Bibes, M.; Jacquet, E.; Bouzehouane, K.; Fusil, S.; Barthélémy, A. *Nat. Mater.* **2008**, *7*, 621.

45. Copie, O.; Garcia, V.; Bödefeld, C.; Carrétéro, C.; Bibes, M.; Herranz, G.; Jacquet, E.; Maurice, J.-L.; Vinter, B.; Fusil, S.; Bouzehouane, K.; Jaffrès, H.; Barthélémy, A. *Phys. Rev. Lett.* **2009**, *102*, 216804.

46. Sing, M.; Berner, G.; Goß, K.; Müller, A.; Ruff, A.; Wetscherek, A.; Thiel, S.; Mannhart, J.; Pauli, S.; Schneider, C.; Willmott, P.; Gorgoi, M.; Schäfers, F.; Claessen, R. *Phys. Rev. Lett.* **2009**, *102*, 176805.

47. Janicka, K.; Velev, J.; Tsymbal, E. *Phys. Rev. Lett.* **2009**, *102*, 106803.

48. Fix, T.; Schoofs, F.; MacManus-Driscoll, J.; Blamire, M. *Phys. Rev. Lett.* **2009**, *103*, 166802.

49. Dubroka, A.; Rössle, M.; Kim, K. W.; Malik, V. K.; Schultz, L.; Thiel, S.; Schneider, C. W.; Mannhart, J.; Herranz, G.; Copie, O.; Bibes, M.; Barthélémy, A.; Bernhard, C. *Phys. Rev. Lett.* **2010**, *104*, 156807.

50. Delugas, P.; Filippetti, A.; Fiorentini, V.; Bilc, D. I.; Fontaine, D.; Ghosez, P. *Phys. Rev. Lett.* **2011**, *106*, 166807.

51. Pallecchi, I.; Codda, M.; Galleani d'Agliano, E.; Marré, D.; Caviglia, A. D.; Reyren, N.; Gariglio, S.; Triscone, J.-M. *Phys. Rev. B* **2010**, *81*, 085414.

52. Huang, Z.; Wang, X. R.; Liu, Z. Q.; Lü, W. M.; Zeng, S. W.; Annadi, A.; Tan, W. L.; Qiu, X. P.; Zhao, Y. L.; Salluzzo, M.; Coey, J. M. D.; Venkatesan, T. *Phys. Rev. B* **2013**, *88*, 161107.

53. Brinks, P.; Siemons, W.; Kleibeuker, J. E.; Koster, G.; Rijnders, G.; Huijben, M. *Appl. Phys. Lett.* **2011**, *98*, 242904.

54. Schooley, J. F.; Hosler, W. R.; Cohen, M. L. *Phys. Rev. Lett.* **1964**, *12*, 474.

55. Reyren, N.; Gariglio, S.; Caviglia, A. D.; Jaccard, D.; Schneider, T.; Triscone, J.-M. *Appl. Phys. Lett.* **2009**, *94*, 112506.

56. Richter, C.; Boschker, H.; Dietsche, W.; Fillis-Tsirakis, E.; Jany, R.; Loder, F.; Kourkoutis, L. F.; Muller, D. A.; Kirtley, J. R.; Schneider, C. W.; Mannhart, J. *Nature* **2013**, 502, 528.

57. Bell, C.; Harashima, S.; Hikita, Y.; Hwang, H. Y. *Appl. Phys. Lett.* **2009**, *94*, 222111.

58. Popovic, Z. S.; Satpathy, S.; Martin, R. M. *Phys. Rev. Lett.* **2008**, *101*, 256801.

59. Wong, F. J.; Chopdekar, R. V.; Suzuki, Y. *Phys. Rev. B* **2010**, *82*, 165413.

60. Hernandez, T.; Bark, C. W.; Felker, D. A.; Eom, C. B.; Rzchowski, M. S. *Phys. Rev. B* **2012**, *85*, 161407.

61. Ben Shalom, M.; Ron, A.; Palevski, A.; Dagan, Y. *Phys. Rev. Lett.* **2010**, *105*, 206401.

62. Seo, S. S. A.; Marton, Z.; Choi, W. S.; Hassink, G. W. J.; Blank, D. H. A.; Hwang, H. Y.; Noh, T. W.; Egami, T.; Lee, H. N. *Appl. Phys. Lett.* **2009**, *95*, 082107.

63. A. Joshua, S. Pecker, J. Ruhman, E. Altman; S. Ilani, *Nat. Commun.* **2012**, *3*, 1129.

64. Guduru, V. K.; Granados del Aguila, A.; Wenderich, S.; Kruize, M. K.; McCollam, A.; Christianen, P. C. M.; Zeitler, U.; Brinkman, A.; Rijnders, G.; Hilgenkamp, H.; Maan, J. C. *Appl. Phys. Lett.* **2013**, *102*, 051604.

65. Liu, Z. Q.; Lü, W. M.; Wang, X.; Huang, Z.; Annadi, A.; Zeng, S. W.; Venkatesan, T.; Ariando. *Phys. Rev. B* **2012**, *85*, 155114.

66. Bark, C. W.; Felker, D. A.; Wang, Y.; Zhang, Y.; Jang, H. W.; Folkman, C. M.; Park, J. W.; Baek, S. H.; Zhou, H.; Fong, D. D.; Pan, X. Q.; Tsymbal, E. Y.; Rzchowski, M. S.; Eom, C. B. *Proc. Natl. Acad. Sci. USA* **2011**, *108*, 4720.

67. Migita, S. S.; Kasai, Y. *J. Low Temp. Phys.* **1996**, *105*, 1337.

68. Takahashi, R.; Matsumoto, Y.; Ohsawa, T.; Lippmaa, M.; Kawasaki, M. *J. Cryst. Growth* **2002**, *234*, 505.

69. Koster, G.; Rijnders, G. J. H. M.; Blank, D. H. A.; Rogalla, H. *Appl. Phys. Lett.* **1999**, *74*, 3729.

70. Pentcheva, R.; Pickett, W. *Phys. Rev. B* **2006**, *74*, 035112.

71. Park, M.; Rhim, S.; Freeman, A. *Phys. Rev. B* **2006**, *74*, 205416.

72. Zhang, L.; Zhou, X.-F.; Wang, H.-T.; Xu, J.-J.; Li, J.; Wang, E. G.; Wei, S.-H. *Phys. Rev. B* **2010**, *82*, 125412.

73. Rubano, A.; Günter, T.; Fink, T.; Paparo, D.; Marrucci, L.; Cancellieri, C.; Gariglio, S.; Triscone, J.-M.; Fiebig, M. *Phys. Rev. B* **2013**, *88*, 035405.

74. Janicka, K.; Velev, J. P.; Tsymbal, E. Y. *J. Appl. Phys.* **2008**, *103*, 07B508.

75. Ben Shalom, M.; Tai, C. W.; Lereah, Y.; Sachs, M.; Levy, E.; Rakhmilevitch, D.; Palevski, A.; Dagan, Y. *Phys. Rev. B* **2009**, *80*, 140403(R).
76. Seri, S.; Klein, L. *Phys. Rev. B* **2009**, *80*, 180410.
77. Fitzsimmons, M. R.; Hengartner, N. W.; Singh, S.; Zhernenkov, M.; Bruno, F. Y.; Santamaria, J.; Brinkman, A.; Huijben, M.; Molegraaf, H. J. A.; de la Venta, J.; Schuller, I. K. *Phys. Rev. Lett.* **2011**, *107*, 217201.
78. Fix, T.; MacManus-Driscoll, J. L.; Blamire, M. G. *Appl. Phys. Lett.* **2009**, *94*, 172101.
79. Millis, A. J. *Nat. Phys.* **2011**, *7*, 749.
80. Gariglio, S.; Triscone, J.-M.; Caviglia, A. *Physics* **2011**, *4*, 59.
81. Michaeli, K.; Potter, A. C.; Lee, P. A. *Phys. Rev. Lett.* **2012**, *108*, 117003.
82. Caprara, S.; Peronaci, F.; Grilli, M. *Phys. Rev. Lett.* **2012**, *109*, 196401.
83. Salman, Z.; Ofer, O.; Radovic, M.; Hao,. H.; Ben Shalom, M.; Chow, K. H.; Dagan, Y.; Hossain, M. D.; Levy, C. D. P.; MacFarlane, W. A.; Morris, G. M.; Patthey, L.; Pearson, M. R.; Saadaoui, H.; Schmitt, T.; Wang, D.; Kiefl, R. F. *Phys. Rev. Lett.* **2012**, *109*, 257207.
84. Kalisky, B.; Bert, J. A.; Klopfer, B. B.; Bell, C.; Sato, H. K.; Hosoda, M.; Hikita, Y.; Hwang, H. Y.; Moler, K. A. *Nat. Commun.* **2012**, *3*, 922.
85. Lee, J.-S.; Xie, Y. W.; Sato, H. K.; Bell, C.; Hikita, Y.; Hwang, H. Y.; Kao, C.-C. *Nat. Mater.* **2013**, *12*, 703.
86. Lee, J.-S.; Xie, Y. W.; Sato, H. K.; Bell, C.; Hikita, Y.; Hwang, H. Y.; Kao, C.-C. *Nat. Mater.* **2013**, *12*, 703.
87. Pavlenko, N.; Kopp, T.; Tsymbal, E. Y.; Sawatzky, G. A.; Mannhart, J. *Phys. Rev. B* **2012**, *85*, 020407(R).
88. Pavlenko, N.; Kopp, T.; Tsymbal, E. Y.; Mannhart, J.; Sawatzky, G. A. *Phys. Rev. B* **2012**, *86*, 064431.
89. Salluzzo, M.; Gariglio, S.; Stornaiuolo, D.; Sessi, V.; Rusponi, S.; Piamonteze, C.; De Luca, G. M.; Minola, M.; Marr, D.; Gadaleta, A.; Brune, H.; Nolting, F.; Brookes, N. B.; Ghiringhelli, G. *arXiv* **2013**, *1305.2226*.
90. Park, J.; Cho, B.-G.; Kim, K. D.; Koo, J.; Jang, H.; Ko, K.-T.; Park, J.-H.; Lee, K.-B.; Kim, J.-Y.; Lee, D. R.; Burns, C. A.; Seo, S. S. A.; Lee, H. N. *Phys. Rev. Lett.* **2013**, *110*, 017401.
91. Liu, Z. Q.; Lü, W. M.; Lim, S. L.; Qiu, X. P.; Bao, N. N.; Motapothula, M.; Yi, J. B.; Yang, M.; Dhar, S.; Venkatesan, T.; Ariando. *Phys. Rev. B* **2013**, *87*, 220405(R).
92. Herranz, G.; Basletic, M.; Bibes, M.; Carretero, C.; Tafra, E.; Jacquet, E.; Bouzehouane, K.; Deranlot, C.; Hamzic, A.; Broto, J.-M.; Barthelemy, A.; Fert, A. *Phys. Rev. Lett.* **2007**, *98*, 216803.
93. Willmott, P. R.; Pauli, S. A.; Herger, R.; Schleputez, C. M.; Martoccia, D.; Patterson, B. D.; Delley, B.; Clarke, R.; Kumah, D.; Cionca, C.; Yacoby, Y. *Phys. Rev. Lett.* **2007**, *99*, 155502.
94. Yoshimatsu, K.; Yasuhara, R.; Kumigashira, H.; Oshima, M. *Phys. Rev. Lett.* **2008**, *101*, 026802.
95. Segal, Y.; Ngai, J. H.; Reiner, J. W.; Walker, F. J.; Ahn, C. H. *Phys. Rev. B* **2009**, *80*, 241107.
96. Salluzzo, M.; Cezar, J.; Brookes, N.; Bisogni, V.; De Luca, G.; Richter, C.; Thiel, S.; Mannhart, J.; Huijben, M.; Brinkman, A.; Rijnders, G.; Ghiringhelli, G. *Phys. Rev. Lett.* **2009**, *102*, 166804.
97. Kalabukhov, A.; Boikov, Y.; Serenkov, I.; Sakharov, V.; Popok, V.; Gunnarsson, R.; Börjesson, J.; Ljustina, N.; Olsson, E.; Winkler, D.; Claeson, T. *Phys. Rev. Lett.* **2009**, *103*, 146101.

98. Savoia, A.; Paparo, D.; Perna, P.; Ristic, Z.; Salluzzo, M.; Miletto Granozio, F.; Scotti di Uccio, U.; Richter, C.; Thiel, S.; Mannhart, J.; Marrucci, L. *Phys. Rev. B* **2009,** *80,* 075110.

99. Zhong, Z.; Xu, P. X.; Kelly, P. J. *Phys. Rev. B* **2010,** *82,* 165127.

100. Singh-Bhalla, G.; Bell, C.; Ravichandran, J.; Siemons, W.; Hikita, Y.; Salahuddin, S.; Hebard, A. F.; Hwang, H. Y.; Ramesh, R. *Nat. Phys.* **2010,** *7,* 80.

101. Takizawa, M.; Tsuda, S.; Susaki, T.; Hwang, H. Y.; Fujimori, A. *Phys. Rev. B* **2011,** *84,* 245124.

102. Yamamoto, R.; Bell, C.; Hikita, Y.; Hwang, H. Y.; Nakamura, H.; Kimura, T.; Wakabayashi, Y. *Phys. Rev. Lett.* **2011,** *107,* 036104.

103. Qiao, L.; Droubay, T. C.; Varga, T.; Bowden, M. E.; Shutthanandan, V.; Zhu, Z.; Kaspar, T. C.; Chambers, S. A. *Phys. Rev. B* **2011,** *83,* 085408.

104. Cantoni, C.; Gazquez, J.; Miletto Granozio, F.; Oxley, M. P.; Varela, M.; Lupini, A. R.; Pennycook, S. J.; Aruta, C.; di Uccio, U. S.; Perna, P.; Maccariello, D. *Adv. Mater. (Deerfield Beach, Fla.)* **2012,** *24,* 3952.

105. Huang, B.; Chiu, Y.; Huang, P.; Wang, W.; Tra, V. T.; Yang, J. C.; He, Q.; Lin, J. Y.; Chang, C. S.; Chu, Y. H. *Phys. Rev. Lett.* **2012,** *109,* 246807.

106. Vonk, V.; Huijben, J.; Kukuruznyak, D.; Stierle, A.; Hilgenkamp, H.; Brinkman, A.; Harkema, S. *Phys. Rev. B* **2012,** *85,* 045401.

107. Slooten, E.; Zhong, Z.; Molegraaf, H. J. A.; Eerkes, P. D.; de Jong, S.; Massee, F.; van Heumen, E.; Kruize, M. K.; Wenderich, S.; Kleibeuker, J. E.; Gorgoi, M.; Hilgenkamp, H.; Brinkman, A.; Huijben, M.; Rijnders, G.; Blank, D. H. A.; Koster, G.; Kelly, P. J.; Golden, M. S. *Phys. Rev. B* **2013,** *87,* 085128.

108. Salluzzo, M.; Gariglio, S.; Torrelles, X.; Ristic, Z.; Di Capua, R.; Drnec, J.; Sala, M. M.; Ghiringhelli, G.; Felici, R.; Brookes, N. B. *Adv. Mater.* **2013,** *25,* 2333.

109. Reinle-Schmitt, M. L.; Cancellieri, C.; Li, D.; Fontaine, D.; Medarde, M.; Pomjakushina, E.; Schneider, C. W.; Gariglio, S.; Ghosez, P.; Triscone, J.-M.; Willmott, P. R. *Nat. Commun.* **2012,** *3,* 932.

110. Liu, Z. Q.; Huang, Z.; Lü, W. M.; Gopinadhan, K.; Wang, X.; Annadi, A.; Venkatesan, T.; Ariando. *AIP Adv.* **2012,** *2,* 12147.

111. Kozuka, Y.; Susaki, T.; Hwang, H. Y. *Phys. Rev. Lett.* **2008,** *101,* 096601.

112. Shibuya, K.; Ohnishi, T.; Lippmaa, M.; Oshima, M. *Appl. Phys. Lett.* **2007,** *91,* 232106.

113. Perna, P.; Maccariello, D.; Radovic, M.; Scotti di Uccio, U.; Pallecchi, I.; Codda, M.; Marré, D.; Cantoni, C.; Gazquez, J.; Varela, M.; Pennycook, S. J.; Granozio, F. M. *Appl. Phys. Lett.* **2010,** *97,* 152111.

114. Chen, Y.; Pryds, N.; Kleibeuker, J. E.; Koster, G.; Sun, J.; Stamate, E.; Shen, B.; Rijnders, G.; Linderoth, S. *Nano Lett.* **2011,** *11,* 3774.

115. Campbell, C. T. *Surf. Sci. Rep.* **1997,** *27,* 1.

116. Lee, S. W.; Liu, Y.; Heo, J.; Gordon, R. G. *Nano Lett.* **2012,** *12,* 4775.

117. Delahaye, J.; Grenet, T. *J. Phys. D: Appl. Phys.* **2012,** *45,* 315301.

118. Kan, D.; Terashima, T.; Kanda, R.; Masuno, A.; Tanaka, K.; Chu, S.; Kan, H.; Ishizumi, A.; Kanemitsu, Y.; Shimakawa, Y.; Takano, M. *Nat. Mater.* **2005,** *4,* 816.

119. Enterkin, J. A.; Subramanian, A. K.; Russell, B. C.; Castell, M. R.; Poeppelmeier, K. R.; Marks, L. D. *Nat. Mater.* **2010,** *9,* 245.

120. Li, F.; Wang, Z.; Meng, S.; Sun, Y.; Yang, J.; Guo, Q.; Guo, J. *Phys. Rev. Lett.* **2011,** *107,* 036103.

121. Doennig, D.; Pickett, W. E.; Pentcheva, R. *Phys. Rev. Lett.* **2013,** *111,* 126804.

122. Wang, Z.; Zhong, Z.; Hao, X.; Gerhold, S.; St, B.; Schmid, M.; Jaime, S.; Varykhalov, A.; Franchini, C.; Held, K.; Diebold, U. *arXiv* **2013**, 1309.7042.
123. Herranz, G.; Bergeal, N.; Lesueur, J.; Gazquez, J.; Scigaj, M.; Dix, N.; Sanchez, F.; Fontcuberta, J. *arXiv* **2013**, 1305.2411.
124. Chen, G.; Balents, L. *Phys. Rev. Lett.* **2013**, *110*, 206401.
125. Cen, C.; Thiel, S.; Hammerl, G.; Schneider, C. W.; Andersen, K. E.; Hellberg, C. S.; Mannhart, J.; Levy, J. *Nat. Mater.* **2008**, *7*, 298.
126. Cen, C.; Thiel, S.; Mannhart, J.; Levy, J. *Science* **2009**, *323*, 1026.
127. Cheng, G.; Siles, P. F.; Bi, F.; Cen, C.; Bogorin, D. F.; Bark, C. W.; Folkman, C. M.; Park, J.-W.; Eom, C.-B.; Medeiros-Ribeiro, G.; Levy, J. *Nat. Nanotechnol.* **2011**, *6*, 343.
128. Irvin, P.; Huang, M.; Wong, F. J.; Sanders, T. D.; Suzuki, Y.; Levy, J. *Appl. Phys. Lett.* **2013**, *102*, 103113.
129. Jany, R.; Breitschaft, M.; Hammerl, G.; Horsche, A.; Richter, C.; Paetel, S.; Mannhart, J.; Stucki, N.; Reyren, N.; Gariglio, S.; Zubko, P.; Caviglia, A. D.; Triscone, J.-M. *Appl. Phys. Lett.* **2010**, *96*, 183504.
130. Förg, B.; Richter, C.; Mannhart, J. *Appl. Phys. Lett.* **2012**, *100*, 053506.
131. Park, J. W.; Bogorin, D. F.; Cen, C.; Felker, D. A.; Zhang, Y.; Nelson, C. T.; Bark, C. W.; Folkman, C. M.; Pan, X. Q.; Rzchowski, M. S.; Levy, J.; Eom, C. B. *Nat. Commun.* **2010**, *1*, 94.

CHAPTER 3

MULTIFERROISM AND MAGNETOELECTRIC APPLICATIONS IN BISMUTH FERRITE

LU YOU and JUNLING WANG*

Department of Materials Science and Engineering, Nanyang Technological University, Singapore 639798, Singapore

*Corresponding author. E-mail: jlwang@ntu.edu.sg

CONTENTS

ABSTRACT

Led by the great advances in material syntheses, characterization techniques, and theoretical modeling, the past decade has witnessed the renaissance of multiferroic and magnetoelectric research. Bismuth ferrite, being one of the most studied single-phase multiferroics, represents a rare case of coexistence of electric and magnetic ferroic orders above room temperature, making it most relevant to practical applications which are presented in this chapter.

3.1   INTRODUCTION

Driven by the ever-increasing demand for low power consumption and functional versatility in nonvolatile memory and logic devices, the exploration of single-phase multiferroics with strong magnetoelectric (ME) coupling has become the research forefront of condensed matter physics and material science in recent years [1,2]. Multiferroics, by definition, refer to materials that possess two or more ferroic orders simultaneously, including primarily ferroelectricity, ferromagnetism, and ferroelasticity [3,4]. As a result, multifunctional devices can be achieved by using one single material and, what's more, the cross-couplings between these order parameters open up new physics and potential applications. Among them, the ferroelectric ferromagnets with possible ME coupling appear to be more scientifically and technologically tantalizing, as the modern digital storage technologies indispensably rely on the electric and magnetic controls. However, as pointed out by Spaldin [5], the driving forces for the ferromagnetism (partially filled d orbital) and ferroelectricity (d^0-ness) are mutually exclusive, rendering the single-phase magnetoferroelectric scarce in nature. Thus, from a general sense, multiferroism also covers antiferroic counterparts and other forms of ferroic orders, like ferrotoroidicity. The pioneer work on multiferroics embarked about half a century ago, but the development has been sluggish thereafter (see Ref. [1] for the history of the ME effect). Thanks to the recent advances in sample synthesis techniques [6], including the growths of high-quality single crystals and epitaxial thin films, assisted by the greatly developed theoretical simulations [7,8], a great revival of the multiferroic field has been witnessed for the past decade. Detailed reviews regarding the progress made in the surge of the recent research work on multiferroics or MEs can be found elsewhere [9–12]. Among the limited number of single-phase multiferroics, bismuth ferrite, or $BiFeO_3$ (BFO) plays an unmatchable role in the multiferroic renaissance, given that one-fourth of the publications in this field are relevant to it in the past decade (data from Web of Science). Suffice it to say that BFO is to multiferroics what yttrium-barium copper oxide is to superconductors [13]. Following the ground-breaking *Science* paper in 2003 [14], a flurry of research work has been devoted to unfold all the promising properties of this magic material, part of which are well beyond the scope of ME coupling. The reason why BFO has been most intensively studied is simply because for a very long time it is the only single-phase multiferroic with the robust ferroelectric and antiferromagnetic orders at room temperature [15]. The ferroelectric Curie temperature (T_C) and antiferromagnetic Néel temperature (T_N) are ~1100 and ~643 K [16,17], respectively, rendering it possible for room-temperature applications. The cross-coupling between the ferroic orders further promises BFO for the next-generation of magnetoelectronics, which combines the merits of electrical writing and magnetic reading process [18]. In addition, the recent discoveries of emergent phenomena related to the ferroic domain walls open up new possibilities in potential BFO-based nanoelectronics [19].

With all the fruitful achievements, it is now worthwhile to look back on the fundamental properties of BFO in order to prospect for the future directions. This chapter summarizes some basic physics of the crystal structure, ferroelectricity, magnetism of BFO, and recent developments in ME phenomena in BFO-based heterostructures, aiming to evoke more interest in the exploration for emergent phenomena in BFO and, most importantly, the endeavor of the ultimate goal of multiferroics—a fully electric control of magnetism. Due to the space constraints, however, some aspects like dielectric characteristics and doping effect are not covered. The readers are referred elsewhere [13,20].

3.2 CRYSTAL STRUCTURE AND PHASE TRANSITIONS OF BiFeO$_3$

3.2.1 CRYSTAL STRUCTURE

It is widely accepted that bulk BFO has a typical perovskite structure (ABO_3) with rhombohedral symmetry (space group $R3c$) [21–24], as shown in Figure 3.1, where Bi atoms are located at the eight corners (A-site), Fe atom at the center of the pseudocubic (pc) (B-site), and oxygen atoms at the face centers, forming an octahedron cage. However, some recent reports suggest the real symmetry of bulk BFO is in fact lower, making its crystal structure somewhat controversy [25,26]. The basic unit cell of BFO comprises two vertex-connected simple perovskite units, which can be described in either rhombohedral (a_{rh} = 5.63 Å, α_{rh} = 59.35°) or hexagonal (a_{hex} = 5.58 Å, c_{hex} = 13.87 Å) notations [23,27]. Each perovskite unit is rhombohedrally distorted along [1 1 1]$_{pc}$ direction, with a lattice constant of a_{pc} = 3.965 Å and a rhombohedral angle of ~89.3–89.5° at room temperature [23,28]. Along the three-fold [1 1 1]$_{pc}$ direction, the Fe^{3+} cation is displaced from the center of the oxygen octahedron by about ~0.26 Å and Bi^{3+} cation is shifted by ~0.67 Å away from its centrosymmetric position between two octahedron centers [24,28]. Both of these off-center displacements contribute to the large spontaneous polarization along [1 1 1]$_{pc}$ direction. In addition to the atomic displacement, the two neighbor octahedra along the polar axis rotate by ~11–14° clockwise and counterclockwise around the same axis, respectively [23,28,29]. Oxygen octahedral rotation, sometimes referred to as the antiferrodistortive order, is commonly observed in perovskite oxides due to the imperfect atomic packing. The oxygen octahedral rotation pattern in BFO, under Glazer's tilt system [30], is $c^-c^-c^-$, consistent with the $R3c$ symmetry. Generally, the Goldschmidt tolerance factor [31] is used to describe how well the ionic radii match, and thus the stability of the perovskite structure. The Goldschmidt tolerance factor is given by,

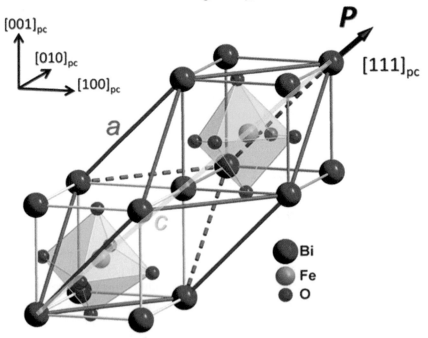

FIGURE 3.1   The crystal structure of BFO with two connected perovskite units distorted along [1 1 1]$_{pc}$ direction. The rhombohedral unit cell is outlined in pink. The yellow line represents the c axis in hexagonal notation.

$$t = \frac{r_A + r_B}{\sqrt{2}(r_B + r_O)}$$

where r_A, r_B, and r_O are ionic radius of A-site, B-site, and oxygen ions, respectively. Using Shannon's effective ionic radii [32], $t = 0.88$ can be obtained for BFO. The ratio equals to 1 for an ideal cubic perovskite with no octahedra rotation. Smaller values suggest that the oxygen octahedra have to tilt to fit into the shrinking space due to smaller A-site ions, causing the Fe–O–Fe bonds to be buckled and staggered. The resulting Fe–O–Fe bond angle in BFO is found to be around 154–156°, which affects both the super-exchange coupling between Fe atoms and the orbital overlap between Fe and O. Therefore, the ME coupling and transport properties are closely linked to its structural characteristics, especially oxygen octahedral rotation, in BFO.

3.2.2   TEMPERATURE AND PRESSURE-DRIVEN PHASE TRANSITIONS

Early studies of bulk BFO concentrated on not only the structural charac-teristics but also the temperature-dependent phase transitions. However, probably due to the poor sample purity and quality, the results are scat-tered. A relatively complete study was carried out by Polomska et al., who mapped out the temperature-composition phase diagram of La-doped BFO using electrical, magnetic, and thermal analyses [33]. Above room temper-ature they identified three transition points, two of which coincides with the T_N and T_C. Other reports using powder X-ray and neutron diffractions confirmed lattice anomalies at these two transitions [24,27,29]. While the small kinks of structural parameters around T_N can be attributed to the spin–lattice coupling, the structural changes at T_C were rarely determined as BFO became increasingly unstable at high temperature. Thanks to the aforemen-tioned revived interest in this field, the high-temperature phase transitions of BFO were revisited recently by several groups using various methods. The results are summarized in Table 3.1. Consistency can be seen in α–β phase transition temperature, namely the T_C around 820–830°C, at which dramatic volume contraction takes place, featuring a first-order transition. However, it is highly controversial regarding the exact structure and symmetry of the β phase. Among the divergent conclusions, two relatively complete and detailed studies performed by Palai et al. [34] and Arnold et al. [35] reached agreement that the β phase is orthorhombic with space group *Pbnm*. Besides,

TABLE 3.1 Possible Structural Phase Transitions Reported in the Literature.

Sample	Technique	Transition temperature (α–β) (°C)	Symmetry and space group (β Phase)	Transition temperature (β–γ)	Symmetry and space group (γ Phase)	References
Single crystal	Raman scattering	810	Cubic, $Pm\bar{3}m$	Not observed	N.A.	[40]
Powder	X-ray diffraction	820	Monoclinic, $P2_1/m$	Not observed till decomposition at 930°C	N.A.	[37]
N.A.	First-principles simulation	800	Tetragonal, $I4/mcm$	1167	Cubic, $Pm\bar{3}m$	[41]
Powder and single crystal	X-ray diffraction	820	Orthorhombic, $Pbnm$	925°C (decomposition at 933°C)	Cubic, $Pm3m$	[34]
Powder	X-ray diffraction	830	Rhombohedral, $R\bar{3}c$	925°C (decomposition at 940°C)	N.A.	[36,42]
Powder	Neutron diffraction	820–830	Orthorhombic, $Pbnm$	930°C (decomposition at 940°C)	Orthorhombic, $Pbnm$	[35,38]

they both observed subsequent β–γ structural transition at around 925°C, which was also identified by Selbach and coworkers [36]. Though they both agreed the β–γ transition was weakly first-order and accompanied by an insulator–metal transition decomposes and even melts far before approaching such high temperature. Partial support for this hypothetical phase can be drawn from the extrapolation of the lattice parameters of the β phase, which converges above 1000°C if a second-order-like phase transition is assumed [37,38]. The conflicts in the literature probably result from the decomposition of BFO at high temperature, especially above T_C. The appearance of secondary phases as well as the melting of the sample complicates the structural analyses. Thus, it is not surprising that some research groups didn't see the β–γ transition. Lastly, it should be noted that the surface layer of BFO may show unique phase transition that is absent in bulk [39].

In addition, phase transitions of BFO at high pressure have been studied by different groups in the past few years using various techniques. Early work by Gavriliuk et al. [43–47] suggested an insulator–metal transition around 45–55 GPa, which is later supported by first-principles calculations [48]. They did not find any phase transitions at low pressure region, whereas another theoretical study indicated a structural transition from *R3c* to *Pnma* at 13 GPa [49]. Haumont et al. [50,51] soon confirmed this transition to orthorhombic phase, however, at a lower pressure of 10 GPa. Additionally, they discovered an intermediate transition from *R3c* to monoclinic *C2/m* at 3.5 GPa. Similar transition was also reported by Belik and coworkers [52], who observed transformation of *R3c* to two consecutive orthorhombic phases at 4 and 7 GPa. Finally, Guennou et al. [53] performed the most detailed investigation that mapped out the complete phase transition sequence in the range of 0–60 GPa. Consistent with previous results, initial *R3c* phase became unstable above 4 GPa and experienced a narrow phase sequence before entering the *Pnma* phase region at 11 GPa. In addition to the low-pressure phase transitions, they were able to identify two high-pressure transitions, with the latter one coinciding with the aforementioned insulator–metal transition. Looking throughout all the results, it is probably safe to bet that BFO remains the *R3c* phase below ~4 GPa and becomes the orthorhombic *Pnma* phase above ~10 GPa. In between may exist multiple competing phases that appear concurrently or in sequence, as evidenced by the large unit cells and complex domain structures. Lastly, the insulator–metal transition takes place at 40–50 GPa. However, symmetry breaking persists in the paramagnetic metallic phase. Similar to the high-temperature phase transitions, the parent cubic phase remains hitherto elusive under high pressure.

In view of the aforementioned literature reports on the temperature/ pressure phase transitions of BFO, a schematic phase diagram for BFO is proposed, first by Scott and Catalan, and later updated by Guennou et al., as shown in Figure 3.2 [53]. It should be noticed that owing to limited experimental data, this phase diagram is just a priori hypothesis, with enormous disagreements in literature. In the proposed phase diagram, solid points are experimental data, while the dash lines are possible phase boundaries. The ground state at ambient condition is rhombohedral, and both high temperature and pressure drive the lattice toward orthorhombic symmetries via first-order transitions. Insulator–metal transition can be trigger by even higher temperature or pressure. Though there are some reports in favor of metallic cubic phase, most studies suggest it is still symmetry breaking. Since most of the research work was performed by fixing either temperature or pressure while varying the other one, there is vast intermediate nonambient P–T area remaining unexplored and worth further investigations.

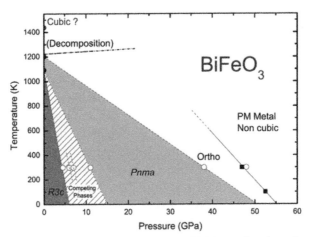

FIGURE 3.2 A possible structural phase diagram of BFO as a function of temperature and pressure (reprinted figure with permission from Ref. [53]). (Reprinted with permission from Guennou, M.; Bouvier, P.; Chen, G. S.; Dkhil, B.; Haumont, R.; Garbarino, G.; Kreisel, J. Multiple high-pressure phase transitions in BiFeO$_3$. *Phys. Rev. B* 2011, *84*, 174107. © 2011 by the American Physical Society.)

3.2.3 BiFeO$_3$ THIN FILMS: STRAIN-INDUCED LOW SYMMETRY PHASES

The studies on high-quality BFO thin films all began with the epitaxial growth on (0 0 1)-oriented SrTiO$_3$ (STO) substrates. The strain resulting from the lattice mismatch between the BFO film and STO substrate was believed to cause significant changes in the structures and lattice parameters of epitaxial BFO thin films, compared to the rhombohedral bulk [14]. A monoclinic structure of the BFO thin films on (0 0 1)-oriented STO was found, based on preliminary X-ray diffraction (XRD) data [54]. Further work by synchrotron XRD determined the structure more precisely to be a M_A-type monoclinic [55]. Thickness-dependent studies verified the M_A phase in thick-film range but revealed tetragonal structure for thin films below 50–70 nm [56,57]. The discrepancy was recently resolved by Daumont et al. [58] using high-resolution reciprocal space mapping on BFO samples with probably better quality. It was found that ultrathin films (below 18 nm) were fully coherent with the substrate and appeared "metrically tetragonal" except for the uncertainty of the oxygen octahedral rotations. Thicker film started to show up twinning domains mediated by in-plane shear vector to relax the strain. The twinning structure formed defect-free 71° domain walls and maintained the out-of-plane lattice parameter and the monoclinic angle almost constant up to 100 nm. An out-of-plane version of such twinning structure was also reported in BFO films grown by sputtering [59]. The symmetry lowering of BFO film grown on (0 0 1)-oriented STO fits an intuitive picture of elastic deformation: the in-plane biaxial compressive strain causes the parent rhombohedral lattice to shrink in plane and expand out of plane while preserving the out-of-plane rhombohedral angle, rendering monoclinic M_A-phase with Cc space group that permits oxygen octahedral rotation.

Thanks to the great advance in crystal growth, a wide spectrum of single-crystal substrates became commercially available in recent years, which arouses a wave of "strain engineering" work in complex oxides, especially with perovskite structure [60]. By applying the concept of strain engineering to BFO, a rich phase diagram emerges. In the systematic work by Chen and coworkers [61], epitaxial (0 0 1)$_{pc}$-oriented BFO thin films were grown on a series of single-crystalline perovskite substrates, with the in-plane misfit strain spanning from highly compressive (−4.4%) on LaAlO$_3$ (LAO) to slightly tensile (+0.6%) on KTaO$_3$. A linear relationship was found between the magnitude of the biaxial strain and the tetragonality (c/a ratio) under the strain ranging from +0.6% to −2.8%; that is to say, the BFO lattice undergoes linear elastic deformation that fits Poisson's effect with a ratio of ~0.49. Depending on the compressive or tensile strain, the lattice can adopt either M_A or M_B monoclinic phase, both of which are still rhombohedral-like because the monoclinic distortion is close to the body diagonal as the bulk. However, when BFO is highly compressed, a sudden jump in the tetragonality was observed, suggesting a possible first-order structural transition. This emergent phase, though with a giant axial ratio above 1.2 [62], possessed neither tetragonal nor M_A or M_B structure. Instead, a monoclinic M_C phase (space group Pm or Pc) was identified based on detailed structural analysis and piezoelectric force microscopy (PFM) measurements [63,64]. Similar M_B–R–M_A–M_C phase sequence driven by strain was also reported by Christen and coworkers, who managed to achieve a true tetragonal structure, nevertheless, by replacing some Bi with Ba [65]. As shown in Figure 3.3, this monoclinic M_C phase differs from the M_A or M_B in such way that, for M_A or M_B, the monoclinic unit cell is obtained by shearing the cubic perovskite cell along the $[1\ 1\ 0]_{pc}$ direction, whereas M_C is obtained by shearing along $[1\ 0\ 0]_{pc}$ direction [66]. However, even larger compressive strain didn't yield a real tetragonal phase at room temperature, though it becomes tetragonal symmetry at elevating temperatures [67,68].

On the other hand, researchers were also searching for possible phase transition toward orthorhombic symmetry in the tensile strain region as theoretically predicted by first-principles calculation [69,70]. However, due to the limited availability of the perovskite substrates with larger lattice compared to BFO, the maximum tensile strain attainable is around +1.4%, way smaller than the predicted threshold of >+5%, though phenomenological calculations suggest a much small value [71–73]. Experimentally, Yang et al. grew BFO thin film on $NdScO_3$ substrate with +1.1% tensile strain and concluded a consequent orthorhombic phase based on the results of the nonlinear optical second harmonic generation and piezoresponse force microscopy. In contrast, Chen et al. claimed that tensile strain as large as +1.4% still rendered BFO a monoclinic M_B phase with 109° stripe domains similar to those typically observed on scandate substrates. The disagreement probably results from the different film thicknesses used in these two studies. Recalling BFO films grown on STO, we observed a structural transition from metrically tetragonal toward monoclinic M_A with increasing film thickness due to the twinning relaxation. The situation could be similar for tensile–strain case, in which the structure of BFO gradually changes from metrically orthorhombic to monoclinic M_B. Further studies are warranted to resolve this problem.

Aside from the widely studied $(0\ 0\ 1)_{pc}$ orientation, BFO has been deposited on substrates with various orientations to change the elastic boundary conditions. Taking the STO substrates as the examples, BFO grown on $(1\ 1\ 1)$-oriented STO showed bulk-like rhombohedral structure because the out-of-plane distortion coincides with the polar direction, whereas $(1\ 1\ 0)$-oriented STO rendered BFO monoclinic M_B structure [54,74]. A subtle insight was further given by growing BFO on STO substrates with low-symmetry planes like (1 2 0) and (1 3 0) [75]. Triclinic phases were found on these BFO films that served as structural bridge between monoclinic M_A and M_B phases with continuously varying lattice parameters.

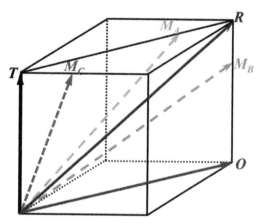

FIGURE 3.3 A simple schematic depicting the possible polar distortions of various phases in BFO. T, R, O, and M represent tetragonal, rhombohedral, orthorhombic, and monoclinic phases, respectively (reprinted figure with permission from Ref. [76]). (Reprinted with permission from Chen, Z.; Qi, Y.; You, L.; Yang, P.; Huang, C. W.; Wang, J.; Sritharan, T.; Chen, L. Large tensile strain induced monoclinic MB phase in BiFeO₃ epitaxial thin films on PrScO₃ substrate. *Phys. Rev. B* 2013, *88*, 054114. © 2013 by the American Physical Society.)

In the above topic concerning the substrate-induced strain engineering on the structures of BFO thin films, it is worth mentioning that the thickness of some BFO films can reach as large as 100 nm without strain relaxation in the out-of-plane lattice parameter, allowing continuous strain tuning with commensurate lattice. The reason, as pointed out above, is because the strain relaxation is mediated by the in-plane shear distortion that largely maintains the in-plane area as well as out-of-plane lattice parameter, assuming volume conservation. The underpinning mechanism should be traced back to the relatively small Goldschmidt tolerance factor of BFO ($t < 0.88$) that was discussed at the very beginning of this chapter. This makes possible the large-degree tilting of oxygen octahedra, with which BFO can accommodate strain at relatively low energy costs. The structural softness of BFO renders a rich phase diagram that may host possible morphotropic phase boundaries (MPBs) highly promising for potential applications [77]. Such a strain-driven MPB has recently been demonstrated in highly compressive BFO films, and a counterpart in the tensile–strain region also appears to be highly possible [78], which deserves more future attention. The coexistences of these competing phases in a single compound without chemical doping also provide an ideal platform for studying the emergent physics and functionalities at the homointerfaces bridging them.

3.3 ELECTRICAL PROPERTIES OF BiFeO₃

3.3.1 BANDGAP AND TRANSPORT BEHAVIORS

BFO is a wide bandgap semiconductor with a broad spread of the reported bandgap values ranging from 2 to 3 eV. BFO nanowires and nanoparticles were usually reported to exhibit smaller bandgaps [79–83] compared to bulk single crystals [34,84], probably due to the surface effect or oxygen defects [83,85]. In thin-film form, most of the previous studies coincided with a direct bandgap within 2.7–2.8 eV [86–92], consistent with the theoretical prediction [93]. A strain-dependent study done recently showed the bandgaps of fully coherent BFO films were barely affected by moderate compressive strains, while it slightly redshifts from 2.82 to 2.75 eV under small tensile strain. However, under large compressive strain, the bandgap showed a large blueshift to 3.1 eV, coinciding with the structural transition to the tetragonal-like phase [94,95]. Such bandgap increase can be understood in terms of the d orbital state of the transition metal ion. Owing to the strongly shrunk in-plane lattice, the Fe–O–Fe bond angle becomes more severely buckled, resulting in less orbital overlapping and thus smaller carrier hopping integral, whereas the out-of-plane lattice becomes so overstretched that one of the oxygen loses bonding with Fe ion, which also leads to a more insulating character along the axial direction.

In high-quality BFO single crystals, the dc resistivity at room temperature can be larger than 10^{10} Ω cm [34,96–98], which decreases as temperature increases, just like any other wide-bandgap semiconductors. As mentioned previously, the bandgap was reported to vanish at the β–γ structural phase transition at ~930°C, signifying the insulator–metal transition, as shown in Figure 3.4a [34, 38]. Alternatively, the insulator–metal transition can be driven by high pressure at ~50 GPa at room temperature [44–47]. The pressure-induced insulator–metal transition was attributed to the high-spin–low-spin crossover in half-filled d orbital of Fe^{3+} ion (Fig. 3.4b) [47]. However, this scenario cannot explain the temperature-driven one as BFO is magnetically disordered at this temperature range. Instead, Catalan and Scott proposed an explanation to reconcile both effects, in which the structural change caused the straightening of the Fe–O–Fe angle and consequently a larger orbital overlapping and bandwidth [13]. However, the high-temperature cubic phase they suggested is still highly debated [38].

The reported values of electrical resistivity of BFO thin films show a large scatter in the literature. This is because the resistivity of the thin film could be affected by a number of factors, such as phase purities, defect concentrations, domain structures, interfacial barriers, and so on. Typically, a resistivity value of at least 10^8–10^9 Ω cm warrants the polarization hysteresis measurements [14,99–103], because higher leakage will overwhelm the true ferroelectric polarization and also shunt the external voltage acting on the film [102]. The leakage problem prevailed in the early BFO thin-film samples, which hindered the determination of the intrinsic polarization. The main reason was probably that the high volatility of Bi posed a great challenge of controlling the chemical stoichiometry in thin films irrespective of the film-growth methods. It seemed natural to use Bi-excess targets to compensate the Bi loss during film deposition as widely reported in the literature. However, most people ignored the fact that in physical vapor depositions, the ablation/sputtering yield of Bi was also much higher than Fe, rendering the films Bi excess. Fortunately, it is possible to tune the film compositions by varying temperature or oxygen pressure. Thus, the optimization of the BFO thin films is usually accomplished by reaching the balance between the material yields (target composition, ablation power) and the thermodynamic environment (substrate temperature, oxygen partial pressure) [104–108]. In other words, no matter whether Bi-rich or Bi-poor target is used, stoichiometric films can be acquire by choosing proper substrate temperature and oxygen pressure. Thus, a flurry of studies that reported enhanced resistivity and ferroelectric polarization are probably the consequences of tuning Bi/Fe ratios in the targets, which makes the films more stoichiometric and resistive [109–114]. In addition to the film compositions and defect chemistry, the leakage mechanisms of BFO thin films are closely linked with the electrode materials, which form different interfacial contacts with the BFO films. Depending on the magnitudes of the interfacial barriers, the dominant leakage mechanism can be driven from interface-limited to bulk-limited [113,115–118]. Lastly, it is worth pointing out that the ferroelectric domain structures also play a role in determining the transport behavior of the BFO films due to the emergent conductivity arising from the domain walls [119].

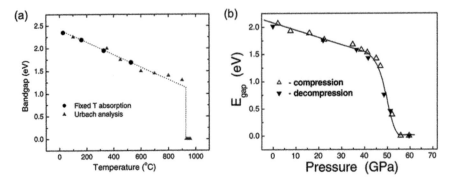

FIGURE 3.4   Bandgap evolution of BFO as a function of (a) temperature and (b) hydrostatic pressure (reprinted figures with permission from Refs. [34,47]). (a: Reprinted with permission from Palai, R.; Katiyar, R. S.; Schmid, H.; Tissot, P.; Clark, S. J.; Robertson, J.; Redfern, S. A. T.; Catalan, G.; Scott, J. F. β phase and γ–β metal-insulator transition in multiferroic BiFeO3. Phys. Rev. B 2008, 77, 014110. © 2008 by the American Physical Society; b: Reprinted with permission from Gavriliuk, A. G.; Struzhkin, V. V.; Lyubutin, I. S.; Ovchinnikov, S. G.; Hu, M. Y.; Chow, P. Another mechanism for the insulator-metal transition observed in Mott insulators. *Phys. Rev.* B2008, *77,* 155112. © 2008 by the American Physical Society.)

3.3.2   FERROELECTRICITY

BFO is a robust ferroelectric with T_C as high as ca. 1103 K. Although high T_C does not necessarily imply large ferroelectric polarization, empirically, it is usually associated with a sizable value. However, the earliest polarization measurement in record performed by Teague et al. on single crystal showed a small polarization value of 3.5 µC cm^{-2} along $[1\ 0\ 0]_{pc}/[0\ 1\ 2]_{hex}$ direction [17], rendering a total polarization of 6 µC cm^{-2}, given that the polar axis in bulk BFO lies in the $[1\ 1\ 1]_{pc}/[0\ 0\ 1]_{hex}$ direction. A large remanent polarization had never been achieved until the *Science* paper published in 2003, which reported an unprecedented polarization value of ca. 60 µC cm^{-2} in (0 0 1)$_{pc}$-oriented epitaxial thin films, with more than one order of magnitude enhancement (Fig. 3.5a). Subsequent studies on high-quality single crystals [96,97] and ceramics [120] revealed similar polarization value to that of the thin films (Fig. 3.5b), confirming the intrinsic nature of the large polarization of BFO. Orientation-dependent measurements further substantiated that the spontaneous polarization was lying along the body diagonal of the pc cell with a value of ca. 100 µC cm^{-2} [49,54,121–123], placing BFO as the ferroelectric material with the largest polarization ever measured (note that its supertetragonal polymorph has an even higher polarization of ca. 150 μC cm^{-2}) [124–126].

Initially, it was proposed that the large polarization was enhanced by the epitaxial strain, just as those reported strain effects for $BaTiO_3$ or $SrTiO_3$ [127,128]. However, similar polarization value measured from bulk BFO negated such explanation. We should note that the driving force of the ferroelectricity in BFO is the chemically active $6s$ lone pair of the A-site Bi^{3+}, in striking contrast to those with displacive B-site cations. Besides, there is no direct connection between the c/a ratio and the magnitude of the polarization, because the c/a ratio is a metrical measurement of the external shape of the lattice, whereas the polarization is determined by the internal displacement of the ions. As a consequence, the sensitivity of polarization to c/a ratio or epitaxial strain turns out not as large as initially thought [129–132]. Nevertheless, as discussed above, epitaxial strain did modify the crystal structures of the BFO films, which should also affect the ferroelectric property as the lattice and polarization degrees of freedom are usually coupled.

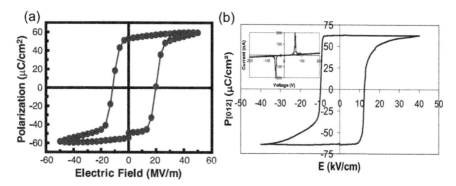

FIGURE 3.5   (a) First fully saturated polarization-electric field (P–E) hysteresis loop of BFO thin film obtained in 2003 (adapted from Ref. [14]. Reprinted with permission from AAAS) and (b) P–E hysteresis loop of high-quality single-crystal BFO measured in 2007. Both polarization loops were measured along $[1\ 0\ 0]_{pc}/[0\ 1\ 2]_{hex}$ direction (reprinted with permission from Ref. [96]). (a: Adapted from Wang, J.; Neaton, J. B.; Zheng, H.; Nagarajan, V.; Ogale, S. B.; Liu, B.; Viehland, D.; Vaithyanathan, V.; Schlom, D. G.; Waghmare, U. V.; Spaldin, N. A.; Rabe, Wuttig, M.; Ramesh, R. Epitaxial BiFeO$_3$ Multiferroic Thin Film Heterostructures. *Science* 2003, *299*, 1719. Reprinted with permission from AAAS; b: Reprinted from Lebeugle, D.; Colson, D.; Forget, A.; Viret, M. Very large spontaneous electric polarization in BiFeO$_3$ single crystals at room temperature and its evolution under cycling fields. *Appl. Phys. Lett.* 2007, *91*, 022907, with the permission of AIP Publishing.)

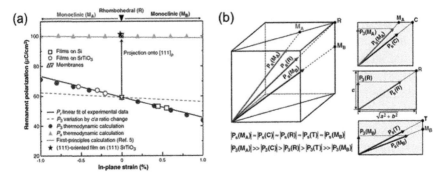

FIGURE 3.6 Strain dependence of polarization in $(0\ 0\ 1)_{pc}$-oriented epitaxial BFO films (reprinted figure with permission from Ref. [133]). (Reprinted with permission from Jang, H. W.; Baek, S. H.; Ortiz, D.; Folkman, C. M.; Das, R. R.; Chu, Y. H.; Shafer, P.; Zhang, J. X.; Choudhury, S.; Vaithyanathan, V.; Chen, Y. B.; Felker, D. A.; Biegalski, M. D.; Rzchowski, M. S.; Pan, X. Q.; Schlom, D. G.; Chen, L. Q.; Ramesh, R.; Eom, C. B. Strain-Induced Polarization Rotation in Epitaxial (001) BiFeO3 Thin Films. *Phys. Rev. Lett.* 2008, *101*, 107602. © 2008 by the American Physical Society.)

In order to understand the strain effects on the polarization of BFO thin films, a thermodynamic analysis and direct measurements of the remnant polarization were carried out by Jang et al. on $(0\ 0\ 1)_{pc}$-oriented epitaxial BFO films as well as free-standing membranes [133]. As discussed previously, the biaxial strain in the $(0\ 0\ 1)_{pc}$ plane would distort the BFO lattice from ideal rhombohedral phase to monoclinic M_A or M_B phase for compressive or tensile strain, respectively. Accompanying the symmetry breaking was the corresponding polarization rotation within the monoclinic symmetry plane, that is, $(1\ 1\ 0)_{pc}$. Specifically, in-plane contraction resulted in $P_z > P_x = P_y$, whereas in-plane extension led to $P_z < P_x = P_y$, as detailed in Figure 3.6. It is noteworthy, however, that the absolute value of the spontaneous polarization (P_s) was barely affected by the epitaxial strain. The polarization rotation scenario was also supported by the theoretical simulations [134–136] and further tested experimentally on a wide spectrum of different substrates or via a piezoelectric substrate [137,138]. The calculated coefficient of P_z change according to the applied strain varied from 4–5 μC cm^{-2} %$^{-1}$ to 12 μC cm^{-2} %$^{-1}$, depending on whether the oxygen octahedral tilt was taken into account. This highlighted the competition between polar and antiferrodistortive orders in BFO. Such polarization rotation path driven by strain brings the reminiscence of similar behaviors reported for ferroelectric relaxors, in which the external stimulus is electric field instead [139,140]. More insight was provided in in-plane polarization studies using planar electrodes, where the variation of in-plane polarization components correlated well with the sign and magnitude of the epitaxial strain [76,141,142]. Beyond the relatively small-strain-induced isosymmetric structural changes, large compressive strain can trigger aforementioned first-order structural transition toward a supertetragonal phase with giant c/a ratio. The polarization, in this case, jumps out from the $(1\ 1\ 0)_{pc}$ plane into $(1\ 0\ 0)_{pc}$ plane, together with a large increment of the total polarization [61,65]. The finite in-plane polarization component of this supertetragonal phase was determined by planar polarization measurements as well as PFM studies, suggesting indeed monoclinic symmetry instead of a true tetragonal structure [63,64,68,143–147]. The out-of-plane polarization of the supertetragonal phase was seldom reported [62], because of, first, the limited thickness that can be grown without structural relaxation [148–150], and second, the huge electric field required to switch the polarization may exceed the dielectric breakdown limit of the capacitor. However, from the polarization measurements of the polycrystalline and mixed-phase films [124,126,151], we can still catch a glimpse of the giant polarization in this polymorph, which is approaching the theoretically predicted value [49,125,129,152]. Besides the polarization rotation toward the film normal, the extra enhancement of the total polarization was attributed to the large Fe^{3+} off-centering relative to the negative charge center as evidenced by transmission electron microscopy studies [126,153,154].

The influences of epitaxial strain on the ferroic transition temperatures (Curie temperature T_C and Néel temperature T_N) in BFO thin films have also been studied in the strain-induced M_A–R–M_B phase sequence (Fig. 3.7) [155]. A combination of theoretical and experimental findings showed that the ferroelectric Curie temperature was generally suppressed by biaxial strain despite its sign, whereas the magnetic Néel temperature remains almost unchanged. This behavior contrasts sharply with that of B-site driven ferroelectrics in which Curie temperature is enhanced by strain [127,128]. The underpinning mechanism was related to the close connection between polar and oxygen tilting instabilities. While the compressive strain tends to increase the c/a ratio, it also greatly restrains the oxygen octahedral rotation along the in-plane axes, which destabilizes the ferroelectric order. This situation is qualitatively similar to the biaxial strain effect on $La_{0.7}Sr_{0.3}MnO_3$ thin films, whose Curie temperature decrease for both compressive and tensile strain due to increasing Jahn–Teller splitting [156]. Other reports on partial relaxed BFO films grown on STO substrates also supported the suppression of T_C due to epitaxial strain [157,158].

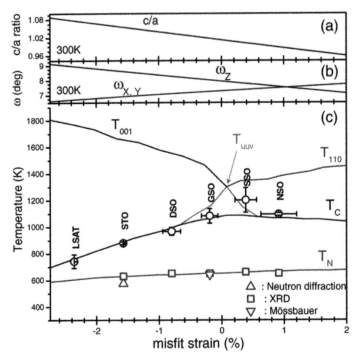

FIGURE 3.7 Strain effects on the ferroelectric Curie temperature and antiferromagnetic Néel temperature of $(0\ 0\ 1)_{pc}$-oriented epitaxial BFO films. Experimental data are shown in symbols and theoretical results are in lines: (a) tetragonality (c/a ratio), (b) antiferrodistortive angles along x, y, and z axes, and (c) transition temperatures as a function of misfit strain (reprinted figure with permission from Ref. [155]). (Reprinted with permission from Infante, I. C.; Lisenkov, S.; Dupé, B.; Bibes, M.; Fusil, S.; Jacquet, E.; Geneste, G.; Petit, S.; Courtial, A.; Juraszek, J.; Bellaiche, L.; Barthélémy, A.; Dkhil, B. Bridging Multiferroic Phase Transitions by Epitaxial Strain in BiFeO$_3$. *Phys. Rev. Lett.* 2010, *105*, 057601. © 2010 by the American Physical Society.)

The giant spontaneous polarization with high Curie temperature as well as the lead-free nature make BFO more advantageous over conventional Pb(Zr,Ti)O$_3$ (PZT) in ferroelectric-based applications due to the ever-increasing environmental concerns. Among the various functions in sensing, actuation, energy harvesting, and information storage, the nonvolatile nature of the ferroelectric polarization is the most promising for the next-generation memory, in which the two logic states, "1" and "0," are represented by two different polarization directions that can be switched by external electric field. In the past few years, the rediscovery of large polarization in BFO has sparked manifold elegant paradigms of nonvolatile memory by exploiting the spin-off effects of ferroelectricity, for example, memristor/switchable diode [159–163], ferroelectric tunnel junction (FTJ) [164], ferroelectric field effect [165–167], ferroelectric photovoltaic effect [86,98,168–173], etc. In addition to the nonvolatility, these memory archetypes all enable non-destructive read-out of the polarization states, thus bypassing the fatigue problem that long plagues ferroelectric-based memories [123,174–176]. However, the reliability of the ferroelectricity at nanoscale, like retention [177–179], imprint [180–183], finite size effect [184–187], and leakage problems, still pose a substantial hurdle toward real devices.

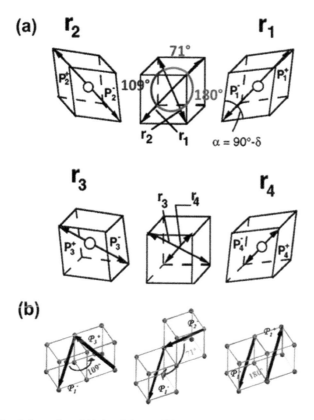

FIGURE 3.8 Schematics of (a) the eight possible domain variants (reprinted from Ref. [188] with the permission of AIP Publishing) and (b) three types of domain walls in rhombohedral ferroelectrics (reprinted figure with permission from Ref. [19]). (a: Reprinted from Streiffer, S. K.; Parker, C. B.; Romanov, A. E.; Lefevre, M. J.; Zhao, L.; Speck, J. S.; Pompe, W.; Foster, C. M.; Bai, G. R. Domain patterns in epitaxial rhombohedral ferroelectric films. I. Geometry and experiments. J. Appl. Phys. 1998, 83, 2742, with the permission of AIP Publishing; b: Reprinted with permission from Catalan, G.; Seidel, J.; Ramesh, R.; Scott, J. F. Domain wall nanoelectronics. Rev. Mod. Phys. 2012, 84, 119. © 2012 by the American Physical Society.)

3.3.3 DOMAINS AND DOMAIN WALLS

In ferroelectric materials, domains, that is, small regions with uniformly oriented polarization within each region, are formed to minimize the free energy in terms of electrical and mechanical boundary conditions. And the boundaries between different domains are called domain walls. The studies on domains and domain walls in BFO are of particular interest owing to the fascinating properties emerging at the (multi)ferroic domain walls, where symmetry-breaking and competition of order parameters take place.

3.3.3.1 DOMAIN VARIANTS AND DOMAIN ENGINEERING

The domain structures in epitaxial rhombohedral ferroelectric thin films were systematically formulated by Streiffer and coworkers in 1998 [188]. Generally, considering the rhombohedral $R3c$ symmetry, the polarization vector can lie along either one of the four body diagonals in the pc cell, as depicted in Figure 3.8. As a consequence, BFO possesses four structural (ferroelastic) variants, or eight polarization (ferroelectric) variants, given two antiparallel directions for each diagonal. Depending on the angles between different polarization variants, three types of domain walls can be identified, namely, 71°, 109°, and 180° walls (note that although these angles are calculated based on an ideal cubic cell, they are still widely used in pc structure for simplification). By evaluating the mechanical and charge compatibility, equilibrium 71° and 109° domain walls correspond to $\{1\,0\,0\}_{pc}$ and $\{1\,0\,1\}_{pc}$ twin boundaries, respectively, while 180° walls may take any planes containing the polar vector. Since the individual domains are energetically degenerate, stripe domain patterns with equal domain width are expected [188]. However, as BFO is prone to decompose at high temperature, film growth is carried out at a temperature much lower than T_c. Thus, the domain pattern forms concomitantly during film growth, which could be thermodynamically nonequilibrium in high energetic deposition process. By tuning the film growth conditions, both stripe-like and mosaic/fractal domain patterns can be obtained [101,189,190].

Due to the fourfold degeneracy of the ferroelastic domains, early BFO thin films grown on (0 0 1)-oriented STO substrates usually exhibited relatively random domain structures with complex switching characteristics [191–193]. To obtain periodic domain patterns with reduced domain variants, symmetry-breaking on the substrate surface is required. Commonly used methods include vicinal substrates, different substrate orientations (which can be deemed as an extreme case of vicinal substrates) and low-symmetry

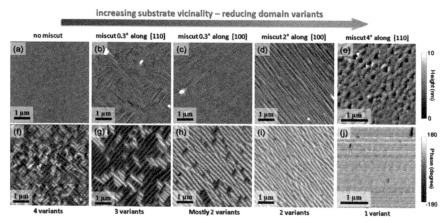

FIGURE 3.9 Domain engineering of BFO thin films using different vicinal STO substrates. (a)–(e) Surface morphologies and (f)–(j) corresponding in-plane PFM images of BFO thin films grown on STO substrates with different miscut directions and angles. With increasing miscut angle, the ferroelectric domain variants are reduced, however, at the expense of increasing surface roughness.

substrates. Miscut substrates are widely employed in various film growths to induce anisotropic properties. Domain engineering in BFO thin films has borrowed this concept to achieve domain selections by vicinal substrates with various miscut angles along different in-plane directions [194–198]. Specifically, taking (0 0 1)-oriented STO as an example, gradual increase of the miscut along [1 0 0]/[0 1 0] direction will reduce the domain variants by half with the extreme case being (1 1 0)-oriented film (miscut angle equals to 45°). If the miscut is along [1 1 0]/[1 −1 0] direction, the domain variants gradually reduce from 4 to 3, and finally a single-domain state is stabilized, with (1 1 1)-oriented film being the extreme case (miscut angle equals to 60°). The evolution of the domain structure with various substrate vicinalities is displayed in Figure 3.9. The mechanism can be understood on the ground of anisotropic strain imposed by the miscut substrates [199], associated with an anisotropic strain relaxation process [200,201]. Orientation-dependent domain structures thus can be considered as substrates with large vicinal angles [196,202]. Besides vicinal substrates, another means to induce anisotropic strain is simply to use low-symmetry substrates, like orthorhombic rare-earth scandates (ReScO$_3$) [203–206]. The anisotropic strain here emphasizes more on the shear strain caused by the monoclinic distortion as well as the clamping of the oxygen octahedral tilt from the substrates, rather than the metrically unequal lattice parameters. Depositing rhombohedral BFO on (1 1 0)$_O$-oriented orthorhombic substrates favors two structural variants, which, however, can have two types of assembly manners (71° and 109° stripe-domain patterns) according to Streiffer' framework (Fig. 3.10) [188]. Along the out-of-plane direction, the global polarizations in 71° and 109° stripe-domain structures correspond to a fully poled state and a fully compensated state, respectively. Therefore, it is possible to control the ferroelectric domain pattern by tuning the electrical boundary condition, as demonstrated by varying the thickness of the bottom electrode [203]. However, later work showed that simply by annealing the scandate substrates, one can switch the 109° stripe-domain pattern to 71° one [207,208], signaling the role of substrate termination on determining the bulk polarization direction [209]. Furthermore, it has also been shown that the fraction of 71° and 109° stripe domains can be regulated by changing the target composition or growth temperature [210], similar to the observation of polarization inversion in BFO films grown on STO [182,183]. All these findings suggest that the preferred polarization orientation of a ferroelectric film is a result of complex interaction between various factors, for example, interface charge discontinuity, strain relaxation gradient (flexoelectric effect), chemical gradient (nonstoichiometry/defect), etc.

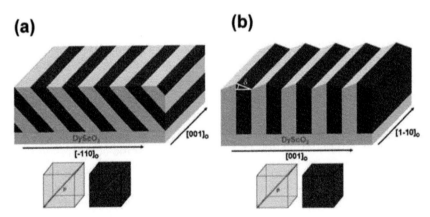

FIGURE 3.10 Schematic of periodic 71° and 109° domain patterns (reprinted with permission from Ref. [203]). (Reprinted with permission from Chu, Y. H.; He, Q.; Yang, C. H.; Yu, P.; Martin, L. W.; Shafer, P.; Ramesh, R. Nanoscale Control of Domain Architectures in BiFeO3 Thin Films. *Nano Lett.* 2009, *9*, 1726. © 2009 American Chemical Society.)

A direct benefit from highly ordered domain structure is the greatly improvement of the polarization switching owing to reduced domain wall pinning [197,211]. Similarly, the dielectric nonlinearity and coercive field can also be reduced by cutting down the number of domain variants in the films [212,213]. Here, the decrease of coercive field in single-domain BFO film appears a bit counter-intuitive, as multidomain structures should facilitate the domain nucleation [214]. This can be explained by that the polarization switching in random-domain BFO thin films is mainly limited by the lateral domain wall motion instead of domain nucleation. Polarization switching in BFO thin films is predominantly 180° reversal under vertical electric field, because no extra elastic energy is incurred in this manner. However, if no elastic constraint is considered, direct 180° switching actually has a larger activation energy compared to 71° or 109° switching. Thus, a relaxation-mediated 180° switching scenario was proposed to explain the switching behavior in single-domain BFO films, and in order to stabilize the 71° or 109° ferroelastic switching, substrate clamping had to be removed [177,215]. In contrast, in-plane electric field is capable of producing ferroelastic switching as the 180° non-ferroelastic switching is forbidden due to the lack of bottom electrodes [141,216,217]. The ferroelastic switching is of great importance in the ME coupling effect of BFO because it switches the magnetic order as well via symmetry breaking (this will be elaborated in the following section) [218,219]. Interestingly, the ferroelastic switching can also be controlled by the in-plane component of the inhomogeneous field emanating from a local probe. Specific domain variants can be generated through different combinations of fast and slow scanning axes [220]. Similar effect was also applied to the strain-driven polymorphic BFO films to produce selective domain variants [221–223].

3.3.3.2   EMERGENT PHENOMENA AT DOMAIN WALLS

With great advance in nanoscale and even atomic-level characterization techniques, exploring exotic functionalities at ferroic domain walls is arising as a burgeoning research field in recent years [19,224–226]. As an active type of topological defects, domain walls host a broad spectrum of new physics and properties that are absent in the bulk, due to intrinsic symmetry-breaking. The study of domain walls in BFO thin films are of particular interest as multiple ferroic order parameters coexist and interact with each other [227,228].

The unique character of domain wall in BFO was first manifested by its enhanced conductivity in otherwise insulating films grown on STO substrates with different orientations [119]. The arising conduction was explained from both intrinsic and extrinsic aspects. Some believed that local structure and symmetry changes lead to modifications of the Fe–O–Fe bond angles, and subsequent reduction in local bandgaps [119,229,230]. However, some others proposed that extrinsic factors like oxygen vacancies or other defects accumulating at the domain walls to compensate the polar disconti-nuity are at the origin of the enhanced conduction [231,232]. One supporting evidence is that the domain wall conductivity can be tuned by the concen-tration of the oxygen vacancies (Fig. 3.11) [233,234]. Further experimental observations of tunable electronic conductions at charged domain walls due to modulated electrostatic doping or domain wall geometry seem to favor the latter mechanism [235–237], which also applies to other ferroelectrics [238–240]. Besides, it is highly possible that the true-domain wall conduc-tion is convoluted with the parasitic switching current induced by domain

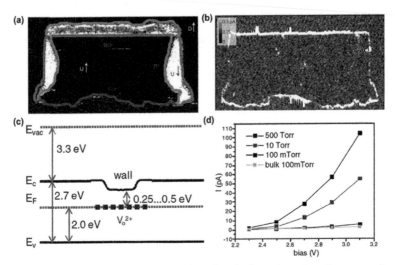

FIGURE 3.11 (a) In-plane PFM image of an electrically written domain pattern showing all three types of domain walls. (b) Corresponding conductive AFM image showing enhanced conductivity at 109° and 180° domain walls (adapted by permission from Macmillan Publishers Ltd: Nature Materials [119], copyright 2009). (c) Band diagram of 109° domain wall with the accumulation of oxygen vacancies. (d) The level of domain wall conduction depends on density of the oxygen vacancies (reprinted figure with permission from Ref. [233]). (a and b: Reprinted by permission from Macmillan Publishers Ltd: Seidel, J.; Martin, L. W.; He, Q.; Zhan, Q.; Chu, Y. H.; Rother, A.; Hawkridge, M. E.; Maksymovych, P.; Yu, P.; Gajek, M.; Balke, N.; Kalinin, S. V.; Gemming, S.; Wang, F.; Catalan, G.; Scott, J. F.; Spaldin, N. A.; Orenstein, J.; Ramesh, R. Conduction at domain walls in oxide multiferroics. *Nat. Mater.* 2009, *8*, 229. © 2009; c and d: Reprinted with permission from Seidel, J.; Maksymovych, P.; Batra, Y.; Katan, A.; Yang, S. Y.; He, Q.; Baddorf, A. P.; Kalinin, S. V.; Yang, C. H.; Yang, J. C.; Chu, Y. H.; Salje, E. K. H.; Wormeester, H.; Salmeron, M.; Ramesh, R. Domain wall conductivity in La-doped BiFeO$_3$. *Phys. Rev. Lett.* 2010, 105, 197603. © 2010 by the American Physical Society.)

wall displacement [241]. In this sense, time-dependent current measurement and in-situ monitor of the domain pattern are highly desirable. It should be mentioned that recently enhanced conductivity has also been observed in the polymorphic phase boundaries in highly strained BFO thin films [223,242].

In addition to electronic conduction, the electrostatic potential steps at the domain walls were also believed to be the source of the abnormal above-bandgap photovoltage discovered in domain-engineered BFO thin films (Fig. 3.12) [207]. Periodic 71° domain walls served as nanoscale photovoltaic cells that connect in series to generate ultrahigh open-circuit voltages, fundamentally in analogous to a tandem solar cell [243]. Results on 109° domain walls were also reported [244,245]. However, Bhatnagar et al. provided a different interpretation of this abnormal photovoltaic effect [208]. By carrying out a detailed temperature-dependent study of the photovoltaic effect in both 71° and 109° stripe-domain BFO films, they concluded that the above-bandgap photovoltages originated from the bulk photovoltaic effect, an intrinsic property of a non-centrosymmetric material. Domain walls, in this case, acted rather as shunts that masked the open-circuit voltage when measured from certain directions [246]. More studies are needed to resolve the discrepancy raised here.

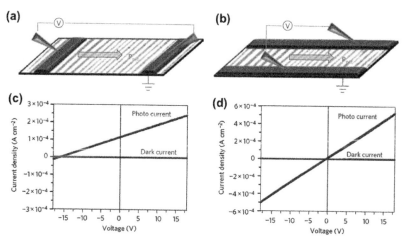

FIGURE 3.12   Abnormal photovoltaic effect in BFO thin films with periodic 71° domain walls: (a) and (b) are schematics of electrode geometry with respect to the domain walls; (c) and (d) are the corresponding light and dark *I–V* measurements based on the device structures shown in (a) and (b) (reprinted with permission from Ref. [207]). (Reprinted by permission from Macmillan Publishers Ltd: Yang, S. Y.; Seidel, J.; Byrnes, S. J.; Shafer, P.; Yang, C. H.; Rossell, M. D.; Yu, P.; Chu, Y. H.; Scott, J. F.; Ager, J. W.; Martin, L. W.; Ramesh, R. Above-bandgap voltages from ferroelectric photovoltaic devices. *Nat. Nanotechnol.* 2010, *5*, 143. © 2010.)

Lastly, enhanced ferromagnetic moments were theoretically predicted at 109° domain walls, as a result of increasing local canting of the antiferro-magnetic spins driven by structural changes [229]. Experimentally, larger exchange bias was indeed observed in CoFe layer when grown on BFO thin films with more 109° domain walls [247], and periodic 109° domain wall arrays also showed significant magnetoresistance below 200 K [248]. Moreover, spontaneous magnetization was also found at the strain-driven polymorphic phase boundaries, which can be electrically written and erased [249]. All these phenomena point to possibly the same origin, that is, spin frustrations due to competing magnetic interactions at the domain bound-aries [250].

3.4 MAGNETIC ORDERS AND MAGNETOELECTRIC COUPLING IN BiFeO$_3$

The essence of a multiferroic material lies in the cross-coupling between the ferroic order parameters, which may lead to additional degrees of control over the functionalities. Magnetoelectric coupling, for example, describes how one can use electric field to manipulate the magnetic order parameters or conversely control the ferroelectricity using magnetic field. This effect can be utilized to achieve the long-sought-after concept of "electric control of magnetism" that holds huge promise in electrical tunable magnetic devices with greatly reduced power consumption. Ever since the discovery of giant ferroelectric polarization in high-quality epitaxial thin films and single crystals of BiFeO$_3$, its magnetic order parameters have become the most intriguing yet controversial research focus throughout the past decade. To date, the detailed magnetic structures of BFO remain inconclusive, espe-cially in epitaxial thin films, where epitaxial strain plays an important role. Next, a broad spectrum of the research work on the magnetic properties of BFO and the related device applications will be reviewed, roughly in a chronological order.

3.4.1 MAGNETIC ORDERS

3.4.1.1 BULK

Pioneer work to unravel the magnetic orders in multiferroic BFO can be traced back to the 1960s. Unlike the ferroelectric property which has been veiled for decades due to the poor sample quality, the magnetic properties of BFO have been characterized with surprisingly fair accuracy at the first beginning. The first neutron diffraction measurement by Kiselev et al. [16] have already revealed a G-type antiferromagnetism with T_N close to 650 K, which turns out to be coincident with most of the following studies (see Ref. [29] and the references therein). Another milestone was built by Sosnowska et al. [15], who found an additional cycloidal spin modulation superimposed on the G-type antiferromagnetism in polycrystalline BFO samples. The period of this incommensurate spin cycloid is about 62 nm, with the propagation vectors along $\langle 1\ 1\ 0 \rangle$ pc axes. The spin rotation plane is defined by the ferroelectric polarization vector and the spin propagation vector. In other words, the antiferromagnetic spins are confined in the cycloid planes, that is, the antiferromagnetic easy planes. These findings were qualitative and

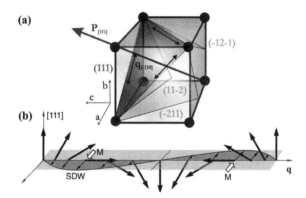

FIGURE 3.13 Magnetic structure of bulk BFO. (a) Given that the ferroelectric polarization is along [1 1 1] direction, there are three degenerate spin propagation vectors q along $\langle 1\ 1\ 0 \rangle$ directions, as depicted by the double arrows in the (1 1 1) plane. The spins are rotating in the planes (antiferromagnetic easy plane) defined by the polarization vector and the spin propagation vectors q, that is, $\{1\ 1\ 2\}$ planes (reprinted figure with permission from Ref. [261]. Copyright 2010 by the American Physical Society.) (b) A schematic drawing of the spin-cycloidal modulation in BFO single crystal. In addition to a spin cycloid, local canting of the magnetic moments out of the spin-cycloid plane produces a spin density wave of the same wavelength (~620 Å). Pseudocubic index is used for crystallographic axes and planes (reprinted figure with permission from Ref. [260]). (a: Reprinted with permission from Lebeugle, D.; Mougin, A.; Viret, M.; Colson, D.; Allibe, J.; Béa, H.; Jacquet, E.; Deranlot, C.; Bibes, M.; Barthélémy, A. Exchange coupling with the multiferroic compound BiFeO$_3$ in antiferromagnetic multidomain films and single-domain crystals. *Phys. Rev. B* 2010, 81, 134411. © 2010 by the American Physical Society; b: Reprinted with permission from Ramazanoglu, M.; Laver, M.; Ratcliff, II, W.; Watson, S. M.; Chen, W. C.; Jackson, A.; Kothapalli, K.; Lee, S.; Cheong, S. W.; Kiryukhin, V. Local Weak Ferromagnetism in Single-Crystalline Ferroelectric BiFeO$_3$. *Phys. Rev. Lett.* 2011, 107, 207206. © 2011 by the American Physical Society.)

quantitatively verified by later studies in high-quality BFO single crystals from different groups [251–253]. A summary of these experimental results done by neutron diffractions can be found in Ref. [254]. Due to the rhombohedral symmetry, the propagation vectors are in fact threefold degenerate for a given polarization direction as illustrated in Figure 3.13a. Therefore, there exist three symmetry-equivalent spin–cycloid planes, that is {1 1 2} planes using pc index. However, some ferroelectric–monodomain single crystals were reported to possess unique spin propagation vector, which can be explained by a symmetry-lowering of the crystal due to residual strain [25,26,251,252]. An important implication of the spin cycloid is the absence of the linear ME effect and weak ferromagnetism in BFO [255]. This conclusion is supported by the magnetization measurements on BFO single crystal, which showed that the weak ferromagnetic signal was completely attributed to small amount of impurities in the sample [97]. However, local spin canting due to Dzyaloshinski–Moriya (DM) interaction [256,257] is indeed permitted by the symmetry of bulk BFO [218,255]. Such delicate superstructure has been theoretically predicted [218,258,259] and recently experimentally observed by Ramazanoglu et al. [260], who discovered a spin canting angle of ca. 1° out of the cycloid plane. As shown in Figure 3.13b, the cycloid magnetic structure consists of two simultaneous waves: an "in-plane" wave with the magnetic dipoles rotating within the cycloid plane; and an "out-of-plane" wave that is normal to the cycloid plane with modulating out-of-plane components of the magnetic dipoles. This "out-of-plane" wave is indeed a spin density wave with an average local magnetization of $0.06\ \mu_B$/Fe. However, there are debates on whether the local canting of the spin cycloid can be considered as the precursor of the weak ferromagnetism expected in BFO if the cycloid modulation is suppressed, for example, by epitaxial strain in thin-film form. As pointed out by theoretical calculation [258], they might be related to two different and competing energies. Specifically, the atomistic simulation suggested that the out-of-plane modulation was inherent to the spin cycloid, whereas the spin-canted weak ferromagnetism driven by the DM interaction was closely linked to the tilting of oxygen octahedra [258]. Consequently, the two different mechanisms proposed lead to opposite configurations of the spin density wave. If the DM interaction is considered, the spin-canting moment M is maximum when the antiferromagnetic vector L is normal to the ferroelectric polarization P, following the right-handed system: $M \sim P \times L$ [218,259,260]. However, the atomisitic simulation indicated a maximum magnitude of canting when the spins are parallel/antiparallel to the polarization [258]. Further detailed investigations are imperative to resolve this discrepancy.

Lastly, we will address the stability of the magnetic ground state of bulk BFO against external perturbations (e.g., thermal fluctuation, magnetic field, strain). Temperature-dependent studies on the magnetic properties of bulk BFO mainly concern two controversies over the magnetic anomalies at low temperature, namely, "magnetic reorientations" and "magnetic anharmonicities" [262]. Seminal work by Sosnowska et al. revealed that the cycloidal magnetic structure barely changed from 78 to 463 K [15]. The lower and upper temperature limits were later expanded to 4 K and Néel temperature, respectively [263,264]. However, in 2008, several magnetic anomalies below room temperature were reported independently by different groups using Raman spectroscopy in BFO single crystals [265–267] (see a review in Ref. [13] for details) These findings pointed to a possible spin reorientation transition at low temperature. Subsequent investigations from the same and different groups, nevertheless, negated the hypothetic spin reorientation scenario below room temperature [268–270]. Instead, a slow growth of the cycloid period with elevating temperature was observed [262,270]. The second debate centered around the anharmonicity of the spin cycloid. Zalesskii and coworkers are the first to reveal a distorted cycloid structure at low temperature using nuclear magnetic resonance measurements, signifying significant anisotropy in the cycloid plane [271–273]. However, this anharmonicity was later proved to be greatly overestimated according to the results obtained from neutron diffractions [262,264,270], which was also supported by the atomistic simulation [258]. With increasing temperature, essentially a circular cycloid can be expected at room temperature. Based on the temperature-dependent studies described above, we can conclude that the ground state of the spin–cycloid structure of bulk BFO is very stable from low temperature up to T_N.

On the other hand, the spin cycloid can be destroyed by a large magnetic field of ~20 T, leading to a homogeneous antiferromagnetic state with possible canting moments and linear ME effect [255,274–281]. An example is provided in Figure 3.14, where the magnetization and polarization exhibit sudden jumps at the critical magnetic field. By extrapolating the magnetic response to zero field, a net magnetization of 0.03 μ_B Fe^{-1} can be deduced, indicating a canted antiferromagnetic state. As a consequence, the linear ME effect previously masked by the spin cycloid can be observed above the critical field. Besides, it can be seen that the critical field decreases with increasing temperature. In contrast to the large magnetic field, a tiny elastic strain (less than 10^{-4}) is able to toggle the magnetic easy plane (spin–cycloid plane) in BFO single crystal, signifying the extreme sensitivity of the magnetic structures to the strain [282]. This finding implies that epitaxial strain (on the order of 10^{-2}) should play a critical role in determining the magnetic ground state in BFO thin films.

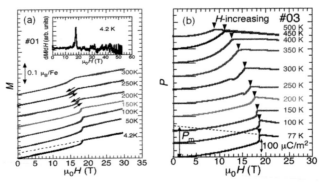

FIGURE 3.14 (a) Magnetization and (b) polarization responses under high-magnetic field. The dash lines in (a) and (b) represent the extrapolations of the M–H and P–H curves above transition field (reprinted figure with permission from Ref. [279]). (Reprinted with permission from Tokunaga, M.; Azuma, M.; Shimakawa, Y. High-Field Study of Strong Magnetoelectric Coupling in Single-Domain Crystals of $BiFeO_3$. *J. Phys. Soc. Jpn.* 2010, *79*, 064713.)

3.4.1.2 EPITAXIAL THIN FILMS

The magnetic structures of epitaxial BFO thin films are much more complicated and controversial than those in single crystals because they are closely related to the film orientation, domain structures, film chemistry, strain state, etc. The debates over the magnetic properties of epitaxial BFO thin films were provoked by the famous *Science* paper, where a saturation magnetization as large as 1 μ_B Fe^{-1} as well as a finite linear ME effect were measured in epitaxial BFO grown on STO substrate at room temperature [14]. These striking results contradicted with the spin–cycloid magnetic structure in bulk BFO. Due to the observed thickness dependence of the magnetization, the enhanced ferromagnetism was attributed to the strain effect. However, following studies by other groups reveal that similar BFO/STO samples exhibited saturation magnetizations 1–2 orders of magnitude smaller (≤ 0.05 μ_B Fe^{-1} by Eerenstein et al. [283] and ≤ 0.02 μ_B Fe^{-1} by Béa et al. [104]), which are consistent with the value estimated from the local canting moment in the single-crystalline BFO as described above. Further investigations attributed the abnormally large ferromagnetic response to the valence change of Fe ions from 3+ to 2+ due to the presence of oxygen vacancies [284] or the formation of ferrimagnetic γ-Fe_2O_3 impurity phases [285]. These systematic studies confirmed the intrinsic weak ferromagnetism of BFO epitaxial thin films, and the release of the latent canting magnetizations was explained based on the destruction of the spin cycloid by epitaxial constraint (a strain of 0.5% is enough to break the cycloid order) [286]. A more detailed neutron diffraction study corroborates this conclusion that the spin cycloid is absent in the BFO/STO sample with −1.5% compressive strain [57]. However, with increasing film thickness and strain relaxation process, bulk-like cycloidal spin modulation can be completely or partially restored, depending on the actual strain state of the films. As most of the studied BFO films are grown on STO substrate, we can have a comparison between different samples with the epitaxial strain ranging from −1.5% (fully strained) to 0 (fully relaxed). For fully strained thin film, Holcomb et al. [287] discovered the absence of spin cycloid and an easy magnetic axis along $\langle 1\ 1\ 2\rangle$ due to the inverse magnetostrictive effect in BFO, while in thick film (strain of −0.28%), an easy magnetic plane is observed, suggesting a recovery of cycloidal modulation. Similar results in thick film were also reported by Ke et al. [288], who studied the magnetic structures of BFO films with engineered ferroelectric domains. Since all of films were considerably relaxed, spin cycloid was detected in all samples. The least relaxed sample (strain of −0.4%) exhibited longer modulation wavelength compared to the single crystal, indicating a trend for the destruction of the spin cycloid by epitaxial strain. All these findings seem to be consistent with previous studies: larger epitaxial strain tends to break the spin–cycloid ground state, and stabilize a homogeneous G-type antiferromagnetism. Nevertheless, exceptional example was also reported by Ratcliff et al. [289] who performed neutron diffraction measurements on 1-μm-thick BFO films grown on STO substrates with different orientations. It was found that (0 0 1)-oriented film, though largely relaxed, still showed commensurate G-type antiferromagnetism with no spin modulation. Besides, both (1 1 0) and (1 1 1) films were found to exhibit modulated magnetic structures that may be different from the single crystal. However, detailed information is still lacking. Based on the findings listed above, we can see that the antiferromagnetic order in BFO is highly susceptible to the epitaxial strain, in accordance with the giant pressure effect observed in the single crystal [282].

More systematic studies of the strain effect on the magnetic properties of BFO were carried out by French research groups, who epitaxially grew BFO thin films onto a series of single-crystalline substrates with the in-plane strains ranging from compressive −2.4% to tensile +0.9% [155,290]. We remind the reader that within this strain range, BFO films adopt a sequential change of crystal symmetry from M_A–R–M_B. Though the ferroelectric Curie temperature was reported to be strongly affected by the strain, the antiferromagnetic Néel temperature shows negligible dependence on the strain [155]. Regarding the central question that whether the bulk spin cycloid is destroyed in thin-film form, they found that at low compressive strain bulk-like spin–cycloidal modulation exists. However, at high strain, the spin cycloid is driven toward pseudo-collinear G-type antiferromagnetism, and at moderate tensile strain, a more complicated cycloidal modulation appears. A complete strain–magnetic phase diagram is illustrated in Figure 3.15 [290]. It should be noted that phase coexistence may occur at the boundaries between different strain states. This picture seems to agree with other reports mentioned previously. However, controversy still rises over the exact orientations of the antiferromagnetic moments in the pseudo-collinear conditions. As indicated in Figure 3.3, the spins of BFO film tend to align along the in-plane $\langle 1\ 1\ 0 \rangle$ pc directions when grown on STO with relatively large compressive strain. This is completely at odds with that reported in Ref. [287], where the magnetic easy axes are along $\langle 1\ 1\ 2 \rangle$ directions with large out-of-plane components. Moreover, in Ref. [290], the film grown on DSO substrate with small compressive strain exhibited coexistence of bulk-like spin cycloid and pseudo-collinear antiferromagnetism, whereas the results from another group showed only pseudo-collinear alignment of the spins with in-plane $\langle 1\ 1\ 0 \rangle$ easy axes [291,292]. Another important conclusion drawn from Ref. [290] is a gradual rotation of the antiferromagnetic spins from complete in-plane orientation toward out-of-plane direction with decreasing compressive strain/increasing tensile strain. This phenomenon is consistent with another recent report, where the angle between the antiferromagnetic axis and the film normal decreases from 66° for BFO/NSO (large tensile strain) to 34° for BFO/DSO (small compressive strain) [73]. Therefore, the exchange coupling between BFO and a ferromagnet with in-plane anisotropy should decrease with increasing tensile strain as demonstrated in Ref. [290].

In summary, general consensus on the magnetic structure can be achieved for bulk/single crystal BFO based on the experiment efforts throughout decades. A spin–cycloid model is confirmed, with three equivalent spin propagation vectors and possible local canting of the antiferromagnetic moments. On the contrary, a clear picture of the magnetic ground states in BFO epitaxial thin films remains elusive. According to the divergent results obtained by different groups, we speculate that the exact magnetic orders depend sensitively on the detailed film growth conditions, which may affect the domain structures and film chemistry significantly. For example, BFO films with order and disorder domain structures may give rise to drastically different magnetic responses and exchange coupling behaviors with a ferromagnet, implying different magnetic structures [247]. What we can conclude from the existing reports at current stage is (1) the intrinsic canting moments in BFO thin films are extremely small and not depending on the strain; (2) the spin–cycloid structure tends to break down to pseudo-collinear spin alignment at high strain state be it compressive or tensile; (3) there is a tendency for the spins to rotate from in-plane to out-of-plane when varying the substrates from large compressive strain to large tensile strain.

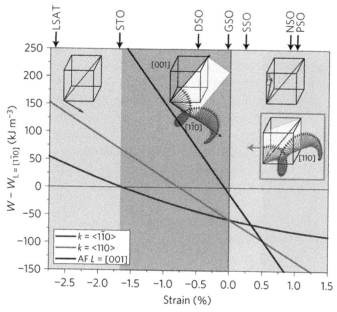

FIGURE 3.15 Magnetic phase diagram of BFO epitaxial films under different strains. Both large compressive and tensile strains stabilize collinear antiferromagnetic orders, however, with their corresponding antiferromagnetic vectors along in-plane [1–10] and out-of-plane [0 0 1] directions, respectively. At small compressive strain, bulk-like "type-1" spin cycloid is restored, with propagation vectors along $\langle 1 -1 0\rangle$ directions. Small tensile strain tends to favor a "type-2" cycloid with propagation vectors along $\langle 1 1 0\rangle$ directions (reprinted with permission from Ref. [290]). (Reprinted by permission from Macmillan Publishers Ltd: Sando, D.; Agbelele, A.; Rahmedov, D.; Liu, J.; Rovillain, P.; Toulouse, C.; Infante, I. C.; Pyatakov, A. P.; Fusil, S.; Jacquet, E.; Carrétéro, C.; Deranlot, C.; Lisenkov, S.; Wang, D.; Le Breton, J. M.; Cazayous, M.; Sacuto, A.; Juraszek, J.; Zvezdin, A. K.; Bellaiche, L.; Dkhil, B.; Barthélémy, A.; Bibes, M. Crafting the magnonic and spintronic response of BiFeO3 films by epitaxial strain. *Nat. Mater.* 2013, *12*, 641. © 2013.)

In order to gain a more comprehensive understanding about the magnetic orders of epitaxial BFO, ferroic domain structures need to be taken into considerations. Preferably, single-domain BFO films grown on different substrates will eliminate the complexity arising from multidomain convolution and reveal the true effect of epitaxial strain on the magnetic orders in BFO. Only after the magnetic orders in BFO epitaxial thin films are fully understood, can we exploit its ME coupling for a better design of device applications.

3.4.2 MAGNETOELECTRIC COUPLING IN BiFeO₃

The most intriguing and promising property of BFO is the intimate connection between ferroelastic, ferroelectric, and antiferromagnetic order parameters, which may have great potential in designing room-temperature multifunctional devices with multiple tunable methods. Electrically controllable spintronic devices, being the most appealing, hinge on the ME coupling effect in BFO itself. Such kind of coupling, though achievable both in single crystal and thin film forms, deserves more investigations, as it forms the cornerstone for the integration of the BFO into real-device applications. To look into the ME coupling effect in BFO, simultaneous characterizations of ferroelectric and antiferromagnetic orders are required. However, such experimental capabilities are limited to several groups so far. We will review on this issue in the following section.

3.4.2.1 BULK (SINGLE CRYSTAL)

In BFO single crystals, the first attempt to explore the electric-field switching effect on the antiferromagnetic order was made by Lebeugle et al. on samples with single ferroelectric domain with unique spin propagation vector [251]. As depicted in Figure 3.16, by applied an electric field, part of the crystal volume switched its polarization by 71°, which was accompanied by a flip of the spin–cycloid plane, as the cycloid plane is determined by the polarization vector and the spin-propagation vector. The physical origin behind this ME coupling effect can be understood based on the antisymmetric DM interaction, which transforms the homogeneous antiferromagnetic state into magnetic spiral in BFO [255]. The inverse effect results in local polarization induced by the magnetization gradients [293]. The interaction between this polarization and the existing ferroelectric polarization leads to a nonzero ME term that accounts for the observed ME coupling effect. At almost the same time, similar experimental results were obtained by Lee et al., however, with equivalent magnetic domains coexisting [252,253]. An additional conclusion they have pointed out is that the populations of magnetic domains after electrical switching are closely linked to the electric field direction. This effect was explained by the piezoelectric strain that lifts the degeneracy among different magnetic domains, as being verified in a later study on the macroscopic uniaxial pressure effect [282].

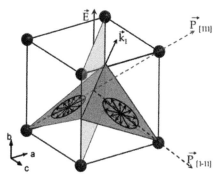

FIGURE 3.16 A schematic showing the electric-field-induced flip of the spin–cycloid plane in ferroelectric–monodomain BFO single crystal. As the cycloid plane is determined by the polarization and spin propagation vectors, electrical switching of the polarization leads to the rotation of the cycloid plane. The spin propagation vector remains unchanged in this case (reprinted with permission from Ref. [251]). (Reprinted with permission from Lebeugle, D.; Colson, D.; Forget, A.; Viret, M.; Bataille, A. M.; Gukasov, A. Electric-field-induced spin flop in BiFeO3 single crystals at room temperature. *Phys. Rev. Lett.* 2008, *100*, 227602. © 2008 by the American Physical Society.)

3.4.2.2 EPITAXIAL THIN FILMS

The studies on the ME coupling effect in epitaxial thin films were carried out even earlier than those reported in single crystals. In 2006, Zhao et al. presented the first demonstration of the electrical switching of antiferromagnetic domains at room temperature [219]. Using combined PFM and photoemission electron microscopy (PEEM) to image the same area of the BFO thin films, they managed to observe a one-to-one correlation between the ferroelectric and antiferromagnetic domains. Furthermore, switching the ferroelectric domains by electric field caused concurrent switching of the antiferromagnetic domains. However, the detailed magnetic structures of the thin films in this study remained ambiguous. They assumed a pseudo-collinear G-type antiferromagnetism according to first-principle calculations, with the antiferromagnetic axis along [1 −1 0] when polarization is along [1 1 1] pc direction. However, we noted that their BFO films were grown on STO substrate with a thickness of 600 nm. The proposed magnetic configuration here disagrees with their following study [287]. Nevertheless, they brought out an important concept that the rotation of the antiferromagnetic easy axis is closely related to the switching path of the ferroelectric polarization, because both of the ferroelectric and antiferromagnetic domains are coupled to the ferroelastic distortion of the BFO lattice. From symmetry and geometry considerations, different polarization switching mechanisms in BFO thin film are shown in Figure 3.17 with their corresponding effects on the antiferromagnetic easy plane. For example, 180° switching won't do anything to the magnetic easy plane, thus, with no flip of spins. In contrast, Both 71° and 109° switching will lead to the rotations of the magnetic easy plane. However, whether the flips of the antiferromagnetic spins can occur depends on their exact orientation in the magnetic easy plane. Therefore, a clear picture of the antiferromagnetic order (especially the relationship between ferroelectric polarization and antiferromagnetic axis/plane) needs to be established first before we make any further step toward the electrical control of the antiferromagnetic order in BFO. Unfortunately, as pointed out above, a consensus is still lacking on this issue, which makes the interpretation of the observed results ambiguous and even contradictive. Let us see another example, where electric-field switching effect on the antiferromagnetic order of ferroelectric monodomain BFO was probed by neutron diffraction as reported by Ratcliff et al. [294]. First, they found that (0 0 1)-oriented single-domain BFO exhibited cycloidal spin structure different from the single crystal. The cycloid plane was found to be perpendicular the polarization with three equivalent propagation directions. This observation seems to be consistent with the antiferromagnetic easy plane proposed in Refs. [219,295], but contradicted to their previous report [289]. Subsequent electrical switching experiments were conducted to change the relative domain populations in the film, thus, leading to modulations of the exchange coupling behavior with a ferromagnetic thin layer. However, the effect looks trivial compared to that observed in single crystal.

To sum up, we can see the electrical switching of the antiferromagnetic orders of BFO is better established in single crystals than in epitaxial thin films. The reason lies in the fact that the magnetic structures of single crystal are more precisely understood with fair agreement being reached in literature reports. Due to DM interaction, cycloidal spin modulations are inherently coupled to the ferroelectric polarizations, leading to a strong ME coupling effect at the atomic level. As a consequence, electrically induced non-180° switching of the polarization should be capable of flipping the cycloid plane, namely, the antiferromagnetic spin directions. On the other hand, research work on epitaxial thin films also reveals intimate connection between the ferroelectric and antiferromagnetic domains. Electrical control of the antiferromagnetic orders in BFO has also been demonstrated. However, the detailed spin configuration in each antiferromagnetic domain is yet to be clarified, which hampers a thorough and comprehensive understanding of the ME switching effect in BFO epitaxial thin films.

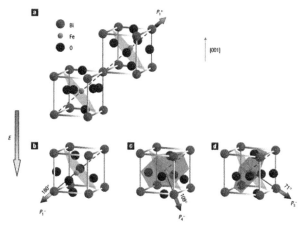

FIGURE 3.17 A cartoon illustrating the responses of the antiferromagnetic easy plane to different polarization switching paths in epitaxial BFO thin films. Though 180° switching won't alter the antiferromagnetic plane, 71° and 109° switching result in rotations of the plane, which may further lead to the flips of the antiferromagnetic spins (reprinted with permission from Ref. [219]). (Reprinted by permission from Macmillan Publishers Ltd: Zhao, T.; Scholl, A.; Zavaliche, F.; Lee, K.; Barry, M.; Doran, A.; Cruz, M. P.; Chu, Y. H.; Ederer, C.; Spaldin, N. A.; Das, R. R.; Kim, D. M.; Baek, S. H.; Eom, C. B.; Ramesh, R. Electrical control of antiferromagnetic domains in multiferroic BiFeO3 films at room temperature. *Nat. Mater.* 2006, *5*, 823. © 2006.)

3.5 EXCHANGE COUPLING AND DEVICE APPLICATIONS IN BiFeO₃-BASED HETEROSTRUCTURES

As discussed in previous sections, electrical controls of the antiferromagnetic orders are viable in both single crystals and epitaxial thin films through the ME coupling effect. However, owing to the local cancelation of the magnetic moments, BFO, be it single crystal or thin films, exhibits negligible ferromagnetism macroscopically, which is not accessible by conventional magnetic reading techniques, thus precluding its direct applications in memory devices. To circumvent this obstacle, Binek and Doudin proposed novel device paradigms by combining the ME effect of a multiferroic and the exchange coupling effect with a thin ferromagnet, in which electrical switching of the antiferromagnetic order in the ME/multiferroic film can effectively change the magnetic state of an adjacent ferromagnetic layer via exchange coupling [18]. This concept was soon employed in ME/multiferroic materials, such as Cr_2O_3 [296,297], $YMnO_3$ [298] as well as $BiFeO_3$ [299]. Figure 3.18 shows a possible device design with a GMR element built on top of a multiferroic material based on this idea. Here, another coupling effect at the ferromagnet–antiferromagnet interface, namely exchange coupling, is introduced in conjunction with the internal ME coupling in MEs/multiferroics. Generally speaking, exchange coupling can manifest itself in two different manners. The first one, termed exchange bias, is a horizontal shift of the magnetization hysteresis loop due to the unidirectional pinning of the ferromagnetic moments by uncompensated antiferromagnetic spins. The second one, is an enhancement of the coercive field resulted from the interaction of the ferromagnetic spins and the fully compensated or unpinned/ reversible antiferromagnetic surface spins. The details regarding the phenomenology and the underpinning mechanisms of exchange coupling are out of the scope of this book chapter and can be found elsewhere [300–303]. In the following section, the exchange coupling effect between BFO and a thin ferromagnetic layer will be discussed, together with its device applications toward the ultimate goal of electrically controllable magnetism.

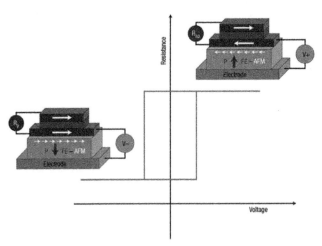

FIGURE 3.18 A possible device design of magnetoresistance random access memory incorporating a GMR element and a multiferroic. A voltage is applied to the multiferroic (antiferromagnetic ferroelectric) layer to switch the polarization. Due to the magnetoelectric coupling effect, the antiferromagnetic spins that attach to the polarization also change their directions. Subsequently, spins in the adjacent ferromagnetic layer also flip through the interfacial exchange coupling, which finally leads to a resistance change of the GMR element (reprinted with permission from Ref. [304]). (Reprinted by permission from Macmillan Publishers Ltd: Bibes, M.; Barthélémy, A. Multiferroics: Towards a magnetoelectric memory. *Nat. Mater.* 2008, 7, 425. © 2008.)

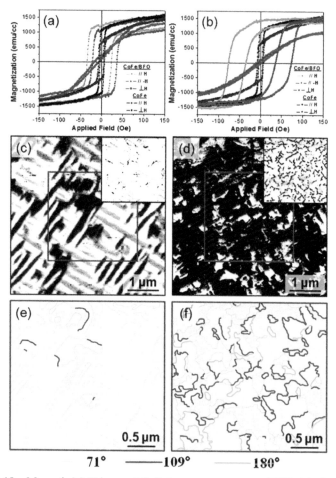

FIGURE 3.19 Magnetic M–H loops of CoFe layer grown on top of BFO thin films with (a) stripe-like (negligible exchange bias) and (b) mosaic-like (significant exchange bias) domain patterns. Corresponding in-plane PFM images of (c) stripe-like and (d) mosaic-like BFO thin films. Out-of-plane PFM images are shown in the insets. (e and f) Populations of different types of domain walls deduced from (c) and (d), respectively (reprinted with permission from Ref. [247]). (Reprinted with permission from Martin, L. W.; Chu, Y. H.; Holcomb, M. B.; Huijben, M.; Yu, P.; Han, S. J.; Lee, D.; Wang, S. X.; Ramesh, R. Nanoscale Control of Exchange Bias with BiFeO3 Thin Films. *Nano Lett.* 2008, *8*, 2050. © American Chemical Society.)

Primitive studies on this topic were launched by growing high-T_C transition metal alloys with soft ferromagnetism onto BFO epitaxial thin films, in which robust room-temperature exchange bias were observed [305–307]. Quantitative analyses of the magnitudes of the exchange bias and the corresponding domain structures of BFO suggested that the strength of the exchange bias is inversely proportional to the domain size of the BFO thin films (note that the ferroelectric and antiferromagnetic domains are coupled) [308], consistent with the random-field theory [309,310]. More specifically, Martin et al. discovered that the magnitudes of the exchange bias field were closely linked to the populations of 109° domain walls in the films [247]. Large exchange bias fields were only observed in mosaic-domain samples in which 109° domain walls prevailed, while stripe-domain samples with ordered 71° domain walls exhibited negligible exchange bias (see Fig. 3.19). This finding seems to be correlated with the enhanced ferromagnetic response observed in the mosaic samples [247] as well as the existence of extraordinary conductivities, even magnetotransport properties at 109° domain walls [119,248]. However, the spintronic devices based on the exchange bias effect of metal–alloy/BFO systems appear extremely fragile upon electrical switching. Both the vertical and planar spin-valve devices built on top of BFO thin films showed irreversible changes in the giant magnetoresistance responses as illustrated in Figure 3.20, which can be attributed to the unfavorable modifications of the domain structures/domain walls under electric cycles [311,312].

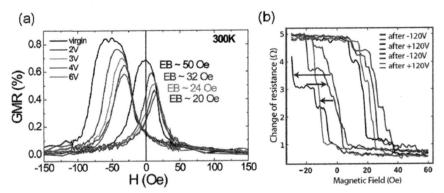

FIGURE 3.20   (a) GMR curves of a spin-valve structure (Au 6 nm/Co 4 nm/Cu 4 nm/CoFeB 4 nm) grown on top of BFO/BFO–Mn bilayer after vertically poling the BFO film at different voltages (reprinted with permission from Ref. [311]. Copyright 2012 American Chemical Society). (b) GMR minor loops of a spin-valve device (Permalloy/Cu/Permalloy/BFO/TbScO$_3$) after repetitive in-plane electrical switching of the BFO domains. Both examples exhibit irreversible changes of the magnitudes of the exchange bias (reprinted with permission from Ref. [312]). (a: Reprinted with permission from Allibe, J.; Fusil, S.; Bouzehouane, K.; Daumont, C.; Sando, D.; Jacquet, E.; Deranlot, C.; Bibes, M.; Barthélémy, A. Room Temperature Electrical Manipulation of Giant Magnetoresistance in Spin Valves Exchange-Biased with BiFeO3. *Nano Lett.* 2012, *12*, 1141. © 2012 American Chemical Society; b: Reproduced with permission from Wang, C.; Ph.D. Thesis: "Characterization of spin transfer torque and magnetization manipulation in magnetic nanostructures," Cornell University, 2012.)

On the other hand, exchange bias is definitely not a must for manifestation of strong exchange coupling at metal–alloy/BFO interface. Enhancement of coercive field and strong magnetic anisotropy of the ferromagnetic layer induced by BFO also signify a nontrivial exchange coupling between the ferromagnetic and antiferromagnetic spins. For instance, Chu et al. presented the first demonstration of dynamic electrical switching of the local ferromagnetism at room temperature by taking advantage of the ME coupling in BFO and the exchange coupling at the CoFe/BFO heterointerface [299]. Again, by using PFM and PEEM, they first found a one-to-one coincidence between the ferroelectric and ferromagnetic domains, mediated by a collinear interfacial spin coupling. Furthermore, the net magnetizations of the CoFe pads can be reversible switched by 90° via electrical switching of the ferroelectric domains of BFO using planar electrodes, as shown in Figure 3.21. The antiferromagnetic structure of BFO used in this report is consistent with Ref. [290], but at odds with their own reports subsequently [287,292].

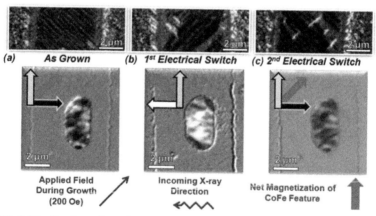

FIGURE 3.21 In-plane ferroelectric domain structures of BFO imaged by PFM and corresponding ferromagnetic domain structures of $Co_{0.9}Fe_{0.1}$ probed by XMCD–PEEM in a coplanar–electrode $Co_{0.9}Fe_{0.1}$/BFO device (a) in the as-grown state, (b) after first electrical switching, and (c) after second electrical switching (reprinted with permission from Ref. [299]). (Reprinted by permission from Macmillan Publishers Ltd: Chu, Y. H.; Martin, L. W.; Holcomb, M. B.; Gajek, M.; Han, S. J.; He, Q.; Balke, N.; Yang, C. H.; Lee, D.; Hu, W.; Zhan, Q.; Yang, P. L.; Fraile-Rodriguez, A.; Scholl, A.; Wang, S. X.; Ramesh, R. Electric-field control of local ferromagnetism using a magnetoelectricmultiferroic. *Nat. Mater.* 2008, *7*, 478. © 2008.)

Following this work, the same group even realized a 180° reversal of the net magnetization of CoFe pads using similar device architectures, but on DSO substrates instead of STO [291]. This time, however, they added in new variables of canted magnetic moments in BFO thin films, which determined the easy axes of the CoFe films grown on top. Through this inter-facial exchange coupling, 180° switching of the in-plane net polarization of BFO film resulted in reversal of the net canted moment, and consequently the flip of the ferromagnetic magnetization of CoFe, as schematically shown in Figure 3.22. Additionally, they found that the magnetic easy axes of the CoFe films were not affected by the magnetic fields applied during growth and an effective coupling field on the order of 10 mT was obtained from micromagnetic simulation, which seemed sufficiently large to toggle the ferromagnetic spins [292]. The absence of exchange bias can be explained by a spin–flop interfacial exchange coupling [313,314] or canted moments driven by DM interaction with a weak barrier for the rotation of oxygen octahedra [292]. However, the antiferromagnetic easy axes observed in BFO/STO and BFO/DSO disagreed with those reported in Ref. [290].

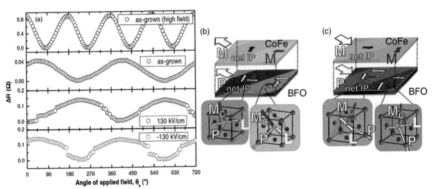

FIGURE 3.22 (a) Anisotropic magnetoresistance (AMR) responses of the CoFe/BFO planar device. From top to bottom: high-field AMR behavior follows a common $\cos^2(\theta_a)$ dependence; low-field AMR behavior of the as-grown CoFe shows a $\cos(\theta_a)$-like dependence; the AMR curve shifts by 180° after electrical switching; the AMR response can be restored by applying a reverse electric field. (b and c) Cartoons depicting the configurations of polarization (P), antiferromagnetic axis (L), and canted moment (M_c) in BFO films, together with the one-to-one relationship of the coupling spins at CoFe/BFO interface (b) in as-grown state and (c) after first electrical switching (reprinted with permission from Ref. [291]). (Reprinted with permission from Heron, J. T.; Trassin, M.; Ashraf, K.; Gajek, M.; He, Q.; Yang, S. Y.; Nikonov, D. E.; Chu, Y. H.; Salahuddin, S.; Ramesh, R. Electric-Field-Induced Magnetization Reversal in a Ferromagnet-MultiferroicHeterostructure. *Phys. Rev. Lett.* 2011, *107*, 217202. © 2011 by the American Physical Society.)

Exchange coupling studies between ferromagnetic metal alloys and BFO were also conducted in single-domain single crystals [315]. Consistent with the results observed in epitaxial thin films with ordered domain structures, uniaxial rather than unidirectional magnetic anisotropy was found, with the easy axis coupled to the cycloidal spin propagation vector, regardless of applied magnetic field directions during growth. The uniaxial anisotropy was explained by the energy minimization of the ferromagnetic spins with wiggling arrangement induced by exchange coupling with the canted moments of the spin cycloid. Switching of the ferroelectric polarization caused the flip of the spin propagation vector, and consequently the rotation of the magnetic easy axis by 90°. Thus, the magnetic anisotropy of the thin ferromagnetic layer can be controlled by the electric field. However, the degree of the electrical control in single-crystal BFO is less deterministic and reproducible compared to the thin film counterparts, because the ferroelectric switching usually leads to multidomain state, and, moreover, the degeneracy of the spin cycloids will be restored after electrical switching. Later on, the different exchange coupling behaviors of thin ferromagnetic layers grown on BFO were discussed and compared between single-crystal and multidomain thin-film samples [261]. In accordance with previous results, the exchange bias was attributed to the imperfection of the antiferromagnetic orders, that is, defects or domain walls, in which pinned/irreversible uncompensated spins could induce unidirectional anisotropy in the ferromagnetic layers by applying in situ magnetic field during growth. As a result, exchange bias was not observed in single-crystal samples with no domain walls, but only in multidomain, more specifically, mosaic-domain samples, where defects and 109° walls prevail. For ferromagnetic thin layers deposited on BFO single crystals, strong uniaxial magnetic anisotropy was found instead, with the easy axis along the cycloid propagation direction. It is worth noting that the situation is very similar to BFO epitaxial thin films with highly ordered 71° stripe domains, in which only uniaxial anisotropy exists with negligible exchange bias. In this case, each stripe domain can be considered as a single-domain unit, which combines to form a global magnetic easy axis in the ferromagnetic layer through exchange coupling. Likewise, the electrical switching behaviors in these samples resemble those observed in BFO single crystals, where the magnetic easy axes of the ferromagnetic layers can be toggled by the polarization switching. However, the detailed coupling mechanisms between the ferromagnetic metal alloys and BFO thin films remain elusive due to the ongoing debates over the actual antiferromagnetic structures in BFO epitaxial thin films on various substrates. For example, both pseudo-collinear and bulk-like spin–cycloid antiferromagnetic orders have been reported for BFO films grown on DSO, in which electrical manipulation of magnetic easy axes of the ferromagnetic layers grown on top are possible.

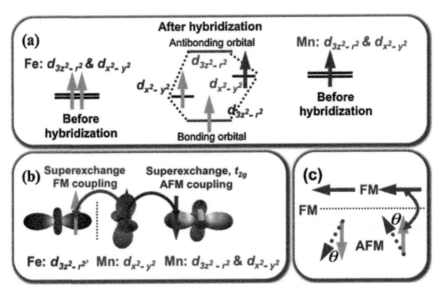

FIGURE 3.23 Proposed model for orbital-reconstruction-induced ferromagnetism state in BFO at the interface with LSMO. (a) Electronic orbital reconstruction between Fe and Mn ions. (b) Proposed coupling mechanism and spin configuration at the interface. (c) Ferromagnetism due to enhanced canting angle of antiferromagnetic spins in BFO via interfacial coupling (reprinted with permission from Ref. [316]). (Reprinted with permission from Yu, P.; Lee, J. S.; Okamoto, S.; Rossell, M. D.; Huijben, M.; Yang, C. H.; He, Q.; Zhang, J. X.; Yang, S. Y.; Lee, M. J.; Ramasse, Q. M.; Erni, R.; Chu, Y. H.; Arena, D. A.; Kao, C. C.; Martin, L. W.; Ramesh, R. Interface ferromagnetism and orbital reconstruction in BiFeO₃- La0.7Sr0.3MnO₃ heterostructures. *Phys. Rev. Lett.* 2010, *105*, 027201. © 2010 by the American Physical Society.)

The ferromagnetic layers discussed above are all polycrystalline or amorphous metal alloys deposited on the perovskite surface of BFO, whose interfaces are essentially incoherent with possible inhomogeneities and disorders. Structurally compatible heteroepitaxial bilayers are believed to exhibit better strength and reliability of the interfacial coupling. The well-studied mixed-valence manganites with colossal magnetoresistance emerge as perfect companions to form coherent interfaces with BFO, where multiple ferroic order parameters coexist and interact. The ferromagnetism in half-metal $La_{0.7}Sr_{0.3}MnO_3$ (LSMO), for example, is mediated by the double exchange coupling between Mn^{3+} and Mn^{4+} ions, whereas superexchange antiferromagnetism is dominant between neighboring Fe^{3+} ions in BFO. When the lattice continuity terminates at the heterointerface, symmetry breaking occurs with inevitable hybridizations between Fe and Mn orbitals via oxygen ligands. Yu et al. reported that such hybridizations

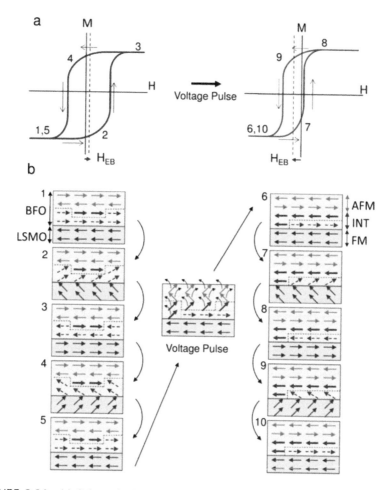

FIGURE 3.24 (a) Schematic drawings of the magnetization hysteresis loops of LSMO before and after electrical switching of the BFO polarization. (b) Proposed model for the electrical switching of the exchange bias. The numbers and arrows in (a) and (b) represent the experimental sequence for the magnetic-field sweeping. At the beginning, the polarization of BFO is pointing downward with the Fe^{3+} ions closer to the $Mn^{3+/4+}$ ions, inducing larger exchange coupling energy. Therefore, more interfacial spins of BFO are unpinned or reversible with the spins of LSMO and less BFO spins are pinned/irreversible, which gives rise to large coercive field and small exchange bias of LSMO. After electrical switching of the BFO polarization, the spins of BFO flip due to its magnetoelectric coupling. Besides, the interfacial exchange coupling is weakened as the Fe^{3+} ions are now pushed away from the interface. Consequently, we see smaller coercive field and larger exchange bias with opposite sign (reprinted with permission from Ref. [319]). (Reprinted with permission from Wu, S. M.; Cybart, S. A.; Yi, D.; Parker, J. M.; Ramesh, R.; Dynes, R. C. Full Electric Control of Exchange Bias Phys. Rev. Lett. 2013, *110*, 067202. © 2013 by the American Physical Society.)

resulted in electronic orbital reconstructions at the interface as shown in Figure 3.23 [316]. Consequently, ferromagnetic superexchange between interfacial Fe and Mn ions induced a stronger canting of the antiferromagnetic moments in BFO, unleashing the ferromagnetism of BFO in the interface region. These pinned/irreversible interfacial spins explained the exchange bias observed in BFO/LSMO heterostructures. This scenario was supported by further structural investigations using high-resolution scanning transmission electron microscopy revealed the suppression of oxygen octahedral tilts and the stabilization of a high-symmetry metallic phase in the interfacial BFO layer adjacent to LSMO film [317]. More intriguingly, the magnitude and the polarity of the exchange bias can be modulated by electrical field, which was explained the electrically induced flip of the antiferromagnetic spins in BFO along with the change in the interfacial exchange energy due to the ionic displacement of the Fe^{3+} relative to $Mn^{3+/4+}$ ions, as schematically depicted in Figure 3.24 [318,319]. Thickness dependences of both the LSMO and BFO layers on the exchange bias effect were also discussed [320]. While the exchange bias field showed conventional inverse proportionality with the thickness of LSMO, an ultrathin limit of 2 nm was found for the BFO layer, below which exchange bias disappeared. This observation is also intuitively understandable, since the antiferromagnetic anisotropy, that is important for exchange bias, decreases with reduced film thickness. Lastly, we should emphasize the critical role of strain in addition to the spin, charge, and orbital degrees of freedom in the BFO/LSMO heterostructures. The strain effect will become a dominant factor at large LSMO thickness and high temperature, where the interfacial exchange coupling is weakened or destroyed by thermal fluctuation. As a result, uniaxial magnetic anisotropy observed in LSMO can be attributed to the lattice distortion imposed by the ferroelastic domains of BFO layer instead of interfacial magnetic coupling [321,322]. To gain better understanding of the physical properties of BFO/LSMO heterostructures with multiple ferroic order parameters, it is crucial to deconvolute their effects at different temperature and dimension scales.

In this section, the coupling phenomena between a ferromagnetic layer and BFO have been elaborated and discussed in detail, which naturally leads to ME devices with electrical tuning capabilities. In general, the heterostructures studied can be divided into two categories depending on the ferromagnetic materials. The first one, being high-T_c ferromagnetic metal alloys grown on BFO thin films or single crystals, exhibits large exchange bias or exchange anisotropy up to room temperature. Although the exchange bias is involved with specific ferroic domain walls, which hitherto cannot be decisively controlled, the electrical manipulation of the magnetic anisotropy of the ferromagnetic layers has been demonstrated in both thin films and single crystals at room temperature. The second group of heterostructures is epitaxial systems, mainly referring to BFO/LSMO bilayers. In addition to the magnetic spins, lattice, charge, and orbital degrees of freedom all play a crucial role at the heterointerface. Exotic exchange bias has been discovered, which can be further modulated by electric field. However, the low operating temperature has posed a critical barrier for future applications.

3.6 OPEN QUESTIONS AND FUTURE DIRECTIONS

In previous sections, the structural, electrical, and magnetic properties of BFO in both bulk (single crystal) and thin-film forms are reviewed and discussed. Over the past decade, the research on this versatile compound proved to be fruitful from both fundamental and application point of views. However, there are still a number of puzzles yet to be unraveled. For example, the exact structures and symmetries of the high-temperature phases remain controversial. Whether a true cubic phase exists or not, and how it may be related to the insulator–metal transition under high temperature and pressure remain to be answered. Besides, the discovery of strain-driven polymorphic phase boundary became another booster for this research field. The rich phase diagram of BFO enclosing a multitude of competing phases enables the control of phase stability by tipping their relative energy scale using external stimuli [77,134,323]. Practices using epitaxial tensile strain or other substrate orientations are strongly encouraged [69,70,78,324,325]. Regarding the ferroelectric property of BFO, although the ferroelectric random-access memory is limited to a niche market, a variety of spin-off memories are flourishing, such as resistive random-access memory, ferroelectric field-effect transistors (ferroelectric FET), FTJ, memory and ferroelectric photovoltaic (FPV) memory, all of which have been demonstrated in BFO-based devices as proofs of concepts. A big advantage of BFO over other ferroelectrics is its additional magnetic degree of freedom. Antiferromagnet itself is emerging as an alternative for generating anisotropic magnetoresistance in spintronic applications [326–330]. By incorporating the ferroelectric and antiferromagnetic aspects of BFO, multi-state information encoding can be realized with the merit of nondestructive read-out [331,332]. A myriad of possibilities exist in the designs of device paradigms and realization of the proofs of concepts, which surely will catch the imaginations of the researchers from all over the world in the foreseeable future [333,334].

The ferroic domain walls in BFO being yet another active element in the design of next-generation nanoelectronics are receiving growing attention from the community. Further work should be focused on deconvolution of the intrinsic (symmetry breaking) and extrinsic (defects) properties, preferably at single-domain-wall limit. Domain wall probably should be treated as an emergent phase with distinct properties compared to the domain itself [335]. Besides, cautions should be devoted to the relative energy scales between different types of domain walls and the impact of the epitaxial strain on them, in which the oxygen octahedral tilt plays a critical role [232,336,337]. In this context, deterministic control of the domain assemblies, both in situ and ex situ, appears highly desirable. In addition to the homointerfaces separating domains, the heteroepitaxial interfaces between BFO and other complex oxides also provide a fertile source for basic research [316,338]. With the aid of atomic-precision film growth techniques, a wide range of oxides can be brought into bonding with BFO with controlled interfacial terminations, where the interplay between spin, charge, lattice, and orbital degrees of freedom at the two dimension is full of surprises (Fig. 3.25) [339–343].

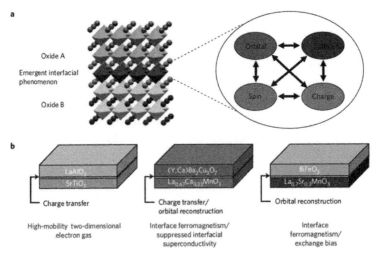

FIGURE 3.25 Schematics showing interplays between degrees of freedom at the complex oxide interfaces (reprinted with permission from Ref. [318]). (Reprinted by permission from Macmillan Publishers Ltd:Wu, S. M.; Cybart, S. A.; Yu, P.; Rossell, M. D.; Zhang, J. X.; Ramesh, R.; Dynes, R. C. Reversible electric control of exchange bias in a multiferroic field-effect device. *Nat. Mater.* 2010, *9*, 756. © 2010.)

As the only hitherto known material with robust room-temperature multiferroism, BFO still holds huge promise for electrically controllable magnetoelectronics with reduced power consumptions. A few of prototypical device architectures based on the ME effect have been theoretically proposed and experimentally demonstrated in recent years. For example, electrical manipulation of the ferromagnetism has been realized in CoFe/BFO heterostructures with planar electrodes. Besides, exchange bias in LSMO/BFO epitaxial thin films has been proved to be electrical switchable. These rudimental results have laid the groundwork for future explorations toward logic and memory applications. However, the existing discrepancies in the antiferromagnetic structures of BFO thin films greatly hamper a comprehensive understanding of its ME coupling as well as the exchange coupling with a ferromagnet, which, in turn, blinds us to designing strategies for devices with better performance. From a fundamental point of view, the physics behind either the ME coupling or the exchange coupling is still rich, thus worth further investigations. In what follows, a series of open questions will be highlighted, jointly with some possible experimental schemes proposed to sort out current discrepancies in this field.

Looking into the literature, general consensus has been reached on the magnetic superstructures in bulk (single crystal) BFO samples, with a spin–cycloid superimposed on the G-type antiferromagnetism. Contrarily, the magnetic orders in BFO epitaxial thin films remain much less explored. Divergent or inconsistent results can be found in the reports published by different groups or the same group. Such inconsistencies could be due to the differences in film growth conditions, which results in dissimilar domain structures, for example, mosaic or striped domain pattern may correspond to disordered or ordered magnetic structures. Furthermore, various characterization techniques have been used in probing the magnetic orders in BFO thin films, for example, neutron diffraction [57,288,289], X-ray linear dichroism [219,287], Mössbauer spectroscopy [155,290], and so on, which differs significantly in sensitivity and resolution. Besides, parasitic effect can also arise from ferroelectric order, as can be seen in the X-ray linear dichroism measurements. To reduce the complexity resulted from multidomain structures, BFO thin films of ferroelectric monodomain are recommended. However, since the domain engineering involves highly vicinal substrates [196,197,200,201], strain relaxation must be carefully monitored to correlate with the observed magnetic structures under different strain state. Multiple probing techniques are required to yield reliable results on the same sample. If strain dependence of the magnetic orders of BFO thin films as reported in Ref. [290] can be confirmed, the design of the device architectures for implementation of BFO thin films in ME or spintronic applications will be greatly facilitated. For example, as the antiferromagnetic easy axis of BFO films can be tuned by strain from in-plane to out-of-plane directions, spintronic building blocks with either in-plane or perpendicular anisotropy can be fabricated on top accordingly to achieve optimized performance.

Likewise, the exploration for dynamic electrical control of antiferromagnetic orders in BFO thin film can also take advantage of the single-domain samples. As mentioned above, the change of the antiferromagnetic orders may have close connection with the polarization switching path. As a result, special attention must be devoted to the delicate control of polarization switching events. Certain combinations of in-plane and out-of-plane electric fields may favor specific switching events in BFO thin films [220]. However, the stability of the domain configurations also needs to be taken into consideration [177,215].

On the other hand, the exchange coupling between BFO and a ferromagnetic layer also plays a vital role in materializing feasible devices. Compared with the low blocking temperature of the exchange bias observed in BFO/LSMO heterostructures, the exchange coupling between BFO and high-T_c metal alloys seems to survive beyond room temperature, which would be highly promising for potential device applications. Different types of exchange anisotropies, that is, unidirectional or uniaxial, must be separately studied and understood. According to the existing reports, unidirectional exchange bias is likely correlated with the topological defects, namely domain walls in the BFO films. This is not surprising because inherent symmetry breaking will occur at the ferroic domain walls, which can be regarded as intrinsic interfaces embedded in the bulk materials. Structural distortion naturally takes place to avoid the lattice discontinuity, which brings about exotic physical properties, like electrical conduction [119], photovoltaic response [207], and magnetoresistance [248], through the interplay between lattice, spin, charge, and orbital degrees of freedom. Hence, it is reasonable to suspect enhanced spin canting or possible spin frustration state with more uncompensated spins at the domain boundaries, especially 109° walls, where additional ferroic disorders may exist [344,345]. Of course, charge defects may also play a role [231,233]. Through domain engineering using orthorhombic substrates [203,210], it is possible for one to create either 71° or 109° domain wall arrays to examine their magnetic properties as well as the exchange coupling effect with a soft ferromagnet.

Lastly, it should be stressed that the ultimate goal of the research on the ME coupling effect of BFO and BFO-based heterostructures is to achieve robust room-temperature control of magnetic properties using electric field. From application point of view, integrated devices require good reliability and ability for miniaturization. Although room-temperature electrical switching of the ferromagnetism has already been realized in planar–electrode devices [291,299], the ability to scale down is lacking. However, in vertical devices, no reliable control of the magnetic functionality has hitherto been reported, which was attribute to the preferred 180° switching under vertical electric field [311]. Such type of switching mechanism should not be able to toggle the antiferromagnetic planes from symmetry consideration [219]. But if we look at the DM interaction that induces local canting moments of the Fe^{3+} ions, it can be expressed as $E_{DM} = -\mathbf{D}(\mathbf{L} \times \mathbf{M})$, where $\mathbf{D}$ is the coupling vector arising from DM interaction, $\mathbf{L}$ is the antiferromagnetic vector of the BFO sublattice and $\mathbf{M}$ is the local canting moment [218]. From this expression, it is clear that $\mathbf{D}$, $\mathbf{L}$, and $\mathbf{M}$ follow a right-handed rule, which means that the direction of the canting is rigidly attached to the antiferromagnetic vector and the DM vector (determined by the rotations of the oxygen octahedra). If during 180° polarization switching $\mathbf{D}$ or $\mathbf{L}$ can be reversed, the canting moment $\mathbf{M}$ will also change its sign. However, the reversal of either $\mathbf{D}$ or $\mathbf{L}$ is considered energetically costly. But if the energy barrier for the reversal can be reduced by the exchange coupling with a proximate ferromagnet, 180° switching of the canting moment may occur, as those seen in the BFO/LSMO heterostructures [318,319]. Besides, as illustrated in Figure 3.26, 180° polarization switching possibly is not a single-step process, there may exist intermediate states during the whole 180° switching events, which might alter the magnetic states of BFO and subsequently the magnetization of the ferromagnetic layer [177]. However, room-temperature manipulation using BFO/metal–alloy devices still remains elusive. Similarly, metal alloy coated on a single-domain BFO thin film is most appropriate to examine the possible electrically induced ferromagnetic switching under vertical device structure. Furthermore, we need to point out an alternative solution for non-180° ferroelastic switching is to fabricate isolated nanoisland capacitors that lift the clamping of the surrounding film [177]. All in all, a myriad of uncertainties and challenges remain in the explorations of room-temperature ME effect in BFO and BFO-based heterostructures, which warrants a lot more research work in the future.

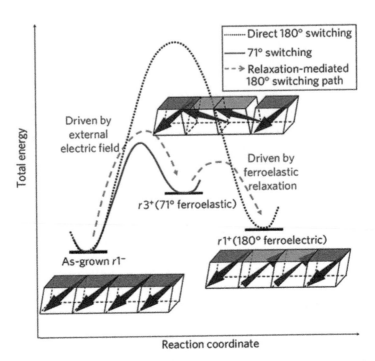

FIGURE 3.26 Schematic showing the possible switching path during a 180° polarization switching. Direct 180° switching is energetically unfavorable due to the large activation barrier. Alternatively, the ferroelectric polarization can be switched first to an intermediate state through 71° ferroelastic switching with less energy barrier, and subsequently due to the instability of ferroelastic domain configuration, relaxes to final 180° ferroelectric domain state (reprinted with permission from Ref. [177]). (Reprinted by permission from Macmillan Publishers Ltd: Baek, S. H.; Jang, H. W.; Folkman, C. M.; Li, Y. L.; Winchester, B.; Zhang, J. X.; He, Q.; Chu, Y. H.; Nelson, C. T.; Rzchowski, M. S.; Pan, X. Q.; Ramesh, R.; Chen, L. Q.; Eom, C. B. Ferroelastic switching for nanoscale non-volatile magnetoelectric devices. *Nat. Mater.* 2010, *9*, 309. © 2010.)

KEYWORDS

- **power consumption**
- **single-phase multiferroics**
- **ferroelectricity**
- **antiferromagnet**
- **magnetoelectric coupling**

REFERENCES

1. Fiebig, M. *J. Phys. D: Appl. Phys.* **2005**, *38*, R123.
2. Spaldin, N. A.; Fiebig, M. *Science* **2005**, *309*, 391.
3. Schmid, H. *Ferroelectrics* **1994**, *162*, 317.
4. Eerenstein, W.; Mathur, N. D.; Scott, J. F. *Nature* **2006**, *442*, 759.
5. Hill, N. A. *J. Phys. Chem. B* **2000**, *104*, 6694.
6. Martin, L. W.; Chu, Y. H.; Ramesh, R. *Mater. Sci. Eng., R* **2010**, *68*, 89.
7. Ederer, C.; Spaldin, N. A. *Curr. Opin. Solid State Mater. Sci.* **2005**, *9*, 128.
8. Silvia, P.; Claude, E. *J. Phys.: Condens. Matter* **2009**, *21*, 303201.
9. Cheong, S. W.; Mostovoy, M. *Nat. Mater.* **2007**, *6*, 13.
10. Ramesh, R.; Spaldin, N. A. *Nat. Mater.* **2007**, *6*, 21.
11. Nan, C.-W.; Bichurin, M. I.; Dong, S.; Viehland, D.; Srinivasan, G. *J. Appl. Phys.* **2008**, *103*, 031101.
12. Wang, K. F.; Liu, J. M.; Ren, Z. F. *Adv. Phys.* **2009**, *58*, 321.
13. Catalan, G.; Scott, J. F. *Adv. Mater.* **2009**, *21*, 2463.
14. Wang, J.; Neaton, J. B.; Zheng, H.; Nagarajan, V.; Ogale, S. B.; Liu, B.; Viehland, D.; Vaithyanathan, V.; Schlom, D. G.; Waghmare, U. V.; Spaldin, N. A.; Rabe, Wuttig, M.; Ramesh, R. Science **2003**, *299*, 1719.
15. Sosnowska, I.; Peterlinneumaier, T.; Steichele, E. *J. Phys. C: Solid State Phys.* **1982**, *15*, 4835.
16. Kiselev, S. V.; Ozerov, R. P.; Zhdanov, G. S. *Sov. Phys. Doklady* **1963**, *7*, 742.
17. Teague, J. R.; Gerson, R.; James, W. J. *Solid State Commun.* **1970**, *8*, 1073.
18. Binek, C.; Doudin, B. *J. Phys.: Condens. Matter* **2005**, *17*, L39.
19. Catalan, G.; Seidel, J.; Ramesh, R.; Scott, J. F. *Rev. Mod. Phys.* **2012**, *84*, 119.
20. Yang, C. H.; Kan, D.; Takeuchi, I.; Nagarajan, V.; Seidel, J. *Phys. Chem. Chem. Phys.* **2012**, *14*, 15953.
21. Michel, C.; Moreau, J.-M.; Achenbach, G. D.; Gerson, R.; James, W. J. *Solid State Commun.* **1969**, *7*, 701.
22. Moreau, J. M.; Michel, C.; Gerson, R.; James, W. J. *J. Phys. Chem. Solids* **1971**, *32*, 1315.
23. Kubel, F.; Schmid, H. *Acta Crystallogr. Sect. B* **1990**, *46*, 698.
24. Palewicz, A.; Przenioslo, R.; Sosnowska, I.; Hewat, A. W. *Acta Crystallogr., Sect. B* **2007**, *63*, 537.
25. Sosnowska, I.; Przeniosło, R.; Palewicz, A.; Wardecki, D.; Fitch, J. *J. Phys. Soc. Jpn.* **2012**, *81*, 044604.
26. Wang, H.; Yang, C.; Lu, J.; Wu, M.; Su, J.; Li, K.; Zhang, J.; Li, G.; Jin, T.; Kamiyama, T.; Liao, F.; Lin, J.; Wu, Y. *Inorg. Chem.* **2013**, *52*, 2388.
27. Bucci, J. D.; Robertson, B. K.; James, W. J. *J. Appl. Crystallogr.* **1972**, *5*, 187.
28. Megaw, H. D.; Darlington, C. N. W. *Acta Crystallogr., Sect. A* **1975**, *31*, 161.
29. Fischer, P.; Polomska, M.; Sosnowska, I.; Szymanski, M. *J. Phys. C: Solid State Phys.* **1980**, *13*, 1931.
30. Glazer, A. *Acta Crystallogr., Sect. B* **1972**, *28*, 3384.
31. Goldschmidt, V. M. *Naturwissenschaften* **1926**, *14*, 477.
32. Shannon, R. D. *Acta Crystallogr., Sect. A* **1976**, *32*, 751.
33. Polomska, M.; Kaczmare, W; Pajak, Z. *Phys. Status Solidi A* **1974**, *23*, 567.
34. Palai, R.; Katiyar, R. S.; Schmid, H.; Tissot, P.; Clark, S. J.; Robertson, J.; Redfern, S. A. T.; Catalan, G.; Scott, J. F. *Phys. Rev. B* **2008**, *77*, 014110.

35. Arnold, D. C.; Knight, K. S.; Morrison, F. D.; Lightfoot, P. *Phys. Rev. Lett.* **2009**, *102*, 027602.

36. Selbach, S. M.; Tybell, T.; Einarsrud, M.-A.; Grande, T. *J. Solid State Chem.* **2010**, *183*, 1205.

37. Haumont, R.; Kornev, I. A.; Lisenkov, S.; Bellaiche, L.; Kreisel, J.; Dkhil, B. *Phys. Rev. B* **2008**, *78*, 134108.

38. Arnold, D. C.; Knight, K. S.; Catalan, G.; Redfern, S. A. T.; Scott, J. F.; Lightfoot, P.; Morrison, F. D. *Adv. Funct. Mater.* **2010**, *20*, 2116.

39. Martí, X.; Ferrer, P.; Herrero-Albillos, J.; Narvaez, J.; Holy, V.; Barrett, N.; Alexe, M.; Catalan, G. *Phys. Rev. Lett.* **2011**, *106*, 236101.

40. Haumont, R.; Kreisel, J.; Bouvier, P.; Hippert, F. *Phys. Rev. B* **2006**, *73*, 132101.

41. Kornev, I. A.; Lisenkov, S.; Haumont, R.; Dkhil, B.; Bellaiche, L. *Phys. Rev. Lett.* **2007**, *99*, 227602.

42. Selbach, S. M.; Tybell, T.; Einarsrud, M.-A.; Grande, T. Adv. Mater. **2008**, *20*, 3692.

43. Gavriliuk, A. G.; Struzhkin, V. V.; Lyubutin, I. S.; Hu, M. Y.; Mao, H. K. *JETP Lett.* **2005**, *82*, 224.

44. Gavriliuk, A. G.; Lyubutin, I. S.; Struzhkin, V. V. *JETP Lett.* **2007**, *86*, 532.

45. Gavriliuk, A. G.; Struzhkin, V.; Lyubutin, I. S.; Trojan, I. A.; Hu, M.; Chow, P. *Mater. Res. Soc. Symp. Proc.* **2006**, *987*, 147.

46. Gavriliuk, A. G.; Struzhkin, V. V.; Lyubutin, I. S.; Troyan, I. A. *JETP Lett.* **2007**, *86*, 197.

47. Gavriliuk, A. G.; Struzhkin, V. V.; Lyubutin, I. S.; Ovchinnikov, S. G.; Hu, M. Y.; Chow, P. *Phys. Rev. B* **2008**, *77*, 155112.

48. Gonzalez-Vazquez, O. E.; Iniguez, J. *Phys. Rev. B* **2009**, *79*, 064102.

49. Ravindran, P.; Vidya, R.; Kjekshus, A.; Fjellvåg, H.; Eriksson, O. *Phys. Rev. B* **2006**, *74*, 224412.

50. Haumont, R.; Kreisel, J.; Bouvier, P. *Phase Transit.* **2006**, *79*, 1043.

51. Haumont, R.; Bouvier, P.; Pashkin, A.; Rabia, K.; Frank, S.; Dkhil, B.; Crichton, W. A.; Kuntscher, C. A.; Kreisel, J. *Phys. Rev. B* **2009**, *79*, 184110.

52. Belik, A. A.; Yusa, H.; Hirao, N.; Ohishi, Y.; Takayama-Muromachi, E. *Chem. Mater.* **2009**, *21*, 3400.

53. Guennou, M.; Bouvier, P.; Chen, G. S.; Dkhil, B.; Haumont, R.; Garbarino, G.; Kreisel, J. *Phys. Rev. B* **2011**, *84*, 174107.

54. Li, J. F.; Wang, J. L.; Wuttig, M.; Ramesh, R.; Wang, N.; Ruette, B.; Pyatakov, A. P.; Zvezdin, A. K.; Viehland, D. *Appl. Phys. Lett.* **2004**, *84*, 5261.

55. Xu, G. Y.; Hiraka, H.; Shirane, G.; Li, J. F.; Wang, J. L.; Viehland, D. *Appl. Phys. Lett.* **2005**, *86*, 032511.

56. Keisuke, S.; Alexander, U.; Volkmar, G.; Heiko, R.; Lutz, B.; Hideo, O.; Toshiyuki, K.; Sadao, U.; Hiroshi, F. *Jpn. J. Appl. Phys.* **2006**, *45*, 7311.

57. Béa, H.; Bibes, M.; Petit, S.; Kreisel, J.; Barthélémy, A. *Philos. Mag. Lett.* **2007**, *87*, 165.

58. Daumont, C. J. M.; Farokhipoor, S.; Ferri, Wojdelstrok, J. C.; Jorge Íñiguez, B., J.; Kooi, B. J.; Noheda, B. *Phys. Rev. B* **2010**, *81*, 144115.

59. Liu, H.; Yang, P.; Yao, K.; Wang, J. *Appl. Phys. Lett.* **2010**, *96*, 023109.

60. Schlom, D. G.; Chen, L. Q.; Eom, C. B.; Rabe, K. M.; Streiffer, S. K.; Triscone, J. M. Annu. Rev. Mater. Res. **2007**, *37*, 589.

61. Chen, Z. H.; Luo, Z. L.; Huang, C. W.; Qi, Y. J.; Yang, P.; You, L.; Hu, C. S.; Wu, T.; Wang, J. L.; Gao, C.; Sritharan, T.; Chen, L. *Adv. Funct. Mater.* **2011**, *21*, 133.

62. Béa, H.; Dupe, B.; Fusil, S.; Mattana, R.; Jacquet, E.; Warot-Fonrose, B.; Wilhelm, F.; Rogalev, A.; Petit, S.; Cros, V.; Anane, A.; Petroff, F.; Bouzehouane, K.; Geneste, G.; Dkhil, B.; Lisenkov, S.; Ponomareva, I.; Bellaiche, L.; Bibes, M.; Barthélémy, A. *Phys. Rev. Lett.* **2009,** *102,* 217603.

63. Chen, Z.; Luo, Z.; Qi, Y.; Yang, P.; Wu, S.; Huang, C.; Wu, T.; Wang, J.; Gao, C.; Sritharan, T.; Chen, L. *Appl. Phys. Lett.* **2010,** *97,* 242903.

64. Damodaran, A. R.; Liang, C.-W.; He, Q.; Peng, C.-Y.; Chang, L.; Chu, Y.-H.; Martin, L. W. *Adv. Mater.* **2011,** *23,* 3170.

65. Christen, H. M.; Nam, J. H.; Kim, H. S.; Hatt, A. J.; Spaldin, N. A. *Phys. Rev. B* **2011,** *83,* 144107.

66. Vanderbilt, D.; Cohen, M. H. *Phys. Rev. B* **2001,** *63,* 094108.

67. Liu, H.-J.; Chen, H.-J.; Liang, W.-I.; Liang, C.-W.; Lee, H.-Y.; Lin, S.-J.; Chu, Y.-H. *J. Appl. Phys.* **2012,** *112.*

68. Beekman, C.; Siemons, W.; Ward, T. Z.; Chi, M.; Howe, J.; Biegalski, M. D.; Balke, N.; Maksymovych, P.; Farrar, A. K.; Romero, J. B.; Gao, P.; Pan, X. Q.; Tenne, D. A.; Christen, H. M. *Adv. Mater.* **2013,** *25,* 5561.

69. Dupé, B.; Prosandeev, S.; Geneste, G.; Dkhil, B.; Bellaiche, L. *Phys. Rev. Lett.* **2011,** *106,* 237601.

70. Yang, Y.; Ren, W.; Stengel, M.; Yan, X. H.; Bellaiche, L. *Phys. Rev. Lett.* **2012,** *109,* 057602.

71. Zeches, R. J.; Rossell, M. D.; Zhang, J. X.; Hatt, A. J.; He, Q.; Yang, C.-H.; Kumar, A.; Wang, C. H.; Melville, A.; Adamo, C.; Sheng, G.; Chu, Y.-H.; Ihlefeld, J. F.; Erni, R.; Ederer, C.; Gopalan, V.; Chen, L. Q.; Schlom, D. G.; Spaldin, N. A.; Martin, L. W.; Ramesh, R. *Science* **2009,** *326,* 977.

72. Huang, C. W.; Chu, Y. H.; Chen, Z. H.; Wang, J.; Sritharan, T.; He, Q.; Ramesh, R.; Chen, L. *Appl. Phys. Lett.* **2010,** *97,* 152901.

73. Yang, J. C.; He, Q.; Suresha, S. J.; Kuo, C. Y.; Peng, C. Y.; Haislmaier, R. C.; Motyka, M. A.; Sheng, G.; Adamo, C.; Lin, H. J.; Hu, Z.; Chang, L.; Tjeng, L. H.; Arenholz, E.; Podraza, N. J.; Bernhagen, M.; Uecker, R.; Schlom, D. G.; Gopalan, V.; Chen, L. Q.; Chen, C. T.; Ramesh, R.; Chu, Y. H. *Phys. Rev. Lett.* **2012,** *109,* 247606.

74. Xu, G.; Li, J.; Viehland, D. *Appl. Phys. Lett.* **2006,** *89,* 222901.

75. Yan, L.; Cao, H.; Li, J. F.; Viehland, D. *Appl. Phys. Lett.* **2009,** *94,* 132901.

76. Chen, Z.; Qi, Y.; You, L.; Yang, P.; Huang, C. W.; Wang, J.; Sritharan, T.; Chen, L. *Phys. Rev. B* **2013,** *88,* 054114.

77. Diéguez, O.; González-Vázquez, O. E.; Wojdeł, J. C.; Íñiguez, J. *Phys. Rev. B* **2011,** *83,* 094105.

78. Lee, J. H.; Chu, K.; Ünal, A. A.; Valencia, S.; Kronast, F.; Kowarik, S.; Seidel, J.; Yang, C.-H. *Phys. Rev. B* **2014,** *89,* 140101.

79. Gao, F.; Yuan, Y.; Wang, K. F.; Chen, X. Y.; Chen, F.; Liu, J.-M.; Ren, Z. F. *Appl. Phys. Lett.* **2006,** *89,* 102506.

80. Luo, J.; Maggard, P. A. *Adv. Mater.* **2006,** *18,* 514.

81. Gao, F.; Chen, X. Y.; Yin, K. B.; Dong, S.; Ren, Z. F.; Yuan, F.; Yu, T.; Zou, Z. G.; Liu, J. M. *Adv. Mater.* **2007,** *19,* 2889.

82. Li, S.; Lin, Y.-H.; Zhang, B.-P.; Wang, Y.; Nan, C.-W. *J. Phys. Chem. C* **2010,** *114,* 2903.

83. Mocherla, P. S. V.; Karthik, C.; Ubic, R.; Ramachandra Rao, M. S.; Sudakar, C. *Appl. Phys. Lett.* **2013,** *103,* 022910.

84. Moubah, R.; Schmerber, G.; Rousseau, O.; Colson, D.; Viret, M. *Appl. Phys. Express* **2012,** *5,* 035802.

85. Clark, S. J.; Robertson, J. *Appl. Phys. Lett.* **2009,** *94,* 202110.

86. Basu, S. R.; Martin, L. W.; Chu, Y. H.; Gajek, M.; Ramesh, R.; Rai, R. C.; Xu, X.; Musfeldt, J. L. *Appl. Phys. Lett.* **2008,** *92,* 091905.

87. Bi, L.; Taussig, A. R.; Kim, H.-S.; Wang, L.; Dionne, G. F.; Bono, D.; Persson, K.; Ceder, G.; Ross, C. A. *Phys. Rev. B* **2008,** *78,* 104106.

88. Hauser, A. J.; Zhang, J.; Mier, L.; Ricciardo, R. A.; Woodward, P. M.; Gustafson, T. L.; Brillson, L. J.; Yang, F. Y. *Appl. Phys. Lett.* **2008,** *92,* 222901.

89. Ihlefeld, J. F.; Podraza, N. J.; Liu, Z. K.; Rai, R. C.; Xu, X.; Heeg, T.; Chen, Y. B.; Li, J.; Collins, R. W.; Musfeldt, J. L.; Pan, X. Q.; Schubert, J.; Ramesh, R.; Schlom, D. G. *Appl. Phys. Lett.* **2008,** *92,* 142908.

90. Kumar, A.; Rai, R. C.; Podraza, N. J.; Denev, S.; Ramirez, M.; Chu, Y.-H.; Martin, L. W.; Ihlefeld, J.; Heeg, T.; Schubert, J.; Schlom, D. G.; Orenstein, J.; Ramesh, R.; Collins, R. W.; Musfeldt, J. L.; Gopalan, V. *Appl. Phys. Lett.* **2008,** *92,* 121913.

91. Allibe, J.; Bougot-Robin, K.; Jacquet, E.; Infante, I. C.; Fusil, S.; Carrétéro, C.; Reverchon, J.-L.; Marcilhac, B.; Creté, D.; Mage, Barthélémy, A.; Bibes, M. *Appl. Phys. Lett.* **2010,** *96,* 182902.

92. Li, W. W.; Zhu, J. J.; Wu, J. D.; Gan, J.; Hu, Z. G.; Zhu, M.; Chu, J. H. *Appl. Phys. Lett.* **2010,** *97,* 121102.

93. Clark, S. J.; Robertson, J. *Appl. Phys. Lett.* **2007,** *90,* 132903.

94. Chen, P.; Podraza, N. J.; Xu, X. S.; Melville, A.; Vlahos, E.; Gopalan, V.; Ramesh, R.; Schlom, D. G.; Musfeldt, J. L. *Appl. Phys. Lett.* **2010,** *96,* 131907.

95. Liu, H. L.; Lin, M. K.; Cai, Y. R.; Tung, C. K.; Chu, Y. H. *Appl. Phys. Lett.* **2013,** *103.*

96. Lebeugle, D.; Colson, D.; Forget, A.; Viret, M. *Appl. Phys. Lett.* **2007,** *91,* 022907.

97. Lebeugle, D.; Colson, D.; Forget, A.; Viret, M.; Bonville, P.; Marucco, J. F.; Fusil, S. *Phys. Rev. B* **2007,** *76,* 024116.

98. Alexe, M.; Hesse, D. *Nat. Commun.* **2011,** *2,* 256.

99. Yun, K. Y.; Noda, M.; Okuyama, M. *Appl. Phys. Lett.* **2003,** *83,* 3981.

100. Wang, J.; Zheng, H.; Ma, Z.; Prasertchoung, S.; Wuttig, M.; Droopad, R.; Yu, J.; Eisenbeiser, K.; Ramesh, R. *Appl. Phys. Lett.* **2004,** *85,* 2574.

101. Chu, Y. H.; Zhan, Q.; Martin, L. W.; Cruz, M. P.; Yang, P. L.; Pabst, G. W.; Zavaliche, F.; Yang, S. Y.; Zhang, J. X.; Chen, L. Q.; Schlom, D. G.; Lin, I. N.; Wu, T. B.; Ramesh, R. *Adv. Mater.* **2006,** *18,* 2307.

102. Eerenstein, W.; Morrison, F. D.; Sher, F.; Prieto, J. L.; Attfield, J. P.; Scott, J. F.; Mathur, N. D. *Philos. Mag. Lett.* **2007,** *87,* 249.

103. Jang, H. W.; Baek, S. H.; Ortiz, D.; Folkman, C. M.; Eom, C. B.; Chu, Y. H.; Shafer, P.; Ramesh, R.; Vaithyanathan, V.; Schlom, D. G. *Appl. Phys. Lett.* **2008,** *92,* 062910.

104. Béa, H.; Bibes, M.; Barthélémy, A.; Bouzehouane, K.; Jacquet, E.; Khodan, A.; Contour, J. P.; Fusil, S.; Wyczisk, F.; Forget, A.; Lebeugle, D.; Colson, D.; Viret, M. *Appl. Phys. Lett.* **2005,** *87,* 072508.

105. Murakami, M.; Fujino, S.; Lim, S.-H.; Salamanca-Riba, L. G.; Wuttig, M.; Takeuchi, I.; Varughese, B.; Sugaya, H.; Hasegawa, T.; Lofland, S. E. *Appl. Phys. Lett.* **2006,** *88,* 112505.

106. You, L.; Chua, N. T.; Yao, K.; Chen, L.; Wang, J. *Phys. Rev. B* **2009,** *80,* 024105.

107. You, L. Ph.D. Thesis, Nanyang Technological University, 2011.

108. Wu, J.; Wang, J.; Xiao, D.; Zhu, J. *ACS Appl. Mat. Interfaces* **2012,** *4,* 1182.

109. Singh, S. K.; Ishiwara, H.; Maruyama, K. *Appl. Phys. Lett.* **2006,** *88,* 262908.

110. Uchida, H.; Ueno, R.; Funakubo, H.; Koda, S. *J. Appl. Phys.* **2006,** *100,* 014106.

111. Hu, G. D.; Cheng, X.; Wu, W. B.; Yang, C. H. *Appl. Phys. Lett.* **2007,** *91*, 112911.
112. Cheng, Z.; Wang, X.; Dou, S.; Kimura, H.; Ozawa, K. *Phys. Rev. B* **2008,** *77*, 092101.
113. Zhu, X. H.; Béa, H.; Bibes, M.; Fusil, S.; Bouzehouane, K.; Jacquet, E.; Barthélémy, A.; Lebeugle, D.; Viret, M.; Colson, D. *Appl. Phys. Lett.* **2008,** *93*, 082902.
114. Kawae, T.; Terauchi, Y.; Tsuda, H.; Kumeda, M.; Morimoto, A. *Appl. Phys. Lett.* **2009,** *94*, 012903.
115. Pabst, G. W.; Martin, L. W.; Chu, Y.-H.; Ramesh, R. *Appl. Phys. Lett.* **2007,** *90*, 072902.
116. Yang, H.; Jain, M.; Suvorova, N. A.; Zhou, H.; Luo, H. M.; Feldmann, D. M.; Dowden, P. C.; DePaula, R. F.; Foltyn, S. R.; Jia, Q. X. *Appl. Phys. Lett.* **2007,** *91*, 072911.
117. Yang, H.; Luo, H. M.; Wang, H.; Usov, I. O.; Suvorova, N. A.; Jain, M.; Feldmann, D. M.; Dowden, P. C.; DePaula, R. F.; Jia, Q. X. *Appl. Phys. Lett.* **2008,** *92*, 192905.
118. Pintilie, L.; Dragoi, C.; Chu, Y. H.; Martin, L. W.; Ramesh, R.; Alexe, M. *Appl. Phys. Lett.* **2009,** *94*, 232902.
119. Seidel, J.; Martin, L. W.; He, Q.; Zhan, Q.; Chu, Y. H.; Rother, A.; Hawkridge, M. E.; Maksymovych, P.; Yu, P.; Gajek, M.; Balke, N.; Kalinin, S. V.; Gemming, S.; Wang, F.; Catalan, G.; Scott, J. F.; Spaldin, N. A.; Orenstein, J.; Ramesh, R. *Nat. Mater.* **2009,** *8*, 229.
120. Shvartsman, V. V.; Kleemann, W.; Haumont, R.; Kreisel, J. *Appl. Phys. Lett.* **2007,** *90*, 172115.
121. Neaton, J. B.; Ederer, C.; Waghmare, U. V.; Spaldin, N. A.; Rabe, K. M. *Phys. Rev. B* **2005,** *71*, 014113.
122. Chu, Y.-H.; Martin, L. W.; Holcomb, M. B.; Ramesh, R. *Mater. Today* **2007,** *10*, 16.
123. Baek, S.-H.; Folkman, C. M.; Park, J.-W.; Lee, S.; Bark, C.-W.; Tybell, T.; Eom, C.-B. *Adv. Mater.* **2011,** *23*, 1621.
124. Yun, K. Y.; Ricinschi, D.; Kanashima, T.; Noda, M.; Okuyama, M. *Jpn. J. Appl. Phys.* **2004,** *43*, L647.
125. Dan, R.; Kwi-Young, Y.; Masanori, O. *J. Phys.: Condens. Matter* **2006,** *18*, L97.
126. Zhang, J. X.; He, Q.; Trassin, M.; Luo, W.; Yi, D.; Rossell, M. D.; Yu, P.; You, L.; Wang, C. H.; Kuo, C. Y.; Heron, J. T.; Hu, Z.; Zeches, R. J.; Lin, H. J.; Tanaka, A.; Chen, C. T.; Tjeng, L. H.; Chu, Y. H.; Ramesh, R. *Phys. Rev. Lett.* **2011,** *107*, 147602.
127. Choi, K. J.; Biegalski, M.; Li, Y. L.; Sharan, A.; Schubert, J.; Uecker, R.; Reiche, P.; Chen, Y. B.; Pan, X. Q.; Gopalan, V.; Chen, L.-Q.; Schlom, D. G.; Eom, C. B. *Science* **2004,** *306*, 1005.
128. Haeni, J. H.; Irvin, P.; Chang, W.; Uecker, R.; Reiche, P.; Li, Y. L.; Choudhury, S.; Tian, W.; Hawley, M. E.; Craigo, B.; Tagantsev, A. K.; Pan, X. Q.; Streiffer, S. K.; Chen, L. Q.; Kirchoefer, S. W.; Levy, J.; Schlom, D. G. *Nature* **2004,** *430*, 758.
129. Ederer, C.; Spaldin, N. A. *Phys. Rev. Lett.* **2005,** *95*, 257601.
130. Lee, H. N.; Nakhmanson, S. M.; Chisholm, M. F.; Christen, H. M.; Rabe, K. M.; Vanderbilt, D. *Phys. Rev. Lett.* **2007,** *98*, 217602.
131. Kim, D. H.; Lee, H. N.; Biegalski, M. D.; Christen, H. M. *Appl. Phys. Lett.* **2008,** *92*, 012911.
132. Damodaran, A. R.; Breckenfeld, E.; Chen, Z.; Lee, S.; Martin, L. W. *Adv. Mater.* **2014.** doi:10.1002/adma.201400254.
133. Jang, H. W.; Baek, S. H.; Ortiz, D.; Folkman, C. M.; Das, R. R.; Chu, Y. H.; Shafer, P.; Zhang, J. X.; Choudhury, S.; Vaithyanathan, V.; Chen, Y. B.; Felker, D. A.; Biegalski, M. D.; Rzchowski, M. S.; Pan, X. Q.; Schlom, D. G.; Chen, L. Q.; Ramesh, R.; Eom, C. B. *Phys. Rev. Lett.* **2008,** *101*, 107602.

134. Dupé, B.; Infante, I. C.; Geneste, G.; Janolin, P. E.; Bibes, M.; Barthélémy, A.; Lisenkov, S.; Bellaiche, L.; Ravy, S.; Dkhil, B. *Phys. Rev. B* **2010**, *81*, 144128.

135. Hatt, A. J.; Spaldin, N. A.; Ederer, C. *Phys. Rev. B* **2010**, *81*, 054109.

136. Wojdel, J. C.; Iniguez, J. *Phys. Rev. Lett.* **2010**, *105*, 046801.

137. Biegalski, M. D.; Kim, D. H.; Choudhury, S.; Chen, L. Q.; Christen, H. M.; Dörr, K. *Appl. Phys. Lett.* **2011**, *98*, 142902.

138. Daumont, C.; Ren, W.; Infante, I. C.; Lisenkov, S.; Allibe, J.; Carrétéro, C.; Fusil, S.; Jacquet, E.; Bouvet, T.; Bouamrane, F.; Prosandeev, S.; Geneste, G.; Dkhil, B.; Bellaiche, L.; Barthélémy, A.; Bibes, M. *J. Phys.: Condens. Matter* **2012**, *24*, 162202.

139. Fu, H.; Cohen, R. E. *Nature* **2000**, *403*, 281.

140. Noheda, B.; Zhong, Z.; Cox, D. E.; Shirane, G.; Park, S. E.; Rehrig, P. *Phys. Rev. B* **2002**, *65*, 224101.

141. You, L.; Liang, E.; Guo, R.; Wu, D.; Yao, K.; Chen, L.; Wang, J. L. *Appl. Phys. Lett.* **2010**, *97*, 062910.

142. Chen, Z.; Zou, X.; Ren, W.; You, L.; Huang, C.; Yang, Y.; Yang, P.; Wang, J.; Sritharan, T.; Bellaiche, L.; Chen, L. *Phys. Rev. B* **2012**, *86*, 235125.

143. Chen, Z.; You, L.; Huang, C.; Qi, Y.; Wang, J.; Sritharan, T.; Chen, L. *Appl. Phys. Lett.* **2010**, *96*, 252903.

144. Mazumdar, D.; Shelke, V.; Iliev, M.; Jesse, S.; Kumar, A.; Kalinin, S. V.; Baddorf, A. P.; Gupta, A. *Nano Lett.* **2010**, *10*, 2555.

145. Chen, W.; Ren, W.; You, L.; Yang, Y.; Chen, Z.; Qi, Y.; Zou, X.; Wang, J.; Sritharan, T.; Yang, P.; Bellaiche, L.; Chen, L. *Appl. Phys. Lett.* **2011**, *99*, 222904.

146. Ko, K.-T.; Jung, M. H.; He, Q.; Lee, J. H.; Woo, C. S.; Chu, K.; Seidel, J.; Jeon, B.-G.; Oh, Y. S.; Kim, K. H.; Liang, W.-I.; Chen, Chu, Y.-H.; Jeong, Y. H.; Ramesh, R.; Park, J.-H.; Yang, C.-H. *Nat. Commun.* **2011**, *2*, 567.

147. Liu, H.-J.; Liang, C.-W.; Liang, W.-I.; Chen, H.-J.; Yang, J.-C.; Peng, C.-Y.; Wang, G.-F.; Chu, F.-N.; Chen, Y.-C.; Lee, H.-Y.; Chang, L.; Lin, S.-J.; Chu, Y.-H. *Phys. Rev. B* **2012**, *85*, 014104.

148. Damodaran, A. R.; Breckenfeld, E.; Choquette, A. K.; Martin, L. W. *Appl. Phys. Lett.* **2012**, *100*, 082904.

149. Damodaran, A. R.; Lee, S.; Karthik, J.; MacLaren, S.; Martin, L. W. *Phys. Rev. B* **2012**, *85*, 024113.

150. Woo, C.-S.; Lee, J. H.; Chu, K.; Jang, B.-K.; Kim, Y.-B.; Koo, T. Y.; Yang, P.; Qi, Y.; Chen, Z.; Chen, L.; Choi, H. C.; Shim, J. H.; Yang, C.-H. *Phys. Rev. B* **2012**, *86*, 054417.

151. Yun, K. Y.; Ricinschi, D.; Kanashima, T.; Okuyama, M. *Appl. Phys. Lett.* **2006**, *89*, 192902.

152. Ricinschi, D.; Yun, K. Y.; Okuyama, M. *Ferroelectrics* **2006**, *335*, 181.

153. Rossell, M. D.; Erni, R.; Prange, M. P.; Idrobo, J. C.; Luo, W.; Zeches, R. J.; Pantelides, S. T.; Ramesh, R. *Phys. Rev. Lett.* **2012**, *108*, 047601.

154. Huang, R.; Ding, H.-C.; Liang, W.-I.; Gao, Y.-C.; Tang, X.-D.; He, Q.; Duan, C.-G.; Zhu, Z.; Chu, J.; Fisher, C. A. J.; Hirayama, T.; Ikuhara, Y.; Chu, Y.-H. *Adv. Funct. Mater.* **2014**, *24*, 793.

155. Infante, I. C.; Lisenkov, S.; Dupé, B.; Bibes, M.; Fusil, S.; Jacquet, E.; Geneste, G.; Petit, S.; Courtial, A.; Juraszek, J.; Bellaiche, L.; Barthélémy, A.; Dkhil, B. *Phys. Rev. Lett.* **2010**, *105*, 057601.

156. Adamo, C.; Ke, X.; Wang, H. Q.; Xin, H. L.; Heeg, T.; Hawley, M. E.; Zander, W.; Schubert, J.; Schiffer, P.; Muller, D. A.; Maritato, L.; Schlom, D. G. *Appl. Phys. Lett.* **2009**, *95*, 112504.

157. Toupet, H.; Le Marrec, F.; Lichtensteiger, C.; Dkhil, B.; Karkut, M. G. *Phys. Rev. B* **2010**, *81*, 140101.

158. Le Marrec, F.; Toupet, H.; Lichtensteiger, C.; Dkhil, B.; Karkut, M. G. *Phase Transit.* **2011**, *84*, 453.

159. Yang, S. Y.; Seidel, J.; Kim, S. Y.; Rossen, P. B.; Yu, P.; Gajek, M.; Chu, Y. H.; Martin, L. W.; Holcomb, M. B.; He, Q.; Maksymovych, P.; Balke, N.; Kalinin, S. V.; Baddorf, A. P.; Basu, S. R.; Scullin, M. L.; Ramesh, R. *Nat. Mater.* **2009**, *8*, 485.

160. Jiang, A. Q.; Wang, C.; Jin, K. J.; Liu, X. B.; Scott, J. F.; Hwang, C. S.; Tang, T. A.; Lu, H. B.; Yang, G. Z. *Adv. Mater.* **2011**, *23*, 1277.

161. Kim, T. H.; Jeon, B. C.; Min, T.; Yang, S. M.; Lee, D.; Kim, Y. S.; Baek, S.-H.; Saenrang, W.; Eom, C.-B.; Song, T. K.; Yoon, J.-G.; Noh, T. W. *Adv. Funct. Mater.* **2012**, *22*, 4962.

162. Tsurumaki, A.; Yamada, H.; Sawa, A. *Adv. Funct. Mater.* **2012**, *22*, 1040.

163. Hong, S.; Choi, T.; Jeon, J. H.; Kim, Y.; Lee, H.; Joo, H.-Y.; Hwang, I.; Kim, J.-S.; Kang, S.-O.; Kalinin, S. V.; Park, B. H. *Adv. Mater.* **2013**, *25*, 2339.

164. Yamada, H.; Garcia, V.; Fusil, S.; Boyn, S.; Marinova, M.; Gloter, A.; Xavier, S.; Grollier, J.; Jacquet, E.; Carrétéro, C.; Deranlot, C.; Bibes, M.; Barthélémy, A. *ACS Nano* **2013**, *7*, 5385.

165. Crassous, A.; Bernard, R.; Fusil, S.; Bouzehouane, K.; Le Bourdais, D.; Enouz-Vedrenne, S.; Briatico, J.; Bibes, M.; Barthélémy, A.; Villegas, J. E. *Phys. Rev. Lett.* **2011**, *107*, 247002.

166. Yamada, H.; Marinova, M.; Altuntas, P.; Crassous, A.; Begon-Lours, L.; Fusil, S.; Jacquet, E.; Garcia, V.; Bouzehouane, K.; Gloter, A.; Villegas, J. E.; Barthelemy, A.; Bibes, M. *Sci. Rep.* **2013**, *3*, 2834.

167. Kim, Y.-M.; Morozovska, A.; Eliseev, E.; Oxley, M. P.; Mishra, R.; Selbach, S. M.; Grande, T.; Pantelides, S. T.; Kalinin, S. V.; Borisevich, A. Y. *Nat. Mater.* **2014** (advance online publication).

168. Choi, T.; Lee, S.; Choi, Y. J.; Kiryukhin, V.; Cheong, S. W. *Science* **2009**, *324*, 63.

169. Yang, S. Y.; Martin, L. W.; Byrnes, S. J.; Conry, T. E.; Basu, S. R.; Paran, D.; Reichertz, L.; Ihlefeld, J.; Adamo, C.; Melville, A.; Chu, Y.-H.; Yang, C.-H.; Musfeldt, J. L.; Schlom, D. G.; Ager, J. W.; Ramesh, R. *Appl. Phys. Lett.* **2009**, *95*.

170. Ji, W.; Yao, K.; Liang, Y. C. *Adv. Mater.* **2010**, *22*, 1763.

171. Lee, D.; Baek, S. H.; Kim, T. H.; Yoon, J. G.; Folkman, C. M.; Eom, C. B.; Noh, T. W. *Phys. Rev. B* **2011**, *84*, 125305.

172. Yi, H. T.; Choi, T.; Choi, S. G.; Oh, Y. S.; Cheong, S. W. *Adv. Mater.* **2011**, *23*, 3403.

173. Guo, R.; You, L.; Zhou, Y.; Shiuh Lim, Z.; Zou, X.; Chen, L.; Ramesh, R.; Wang, J. *Nat. Commun.* **2013**, *4*, 2788.

174. Tagantsev, A. K.; Stolichnov, I.; Colla, E. L.; Setter, N. *J. Appl. Phys.* **2001**, *90*, 1387.

175. Lou, X. J. *J. Appl. Phys.* **2009**, *105*, 024101.

176. Zou, X.; You, L.; Chen, W.; Ding, H.; Wu, D.; Wu, T.; Chen, L.; Wang, J. *ACS Nano* **2012**, *6*, 8997.

177. Baek, S. H.; Jang, H. W.; Folkman, C. M.; Li, Y. L.; Winchester, B.; Zhang, J. X.; He, Q.; Chu, Y. H.; Nelson, C. T.; Rzchowski, M. S.; Pan, X. Q.; Ramesh, R.; Chen, L. Q.; Eom, C. B. *Nat. Mater.* **2010**, *9*, 309.

178. Baek, S. H.; Eom, C. B. *Phil. Trans. Res. Soc. A* **2012**, *370*, 4872.

179. Gao, P.; Nelson, C. T.; Jokisaari, J. R.; Zhang, Y.; Baek, S.-H.; Bark, C. W.; Wang, E.; Liu, Y.; J. Li, Eom, C.-B.; Pan, X. *Adv. Mater.* **2012**, *24*, 1106.

180. Folkman, C. M.; Baek, S. H.; Nelson, C. T.; Jang, H. W.; Tybell, T.; Pan, X. Q.; Eom, C. B. *Appl. Phys. Lett.* **2010**, *96*, 052903.

181. Lee, D.; Jeon, B. C.; Baek, S. H.; Yang, S. M.; Shin, Y. J.; Kim, T. H.; Kim, Y. S.; Yoon, J.-G.; Eom, C. B.; Noh, T. W. *Adv. Mater.* **2012**, *24*, 6490.

182. Jeon, B. C.; Lee, D.; Lee, M. H.; Yang, S. M.; Chae, S. C.; Song, T. K.; Bu, S. D.; Chung, J.-S.; Yoon, J.-G.; Noh, T. W. *Adv. Mater.* **2013**, *25*, 5643.

183. Lee, D.; Jeon, B. C.; Yoon, A.; Shin, Y. J.; Lee, M. H.; Song, T. K.; Bu, S. D.; Kim, M.; Chung, J.-S.; Yoon, J.-G.; Noh, T. W. *Adv. Mater.* **2014**, *26*, 5005.

184. Chu, Y. H.; Zhao, T.; Cruz, M. P.; Zhan, Q.; Yang, P. L.; Martin, L. W.; Huijben, M.; Yang, C. H.; Zavaliche, F.; Zheng, H.; Ramesh, R. *Appl. Phys. Lett.* **2007**, *90*, 072907.

185. Maksymovych, P.; Huijben, M.; Pan, M.; Jesse, S.; Balke, N.; Chu, Y.-H.; Chang, H. J.; Borisevich, A. Y.; Baddorf, A. P.; Rijnders, G.; Blank, D. H. A.; Ramesh, R.; Kalinin, S. V. *Phys. Rev. B* **2012**, *85*, 014119.

186. Rault, J. E.; Ren, W.; Prosandeev, S.; Lisenkov, S.; Sando, D.; Fusil, S.; Bibes, M.; Barthélémy, A.; Bellaiche, L.; Barrett, N. *Phys. Rev. Lett.* **2012**, *109*, 267601.

187. Vilas, S.; Dipanjan, M.; Stephen, J.; Sergei, K.; Arthur, B.; Arunava, G. *New J. Phys.* **2012**, *14*, 053040.

188. Streiffer, S. K.; Parker, C. B.; Romanov, A. E.; Lefevre, M. J.; Zhao, L.; Speck, J. S.; Pompe, W.; Foster, C. M.; Bai, G. R. *J. Appl. Phys.* **1998**, *83*, 2742.

189. Catalan, G.; Béa, H.; Fusil, S.; Bibes, M.; Paruch, P.; Barthélémy, A.; Scott, J. F. *Phys. Rev. Lett.* **2008**, *100*, 027602.

190. Béa, H.; Ziegler, B.; Bibes, M.; Barthélémy, A.; Paruch, P. *J. Phys.: Condens. Matter* **2011**, *23*, 142201.

191. Zavaliche, F.; Das, R. R.; Kim, D. M.; Eom, C. B.; Yang, S. Y.; Shafer, P.; Ramesh, R. *Appl. Phys. Lett.* **2005**, *87*, 252902.

192. Zavaliche, F.; Shafer, P.; Ramesh, R.; Cruz, M. P.; Das, R. R.; Kim, D. M.; Eom, C. B. *Appl. Phys. Lett.* **2005**, *87*, 182912.

193. Zavaliche, F.; Yang, S. Y.; Zhao, T.; Chu, Y. H.; Cruz, M. P.; Eom, C. B.; Ramesh, R. *Phase Transit.* **2006**, *79*, 991.

194. Das, R. R.; Kim, D. M.; Baek, S. H.; Eom, C. B.; Zavaliche, F.; Yang, S. Y.; Ramesh, R.; Chen, Y. B.; Pan, X. Q.; Ke, X.; Rzchowski, M. S.; Streiffer, S. K. *Appl. Phys. Lett.* **2006**, *88*, 242904.

195. Chen, Y. B.; Katz, M. B.; Pan, X. Q.; Das, R. R.; Kim, D. M.; Baek, S. H.; Eom, C. B. *Appl. Phys. Lett.* **2007**, *90*, 072907.

196. Chu, Y. H.; Cruz, M. P.; Yang, C. H.; Martin, L. W.; Yang, P. L.; Zhang, J. X.; Lee, K.; Yu, P.; Chen, L. Q.; Ramesh, R. *Adv. Mater.* **2007**, *19*, 2662.

197. Jang, H. W.; Ortiz, D.; Baek, S. H.; Folkman, C. M.; Das, R. R.; Shafer, P.; Chen, Y.; Nelson, C. T.; Pan, X.; Ramesh, R.; Eom, C. B. *Adv. Mater.* **2009**, *21*, 817.

198. You, L.; Yasui, S.; Ehara, Y.; Zou, X.; Ding, H.; Chen, Z.; Chen, W.; Chen, L.; Funakubo, H.; Wang, J. *Appl. Phys. Lett.* **2012**, *100*, 102901.

199. Winchester, B.; Wu, P.; Chen, L. Q. *Appl. Phys. Lett.* **2011**, *99*, 052903.

200. Sichel, R. J.; Grigoriev, A.; Do, D.-H.; Baek, S.-H.; Jang, H.-W.; Folkman, C. M.; Eom, C.-B.; Cai, Z.; Evans, P. G. *Appl. Phys. Lett.* **2010**, *96*, 051901.

201. Kim, T. H.; Baek, S. H.; Jang, S. Y.; Yang, S. M.; Chang, S. H.; Song, T. K.; Yoon, J.-G.; Eom, C. B.; Chung, J.-S.; Noh, T. W. *Appl. Phys. Lett.* **2011**, *98*, 022904.

202. Zhang, J. X.; Li, Y. L.; Choudhury, S.; Chen, L. Q.; Chu, Y. H.; Zavaliche, F.; Cruz, M. P.; Ramesh, R.; Jia, Q. X. J. Appl. Phys. **2008**, *103*, 094111.

203. Chu, Y. H.; He, Q.; Yang, C. H.; Yu, P.; Martin, L. W.; Shafer, P.; Ramesh, R. *Nano Lett.* **2009**, *9*, 1726.

204. Folkman, C. M.; Baek, S. H.; Jang, H. W.; Eom, C. B.; Nelson, C. T.; Pan, X. Q.; Li, Y. L.; Chen, L. Q.; Kumar, A.; Gopalan, V.; Streiffer, S. K. *Appl. Phys. Lett.* **2009**, *94*, 251911.

205. Johann, F.; Morelli, A.; Biggemann, D.; Arredondo, M.; Vrejoiu, I. *Phys. Rev. B* **2011**, *84*, 094105.

206. Chen, Z. H.; Damodaran, A. R.; Xu, R.; Lee, S.; Martin, L. W. *Appl. Phys. Lett.* **2014**, *104*, 182908.

207. Yang, S. Y.; Seidel, J.; Byrnes, S. J.; Shafer, P.; Yang, C. H.; Rossell, M. D.; Yu, P.; Chu, Y. H.; Scott, J. F.; Ager, J. W.; Martin, L. W.; Ramesh, R. *Nat. Nanotechnol.* **2010**, *5*, 143.

208. Bhatnagar, A.; Roy Chaudhuri, A.; Heon Kim, Y.; Hesse, D.; Alexe, M. *Nat. Commun.* **2013**, *4*, 2678.

209. Yu, P.; Luo, W.; Yi, D.; Zhang, J. X.; Rossell, M. D.; Yang, C.-H.; You, L.; Singh-Bhalla, G.; Yang, S. Y.; He, Q.; Ramasse, Q. M.; Erni, R.; Martin, L. W.; Chu, Y. H.; Pantelides, S. T.; Pennycook, S. J.; Ramesh, R. *Proc. Nat. Acad. Sci. USA* **2012**, *109*, 9710.

210. Guo, R.; You, L.; Motapothula, M.; Zhang, Z.; Breese, M. B. H.; Chen, L.; Wu, D.; Wang, J. *AIP Adv.* **2012**, *2*, 042104.

211. Park, J. W.; Baek, S. H.; Wu, P.; Winchester, B.; Nelson, C. T.; Pan, X. Q.; Chen, L. Q.; Tybell, T.; Eom, C. B. *Appl. Phys. Lett.* **2010**, *97*, 212904.

212. Ihlefeld, J. F.; Folkman, C. M.; Baek, S. H.; Brennecka, G. L.; George, M. C.; Carroll, J. F.; Eom, C. B. *Appl. Phys. Lett.* **2010**, *97*, 262904.

213. Shelke, V.; Mazumdar, D.; Srinivasan, G.; Kumar, A.; Jesse, S.; Kalinin, S.; Baddorf, A.; Gupta, A. *Adv. Mater.* **2011**, *23*, 669.

214. Jesse, S.; Rodriguez, B. J.; Choudhury, S.; Baddorf, A. P.; Vrejoiu, I.; Hesse, D.; Alexe, M.; Eliseev, E. A.; Morozovska, A. N.; Zhang, J.; Chen, L.-Q.; Kalinin, S. V. *Nat. Mater.* **2008**, *7*, 209.

215. Cruz, M. P.; Chu, Y. H.; Zhang, J. X.; Yang, P. L.; Zavaliche, F.; He, Q.; Shafer, P.; Chen, L. Q.; Ramesh, R. *Phys. Rev. Lett.* **2007**, *99*, 217601.

216. Shafer, P.; Zavaliche, F.; Chu, Y.-H.; Yang, P.-L.; Cruz, M. P.; Ramesh, R. *Appl. Phys. Lett.* **2007**, *90*, 072907.

217. Balke, N.; Gajek, M.; Tagantsev, A. K.; Martin, L. W.; Chu, Y.-H.; Ramesh, R.; Kalinin, S. V. Adv. Funct. Mater. **2010**, *20*, 3466.

218. Ederer, C.; Spaldin, N. A. *Phys. Rev. B* **2005**, *71*, 060401(R).

219. Zhao, T.; Scholl, A.; Zavaliche, F.; Lee, K.; Barry, M.; Doran, A.; Cruz, M. P.; Chu, Y. H.; Ederer, C.; Spaldin, N. A.; Das, R. R.; Kim, D. M.; Baek, S. H.; Eom, C. B.; Ramesh, R. *Nat. Mater.* **2006**, *5*, 823.

220. Balke, N.; Choudhury, S.; Jesse, S.; Huijben, M.; Chu, Y. H.; Baddorf, A. P.; Chen, L. Q.; Ramesh, R.; Kalinin, S. V. *Nat. Nanotechnol.* **2009**, *4*, 868.

221. Vasudevan, R. K.; Liu, Y.; Li, J.; Liang, W.-I.; Kumar, A.; Jesse, S.; Chen, Y.-C.; Chu, Y.-H.; Nagarajan, V.; Kalinin, S. V. *Nano Lett.* **2011**, *11*, 3346.

222. You, L.; Chen, Z.; Zou, X.; Ding, H.; Chen, W.; Chen, L.; Yuan, G.; Wang, J. *ACS Nano* **2012**, *6*, 5388.

223. Kim, K.-E.; Jang, B.-K.; Heo, Y.; Hong Lee, J.; Jeong, M.; Lee, J. Y.; Seidel, J.; Yang, C.-H. *NPG Asia Mater.* **2014**, *6*, e81.

224. Seidel, J. *J. Phys. Chem. Lett.* **2012**, *3*, 2905.

225. Vasudevan, R. K.; Wu, W.; Guest, J. R.; Baddorf, A. P.; Morozovska, A. N.; Eliseev, E. A.; Balke, N.; Nagarajan, V.; Maksymovych, P.; Kalinin, S. V. *Adv. Funct. Mater.* **2013**, *23*, 2592.

226. Yang, J. C.; Huang, Y. L.; He, Q.; Chu, Y. H. *J. Appl. Phys.* **2014**, *116*, 153902.

227. Seidel, J.; Singh-Bhalla, G.; He, Q.; Yang, S.-Y.; Chu, Y.-H.; Ramesh, R. *Phase Transit.* **2012**, *86*, 53.

228. Yang, J. C.; Yeh, C. H.; Chen, Y. T.; Liao, S. C.; Huang, R.; Liu, H. J.; Hung, C. C.; Chen, S. H.; Wu, S. L.; Lai, C. H.; Chiu, Y. P.; Chiu, P. W.; Chu, Y. H. *Nanoscale* **2014**, *6*, 10524.

229. Lubk, A.; Gemming, S.; Spaldin, N. A. *Phys. Rev. B* **2009**, *80*, 104110.

230. Chiu, Y.-P.; Chen, Y.-T.; Huang, B.-C.; Shih, M.-C.; Yang, J.-C.; He, Q.; Liang, C.-W.; Seidel, J.; Chen, Y.-C.; Ramesh, R.; Chu, Y.-H. *Adv. Mater.* **2011**, *23*, 1530.

231. Farokhipoor, S.; Noheda, B. *Phys. Rev. Lett.* **2011**, *107*, 127601.

232. Diéguez, O.; Aguado-Puente, P.; Junquera, J.; Íñiguez, J. *Phys. Rev. B* **2013**, *87*, 024102.

233. Seidel, J.; Maksymovych, P.; Batra, Y.; Katan, A.; Yang, S. Y.; He, Q.; Baddorf, A. P.; Kalinin, S. V.; Yang, C. H.; Yang, J. C.; Chu, Y. H.; Salje, E. K. H.; Wormeester, H.; Salmeron, M.; Ramesh, R. *Phys. Rev. Lett.* **2010**, *105*, 197603.

234. Farokhipoor, S.; Noheda, B. *J. Appl. Phys.* **2012**, *112*, 052002.

235. Maksymovych, P.; Seidel, J.; Chu, Y. H.; Wu, P.; Baddorf, A. P.; Chen, L.-Q.; Kalinin, S. V.; Ramesh, R. *Nano Lett.* **2011**, *11*, 1906.

236. Balke, N.; Winchester, B.; Ren, W.; Chu, Y. H.; Morozovska, A. N.; Eliseev, E. A.; Huijben, M.; Vasudevan, R. K.; Maksymovych, P.; Britson, J.; Jesse, S.; Kornev, I.; Ramesh, L.; Bellaiche, L.; Chen, L. Q.; Kalinin, S. V. *Nat. Phys.* **2012**, *8*, 81.

237. Vasudevan, R. K.; Morozovska, A. N.; Eliseev, E. A.; Britson, J.; Yang, J. C.; Chu, Y. H.; Maksymovych, P.; Chen, L. Q.; Nagarajan, V.; Kalinin, S. V. *Nano Lett.* **2012**, *12*, 5524.

238. Maksymovych, P.; Morozovska, A. N.; Yu, P.; Eliseev, E. A.; Chu, Y.-H.; Ramesh, R.; Baddorf, A. P.; Kalinin, S. V. *Nano Lett.* **2011**, *12*, 209.

239. Schröder, M.; Haußmann, A.; Thiessen, A.; Soergel, E.; Woike, T.; Eng, L. M. *Adv. Funct. Mater.* **2012**, *22*, 3936.

240. Sluka, T.; Tagantsev, A. K.; Bednyakov, P.; Setter, N. *Nat. Commun.* **2013**, *4*, 1808.

241. Stolichnov, I.; Iwanowska, M.; Colla, E.; Ziegler, B.; Gaponenko, I.; Paruch, P.; Huijben, M.; Rijnders, G.; Setter, N. *Appl. Phys. Lett.* **2014**, *104*, 132902.

242. Seidel, J.; Trassin, M.; Zhang, Y.; Maksymovych, P.; Uhlig, T.; Milde, P.; Köhler, D.; Baddorf, A. P.; Kalinin, S. V.; Eng, L. M.; Pan, X.; Ramesh, R. *Adv. Mater.* **2014**, *26*, 4376.

243. Seidel, J.; Fu, D.; Yang, S.-Y.; Alarcón-Lladó, E.; Wu, J.; Ramesh, R.; Ager, III, J. W. *Phys. Rev. Lett.* **2011**, *107*, 126805.

244. Guo, R.; You, L.; Chen, L.; Wu, D.; Wang, J. *Appl. Phys. Lett.* **2011**, *99*, 222904.

245. Seidel, J.; Yang, S. Y.; Alarcón-Lladó, E.; Ager, J. W.; Ramesh, R. *Ferroelectrics* **2012**, *433*, 123.

246. Young, S. M.; Zheng, F.; Rappe, A. M. *Phys. Rev. Lett.* **2012**, *109*, 236601.

247. Martin, L. W.; Chu, Y. H.; Holcomb, M. B.; Huijben, M.; Yu, P.; Han, S. J.; Lee, D.; Wang, S. X.; Ramesh, R. *Nano Lett.* **2008**, *8*, 2050.

248. He, Q.; Yeh, C. H.; Yang, J. C.; Singh-Bhalla, G.; Liang, C. W.; Chiu, P. W.; Catalan, G.; Martin, L. W.; Chu, Y. H.; Scott, J. F.; Ramesh, R. *Phys. Rev. Lett.* **2012**, *108*, 067203.

249. He, Q.; Chu, Y. H.; Heron, J. T.; Yang, S. Y.; Liang, W. I.; Kuo, C. Y.; Lin, H. J.; Yu, P.; Liang, C. W.; Zeches, R. J.; Kuo, W. C.; Juang, J. Y.; Chen, C. T.; Arenholz, E.; Scholl, A.; Ramesh, R. *Nat. Commun.* **2011**, *2*, 225.

250. Escorihuela-Sayalero, C.; Diéguez, O.; Íñiguez, J. *Phys. Rev. Lett.* **2012**, *109*, 247202.

251. Lebeugle, D.; Colson, D.; Forget, A.; Viret, M.; Bataille, A. M.; Gukasov, A. *Phys. Rev. Lett.* **2008**, *100*, 227602.

252. Lee, S.; Choi, T.; Ratcliff, W.; Erwin, R.; Cheong, S. W.; Kiryukhin, V. *Phys. Rev. B* **2008**, *78*, 100101(R).

253. Lee, S.; Ratcliff, W.; Cheong, S. W.; Kiryukhin, V. *Appl. Phys. Lett.* **2008**, *92*, 192906.

254. Sosnowska, I. M. *J. Microsc.* **2009**, *236*, 109.

255. Kadomtseva, A. M.; Zvezdin, A. K.; Popov, Y. F.; Pyatakov, A. P.; Vorob'ev, G. P. *JETP Lett.* **2004**, *79*, 571.

256. Dzyaloshinskii, I. E.; Soviet physics JETP **1960**, *10*, 628.

257. Moriya, T. Phys. Rev. **1960**, *120*, 91.

258. Rahmedov, D.; Wang, D.; Íñiguez, J.; Bellaiche, L. *Phys. Rev. Lett.* **2012**, *109*, 037207.

259. Zvezdin, A. K.; Pyatakov, A. P.; Europhys. Lett. **2012**, *99*, 57003.

260. Ramazanoglu, M.; Laver, M.; Ratcliff, II, W.; Watson, S. M.; Chen, W. C.; Jackson, A.; Kothapalli, K.; Lee, S.; Cheong, S. W.; Kiryukhin, V. *Phys. Rev. Lett.* **2011**, *107*, 207206.

261. Lebeugle, D.; Mougin, A.; Viret, M.; Colson, D.; Allibe, J.; Béa, H.; Jacquet, E.; Deranlot, C.; Bibes, M.; Barthélémy, A. *Phys. Rev. B* **2010**, *81*, 134411.

262. Sosnowska, I.; Przeniosło, R. *Phys. Rev. B* **2011**, *84*, 144404.

263. Sosnowska, I.; Loewenhaupt, M.; David, W. I. F.; Ibberson, R. M. *Physica B* **1992**, *180–181* (Part 1), 117.

264. Przeniosło, R.; Palewicz, A.; Regulski, M.; Sosnowska, I.; Ibberson, R. M.; Knight, K. S. *J. Phys.: Condens. Matter* **2006**, *18*, 2069.

265. Cazayous, M.; Gallais, Y.; Sacuto, A.; de Sousa, R.; Lebeugle, D.; Colson, D. *Phys. Rev. Lett.* **2008**, *101*, 037601.

266. Singh, M. K.; Katiyar, R. S.; Scott, J. F. *J. Phys.: Condens. Matter* **2008**, *20*, 252203.

267. Singh, M. K.; Prellier, W.; Singh, M. P.; Katiyar, R. S.; Scott, J. F. *Phys. Rev. B* **2008**, *77*, 144403.

268. Julia, H.-A.; Gustau, C.; José Alberto, R.-V.; Michel, V.; Dorothée, C.; James, F. S. *J. Phys.: Condens. Matter* **2010**, *22*, 256001.

269. Lu, J.; Günther, A.; Schrettle, F.; Mayr, F.; Krohns, S.; Lunkenheimer, P.; Pimenov, A.; Travkin, V. D.; Mukhin, A. A.; Loidl, A. *Eur. Phys. J. B* **2010**, *75*, 451.

270. Ramazanoglu, M.; Ratcliff, II, W.; Choi, Y. J.; Lee, S.; Cheong, S. W.; Kiryukhin, V. *Phys. Rev. B* **2011**, *83*, 174434.

271. Zalesskii, A. V.; Zvezdin, A. K.; Frolov, A. A.; Bush, A. A. *JETP Lett.* **2000**, *71*, 465.

272. Zalesskii, A. V.; Frolov, A. A.; Zvezdin, A. K.; Gippius, A. A.; Morozova, E. N.; Khozeevc, D. F.; Bush, A. S.; Pokatilov, V. S. *J. Exp. Theor. Phys.* **2002**, *95*, 101.

273. Bush, A. A.; Gippius, A. A.; Zalesskii, A. V.; Morozova, E. N. *JETP Lett.* **2003**, *78*, 389.

274. Popov, Y. F.; Zvezdin, A. K.; Vorobev, G. P.; Kadomtseva, A. M.; Murashev, V. A.; Rakov, D. N. *JETP Lett.* **1993**, *57*, 69.

275. Popov, Y. F.; Kadomtseva, A. M.; Vorob'ev, G. P.; Zvezdin, A. K. *Ferroelectrics* **1994**, *162*, 135.

276. Popov, Y. F.; Kadomtseva, A. M.; Krotov, S. S.; Belov, D. V.; Vorob'ev, G. P.; Makhov, P. N.; Zvezdin, A. K. *Low Temp. Phys.* **2001**, *27*, 478.

277. Ruette, B.; Zvyagin, S.; Pyatakov, A. P.; Bush, A.; Li, J. F.; Belotelov, V. I.; Zvezdin, A. K.; Viehland, D. *Phys. Rev. B* **2004**, *69*, 064114.

278. Wardecki, D.; Przeniosło, R.; Sosnowska, I.; Y. Skourski; Loewenhaupt, M. *J. Phys. Soc. Jpn.* **2008**, *77*, 103709.

279. Tokunaga, M.; Azuma, M.; Shimakawa, Y. *J. Phys. Soc. Jpn.* **2010**, *79*, 064713.

280. Ohoyama, K.; Lee, S.; Yoshii, S.; Narumi, Y.; Morioka, T.; Nojiri, H.; Jeon, G. S.; Cheong, S.-W.; Park, J.-G. *J. Phys. Soc. Jpn.* **2011**, *80*, 125001.

281. Park, J.; Lee, S.-H.; Lee, S.; Gozzo, F.; Kimura, H.; Noda, Y.; Choi, Y. J.; Kiryukhin, V.; Cheong, S.-W.; Jo, Y.; Choi, E. S.; Balicas, L.; Jeon, G. S.; Park, J.-G. *J. Phys. Soc. Jpn.* **2011**, *80*, 114714.

282. Ramazanoglu, M.; Ratcliff, II, W.; Yi, H. T.; Sirenko, A. A.; Cheong, S. W.; Kiryukhin, V. *Phys. Rev. Lett.* **2011**, *107*, 067203.

283. Eerenstein, W.; Morrison, F. D.; Dho, J.; Blamire, M. G.; Scott, J. F.; Mathur, N. D. *Science* **2005**, *307*, 1203a.

284. Wang, J.; Scholl, A.; Zheng, H.; Ogale, S. B.; Viehland, D.; Schlom, D. G.; Spaldin, N. A.; Rabe, K. M.; Wuttig, M.; Mohaddes, L.; Neaton, J.; Waghmare, U.; Zhao, T.; Ramesh, R. *Science* **2005**, *307*, 1203.

285. Béa, H.; Bibes, M.; Fusil, S.; Bouzehouane, K.; Jacquet, E.; Rode, K.; Bencok, P.; Barthélémy, A. *Phys. Rev. B* **2006**, *74*, 020101.

286. Bai, F. M.; Wang, J. L.; Wuttig, M.; Li, J. F.; Wang, N. G.; Pyatakov, A. P.; Zvezdin, A. K.; Cross, L. E.; Viehland, D. *Appl. Phys. Lett.* **2005**, *86*, 032511.

287. Holcomb, M. B.; Martin, L. W.; Scholl, A.; He, Q.; Yu, P.; Yang, C. H.; Yang, S. Y.; Glans, P. A.; Valvidares, M.; Huijben, M.; Kortright, J. B.; Guo, J.; Chu, P. P.; Ramesh, R. *Phys. Rev. B* **2010**, *81*, 134406.

288. Ke, X.; Zhang, P. P.; Baek, S. H.; Zarestky, J.; Tian, W.; Eom, C. B. *Phys. Rev. B* **2010**, *82*, 134448.

289. Ratcliff, W.; Kan, D.; Chen, W.; Watson, S.; Chi, S.; Erwin, R.; McIntyre, G. J.; Capelli, S. C.; Takeuchi, I. *Adv. Funct. Mater.* **2011**, *21*, 1567.

290. Sando, D.; Agbelele, A.; Rahmedov, D.; Liu, J.; Rovillain, P.; Toulouse, C.; Infante, I. C.; Pyatakov, A. P.; Fusil, S.; Jacquet, E.; Carrétéro, C.; Deranlot, C.; Lisenkov, S.; Wang, D.; Le Breton, J. M.; Cazayous, M.; Sacuto, A.; Juraszek, J.; Zvezdin, A. K.; Bellaiche, L.; Dkhil, B.; Barthélémy, A.; Bibes, M. *Nat. Mater.* **2013**, *12*, 641.

291. Heron, J. T.; Trassin, M.; Ashraf, K.; Gajek, M.; He, Q.; Yang, S. Y.; Nikonov, D. E.; Chu, Y. H.; Salahuddin, S.; Ramesh, R. *Phys. Rev. Lett.* **2011**, *107*, 217202.

292. Trassin, M.; Clarkson, J. D.; Bowden, S. R.; Liu, J.; Heron, J. T.; Paull, R. J.; Arenholz, E.; Pierce, D. T.; Unguris, J. *Phys. Rev. B* **2013**, *87*, 134426.

293. Cheong, S.-W.; Mostovoy, M. *Nat. Mater.* **2007**, *6*, 13.

294. Ratcliff, II, W.; Yamani, Z.; Anbusathaiah, V.; Gao, T. R.; Kienzle, P. A.; Cao, H.; Takeuchi, I. *Phys. Rev. B* **2013**, *87*, 140405.

295. Ederer, C.; Spaldin, N. A. *Phys. Rev. B* **2005**, *71*, 060401.

296. Borisov, P.; Hochstrat, A.; Chen, X.; Kleemann, W.; Binek, C. *Phys. Rev. Lett.* **2005**, *94*, 117203.

297. He, X.; Wang, Y.; Wu, N.; Caruso, A. N.; Vescovo, E.; Belashchenko, K. D.; Dowben, P. A.; Binek, C. *Nat. Mater.* **2010**, *9*, 579.

298. Laukhin, V.; Skumryev, V.; Martí, X.; Hrabovsky, D.; Sánchez, F.; García-Cuenca, M. V.; Ferrater, C.; Varela, M.; Lüders, U.; Bobo, J. F.; Fontcuberta, J. *Phys. Rev. Lett.* **2006**, *97*, 227201.

299. Chu, Y. H.; Martin, L. W.; Holcomb, M. B.; Gajek, M.; Han, S. J.; He, Q.; Balke, N.; Yang, C. H.; Lee, D.; Hu, W.; Zhan, Q.; Yang, P. L.; Fraile-Rodriguez, A.; Scholl, A.; Wang, S. X.; Ramesh, R. *Nat. Mater.* **2008**, *7*, 478.

300. Berkowitz, A. E.; Takano, K. *J. Magn. Magn. Mater.* **1999**, *200*, 552.

301. Nogues, J.; Schuller, I. K. *J. Magn. Magn. Mater.* **1999**, *192*, 203.

302. Kiwi, M. *J. Magn. Magn. Mater.* **2001**, *234*, 584.

303. Radu, F.; Zabel, H. In: *Magnetic Heterostructures*; Zabel, H., Bader, S. D., Eds.; Springer: Berlin, 2008; Vol. 227, p 97.

304. Bibes, M.; Barthélémy, A. *Nat. Mater.* **2008**, *7*, 425.

305. Béa, H.; Bibes, M.; Cherifi, S.; Nolting, F.; Warot-Fonrose, B.; Fusil, S.; Herranz, G.; Deranlot, C.; Jacquet, E.; Bouzehouane, K.; Barthélémy, A. *Appl. Phys. Lett.* **2006**, *89*, 242114.

306. Dho, J. H.; Qi, X. D.; Kim, H.; MacManus-Driscoll, J. L.; Blamire, M. G. *Adv. Mater.* **2006**, *18*, 1445.

307. Martin, L. W.; Chu, Y. H.; Zhan, Q.; Ramesh, R.; Han, S. J.; Wang, S. X.; Warusawithana, M.; Schlom, D. G. *Appl. Phys. Lett.* **2007**, *91*, 172513.

308. Béa, H.; Bibes, M.; Ott, F.; Dupé, B.; Zhu, X. H.; Petit, S.; Fusil, S.; Deranlot, C.; Bouzehouane, K.; Barthélémy, A. *Phys. Rev. Lett.* **2008**, *100*, 017204.

309. Malozemoff, A. P. *Phys. Rev. B* **1988**, *37*, 7673.

310. Malozemoff, A. P. *J. Appl. Phys.* **1988**, *63*, 3874.

311. Allibe, J.; Fusil, S.; Bouzehouane, K.; Daumont, C.; Sando, D.; Jacquet, E.; Deranlot, C.; Bibes, M.; Barthélémy, A. *Nano Lett.* **2012**, *12*, 1141.

312. Wang, C.; Ph.D. Thesis, Cornell University, 2012.

313. Stiles, M. D.; McMichael, R. D. *Phys. Rev. B* **1999**, *59*, 3722.

314. Schulthess, T. C.; Butler, W. H. *Phys. Rev. Lett.* **1998**, *81*, 4516.

315. Lebeugle, D.; Mougin, A.; Viret, M.; Colson, D.; Ranno, L. *Phys. Rev. Lett.* **2009**, *103*, 257601.

316. Yu, P.; Lee, J. S.; Okamoto, S.; Rossell, M. D.; Huijben, M.; Yang, C. H.; He, Q.; Zhang, J. X.; Yang, S. Y.; Lee, M. J.; Ramasse, Q. M.; Erni, R.; Chu, Y. H.; Arena, D. A.; Kao, C. C.; Martin, L. W.; Ramesh, R. *Phys. Rev. Lett.* **2010**, *105*, 027201.

317. Borisevich, A. Y.; Chang, H. J.; Huijben, M.; Oxley, M. P.; Okamoto, S.; Niranjan, M. K.; Burton, J. D.; Tsymbal, E. Y.; Chu, Y. H.; Yu, P.; Ramesh, R.; Kalinin, S. V.; Pennycook, S. J. *Phys. Rev. Lett.* **2010**, *105*, 087204.

318. Wu, S. M.; Cybart, S. A.; Yu, P.; Rossell, M. D.; Zhang, J. X.; Ramesh, R.; Dynes, R. C. *Nat. Mater.* **2010**, *9*, 756.

319. Wu, S. M.; Cybart, S. A.; Yi, D.; Parker, J. M.; Ramesh, R.; Dynes, R. C. *Phys. Rev. Lett.* **2013**, *110*, 067202.

320. Huijben, M.; Yu, P.; Martin, L. W.; Molegraaf, H. J. A.; Chu, Y. H.; Holcomb, M. B.; Balke, N.; Rijnders, G.; Ramesh, R. *Adv. Mater.* **2013**, *25*, 4739.

321. You, L.; Lu, C.; Yang, P.; Han, G.; Wu, T.; Luders, U.; Prellier, W.; Yao, K.; Chen, L.; Wang, J. *Adv. Mater.* **2010**, *22*, 4964.

322. You, L.; Wang, B.; Zou, X.; Lim, Z. S.; Zhou, Y.; Ding, H.; Chen, L.; Wang, J. *Phys. Rev. B* **2013**, *88*, 184426.

323. Chen, Z.; Prosandeev, S.; Luo, Z. L.; Ren, W.; Qi, Y.; Huang, C. W.; You, L.; Gao, C.; Kornev, I. A.; Wu, T.; Wang, J.; Yang, P.; Sritharan, T.; Bellaiche, L.; Chen, L. *Phys. Rev. B* **2011**, *84*, 094116.

324. Prosandeev, S.; Kornev, I. A.; Bellaiche, L. *Phys. Rev. Lett.* **2011**, *107*, 117602.

325. Fan, Z.; Wang, J.; Sullivan, M. B.; Huan, A.; Singh, D. J.; Ong, K. P. *Sci. Rep.* **2014**, *4*, 3621.

326. Shick, A. B.; Khmelevskyi, S.; Mryasov, O. N.; Wunderlich, J.; Jungwirth, T. *Phys. Rev. B* **2010**, *81*, 212409.

327. Park, B. G.; Wunderlich, J.; Martí, X.; Holý, V.; Kurosaki, Y.; Yamada, M.; Yamamoto, H.; Nishide, A.; Hayakawa, J.; Takahashi, H.; Shick, A. B.; Jungwirth, T. *Nat. Mater.* **2011**, *10*, 347.

328. Wang, Y. Y.; Song, C.; Cui, B.; Wang, G. Y.; Zeng, F.; Pan, F. *Phys. Rev. Lett.* **2012**, *109*, 137201.

329. Marti, X.; Fina, I.; Frontera, C.; Liu, J.; Wadley, P.; He, Q.; Paull, R. J.; Clarkson, J. D.; Kudrnovský, J.; Turek, I.; Kuneš, J.; Yi, D.; Chu, J. H.; Nelson, C. T.; You, L.; Arenholz, E.; Salahuddin, S.; Fontcuberta, J.; Jungwirth, T.; Ramesh, R. *Nat. Mater.* **2014**, *13*, 367.

330. Wang, Y.; Song, C.; Wang, G.; Miao, J.; Zeng, F.; Pan, F. *Adv. Funct. Mater.* **2014**, doi:10.1002/adfm.201401659.

331. Gajek, M.; Bibes, M.; Fusil, S.; Bouzehouane, K.; Fontcuberta, J.; Barthelemy, A.; Fert, A. *Nat. Mater.* **2007**, *6*, 296.

332. Garcia, V.; Fusil, S.; Bouzehouane, K.; Enouz-Vedrenne, S.; Mathur, N. D.; Barthelemy, A.; Bibes, M. *Nature* **2009**, *460*, 81.

333. Bibes, M. *Nat. Mater.* **2012**, *11*, 354.

334. Garcia, V.; Bibes, M. *Nat. Commun.* **2014**, *5*, 5518.

335. Wojdeł, J. C.; Íñiguez, J. *Phys. Rev. Lett.* **2014**, *112*, 247603.

336. Ren, W.; Yang, Y.; Diéguez, O.; Íñiguez, J.; Choudhury, N.; Bellaiche, L. *Phys. Rev. Lett.* **2013**, *110*, 187601.

337. Wang, Y.; Nelson, C.; Melville, A.; Winchester, B.; Shang, S.; Liu, Z.-K.; Schlom, D. G.; Pan, X.; Chen, L.-Q. *Phys. Rev. Lett.* **2013**, *110*, 267601.

338. Yi, D.; Liu, J.; Okamoto, S.; Jagannatha, S.; Chen, Y.-C.; Yu, P.; Chu, Y.-H.; Arenholz, E.; Ramesh, R. *Phys. Rev. Lett.* **2013**, *111*, 127601.

339. Mannhart, J.; Schlom, D. G. *Science* **2010**, *327*, 1607.

340. Zubko, P.; Gariglio, S.; Gabay, M.; Ghosez, P.; Triscone, J.-M. *Annu. Rev. Condens. Matter Phys.* **2011**, *2*, 141.

341. Hwang, H. Y.; Iwasa, Y.; Kawasaki, M.; Keimer, B.; Nagaosa, N.; Tokura, Y. *Nat. Mater.* **2012**, *11*, 103.

342. Tsymbal, E. Y.; Dagotto, E. R.; Eom, P. C.-B.; Ramesh, R. *Multifunctional Oxide Heterostructures.* Oxford University Press: Oxford, 2012.

343. Yu, P.; Chu, Y.-H.; Ramesh, R. *Mater. Today* **2012**, *15*, 320.

344. Nelson, C. T.; Winchester, B.; Zhang, Y.; Kim, S.-J.; Melville, A.; Adamo, C.; Folkman, C. M.; Baek, S.-H.; Eom, C.-B.; Schlom, D. G.; Chen, L.-Q.; Pan, X. *Nano Lett.* **2011**, *11*, 828.

345. Qi, Y.; Chen, Z.; Huang, C.; Wang, L.; Han, X.; Wang, J.; Yang, P.; Sritharan, T.; Chen, L. *J. Appl. Phys.* **2012**, *111*.

CHAPTER 4

RECENT ADVANCES IN RESISTIVE SWITCHING NANOCRYSTALLINE THIN FILMS DERIVED VIA SOLUTION PROCESSED TECHNIQUES

ADNAN YOUNIS*, SEAN LI, and DEWEI CHU*

School of Materials Science and Engineering, University of New South Wales, Sydney 2052, NSW, Australia

*Corresponding authors. E-mail: a.younis@unsw.edu.au; d.Chu@unsw.edu.au

CONTENTS

ABSTRACT

During the last decade, resistive random-access memories (RRAMs) based on metal oxide thin films have attracted great attention owing to their unique advantages, such as simple structure, faster reading and writing speed, smaller bit size, lower power consumption compared to existing nonvolatile memory technologies, etc. In particular, RRAM exhibits two resistances states (ON and OFF) that can be switched by an external bias or preserved as information and this process is referred as resistive switching (RS) process. The RS processes in metal oxides are usually limited to a small region near the interfaces between the electrodes and the oxides layer; therefore, the understanding of the interfaces and growth of nanometer scale oxide films are extremely desirable. Meanwhile, there is a barrier that the current photo lithography tools for making nanoscale features (top-down approaches) are prohibitively costly and complicated. Therefore, this book chapter will mainly focus to review the recent progress in the development of solution processed novel metal oxide thin films with improved RS performances via appropriate materials design, defect engineering, and interface engineering.

The main body of the chapter comprises two portions. In the first half, after brief introduction of RRAMs, the RS phenomenon which had been highlighted in many reports so far will be reviewed in detail. In the remaining half, a detailed description of various solution chemistry-driven schemes including chemical bath deposition (CBD), sol–gel, electrochemical deposition, and hydrothermal processes for the fabrication of metal oxides RRAMs will be presented.

4.1 INTRODUCTION

Resistive switching (RS) memories which are named as "memristors" are predicted to be fourth fundamental passive circuit element according to Leon Chua in 1971 [1,2] and later on, practically demonstrated by Hewlett-Packard laboratory in 2008 [3], has attracted extensive research attention for their high scalability and versatility. The word "memristor" actually originated as a combination of "memory" and "resistor," which also refers to the fact that memristor is a resistor that can memorize its past history. RS memories take advantages of low power consumption, high scalability, and excellent compatibility with the complementary metal–oxide–semiconductor (CMOS) technology which is extensively utilized for constructing modern integrated circuits including microprocessors, microcontrollers, digital and analog logic circuits (CMOS sensor), etc.

Instead of storing information in transistors (flash memory) or capacitors so-called dynamic random-access memories, RS memories depend on the ability to alter the electrical resistance of certain materials by applying an external voltage or current. RS memories have been regarded as next generation nonvolatile memory devices due to its simple structure and compatibility with CMOS technology [4–6]. In comparison with traditional nonvolatile memories (flash), RS memories have unique advantages, such as much faster writing rate, smaller bit cell size, and lower operating voltages. Specially, the resistive transition induced by applying electric pulses with opposite polarities can be achieved within tenths of nanoseconds at room temperature and the resultant resistance states can be retained for 10 years [7].

So far, a great number of materials have been found to show the resistive switching (RS) characteristics (details will be provided in the forthcoming sections) and almost all or most of the memory cells have a simple capacitor-like "MIM" structure, where "M" represents a kind of metal electrode as well as conducting nonmetals, and "I" stands for an insulator or semi-conductor layer sandwiched by two electrodes. Although, RS memories had gained lot more scientific and commercial interests in recent years, but the involved switching mechanisms are still debatable. It is nearly impossible to describe all the work taken out so far. Therefore, in the following sections, the recent progress in our understanding of the mechanism in RRAM will be described.

4.2 RESISTIVE SWITCHING MECHANISMS

A memristor has a typical capacitor-like structure with an insulating/ RS material layer sandwiched between two metal electrodes, as shown in Figure 4.1a. Usually, in the absence of external potential/bias, a memristor cell is in low-conduction state, which is usually referred as a high-resistance state (HRS). An electrical bias is needed to trigger the soft breakdown of the insulating layer, which switches the memristor conduction state from the HRS to a relatively high-conduction state or low-resistance state (LRS) and this process is called forming process. Subsequently, a reverse transition in cell conduction states can also be achieved by applying appropriate reverse polarity conditions which is known as a reset process. Memristors with two or several stable resistance states are preferable for digital information storage. RS can also be achieved in two ways: (1) unipolar nonpolar switching and (2) bipolar switching. For the unipolar memristor, set and reset voltages are of the same polarity but of different magnitudes, whereas for the bipolar switching, set and reset processes appeared on different polarities, as shown in Figure 4.1b,c.

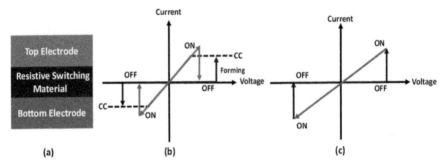

(a)   (b)   (c)

FIGURE 4.1 (a) Schematics of a memristor cell and (b) unipolar switching. The set voltage is always higher than the reset voltage, and the reset current is always higher than the current compliance (CC) during set operation. (c) Bipolar switching. The set process appeared on one polarity and the reset operation requires the opposite polarity.

So far, many materials, including metal sulfides [8], organic compounds [9], metal oxides [10], nonmetal oxides [11], and phase-change materials [12], were used as RS materials. Also, a variety of metals or alloys, such as Au [13], Pt [13], W [14], Al [15], Ni [16], Ag [17], Ti [18], and many others [19–22], were also used as electrode materials and it is believed that the selection of electrode material has strong influence on overall RS device performances. Among all of the reported RS materials, many efforts have been made to clarify the actual switching mechanisms within those materials [23, 24]. The most widely used models to explain switching mechanisms in many material systems could be categorized into two major types, namely, "conductive filament (CF)-type" and "barrier-type" conduction processes.

4.2.1 FILAMENT-TYPE CONDUCTION PROCESS

Filament-type conduction process mainly refers to the formation and rupture of CFs in insulating thin films in the presence or absence of applied potential (Fig. 4.2), which could exhibit either bipolar or unipolar switching behaviors. The span of filament's type conduction process further expands to the valence change (VC) mechanism and electrochemical metallization process (ECM), depending on the chemical compositions of the CFs [24].

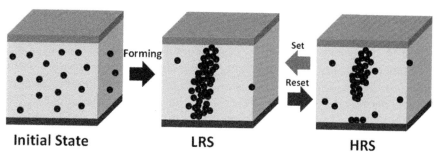

FIGURE 4.2 Schematic illustration of the resistive switching process in the filament-type memristors.

4.2.1.1 VALENCE CHANGE MECHANISMS

On the implication of external bias, the electric field and/or thermal effects may lead to the migration of anions toward one of the electrodes, thus resulting in the formation/annihilation of the CFs. In most metal oxides, the oxygen anion (or the positive-charge-depleted oxygen vacancy) is considered as the mobile species. Structural defects in the insulating layer, such as grain boundaries [25,26] and dislocations [27], where the diffusion energy is low, can easily provide ion diffusion pathway. High-electric field ($>>10^6$ V cm^{-1}) inside the insulating layer can dramatically accelerate the ions'/vacancies' drift velocity by the following relation; $v = \mu E \exp (E / E_0)$, where E_0 is a characteristic field for a particular mobile atom and μ is the ionic mobility [28,29]. Besides, the leakage current can also induce local joule heating effects which could also predominantly accelerate the drift velocity [28–31]. During the forming/set process, defects (e.g., oxygen anion-vacancy pairs) inside the insulating layer can be created through impact ionization [32], and the formation of an oxygen vacancy can be described by

$$O_o \rightarrow V_o^{2+} + O_i^{2-} \tag{4.1}$$

where O_o and O_i^{2-} represents the presence and absence of oxygen atoms on lattice site, whereas V_o^{2+} represents the oxygen vacancy. When the oxygen ions (being negatively charged) move toward the anode, oxygen vacancies start accumulating near cathode and an oxygen-deficient region starts to build up and percolate toward the anode. Meanwhile, the metal cations accommodate at this region and start to trap electrons from the cathode by the following reaction [33]:

$$M^{z+} + ne^- \rightarrow M^{(2-n)+} \tag{4.2}$$

where M^{z+} represents the positive metal cations. The reduced valence states of the metal cations, as a consequence of the electrochemical process, typically turn the oxide into a more conductive phase. Anion motion leads to VCs of the metal (cations), which causes the resistance switching of the metal oxide. The devices which demonstrate these kinds of behavior during resistance switching process are usually named as VC memories [24]. In the following section, some examples will be presented which demonstrated VC mechanism during RS process.

Kwon et al. [34] utilized in situ transmission electron microscopy to observe the formation and annihilation of CFs in a TiO_2 memristor. A complete CF with the diameter of ~15 nm, comprising a Ti_4O_7 Magneli phase, can be observed in Figure 4.3a. Most of the incomplete Ti_4O_7 or Ti_5O_9 CFs were located near the cathode as shown in Figure 4.3b. There observation revealed that once a complete CF was formed, it further blocks the pathway for more filaments to be formed in the adjacent region, as most of the current flowed through the existing CF.

Very recently, a similar type of study had been conducted on Pt/ZnO/Pt memristor system by Chen and coworkers [35] as shown in Figure 4.3. The real-time observation for the growth of CF during set process during in situ transmission electron microscopy (TEM) measurement showed that the CF gradually originated from the cathode in a conical shape and then transformed to a dendritic shape. It formed a columnar structure, once it was connected to the two electrodes. The migration, arrangement, or rearrangement of oxygen vacancies in the presence or absence of electric field was suggested to the formation or annihilation of conducting filaments during set or reset processes.

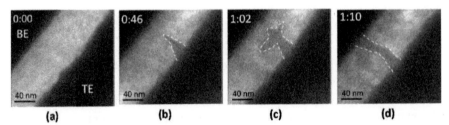

FIGURE 4.3 A series of in-situ TEM images (a–d) showing the forming process in real-time observation. From the TEM images, the dynamic evolution of a CF can be observed. As the voltage increases, the CF grows from a conical shape to a dendritic shape and then reshapes to a column after connecting the two electrodes (reprinted with permission from Ref. [35]).

The typical reset process in RS can usually be attributed to the reverse motion of oxygen ions at the cathode under a reverse bias. But, in reality, the reset mechanism is not as simple and straightforward but is more complicated. Some researchers refer the switching reset process due to the thermal-assisted dissolution/diffusion of the nanoscale filament like a fusing process [33], whereas others attributed it to the thermochemical reaction of the CF with the adjacent oxygen ions [35–38]. In the reset process, the Joule heating caused by the large current density in the CFs could lead to the changes of local elemental composition such that some oxygen vacancies are occupied by the oxygen ions. Once the filament is ruptured, the device is switched from the LRS to HRS [39].

While studying the temperature profile along the CF obtained from computer simulation results, the existence of a hot spot is observed at the center of the CF during the reset process [40–42]. The simulation results were further recently confirmed by in-situ TEM measurements, which demonstrate that the rupture of the CF was actually initiated from the middle of the CF and extended further to the anode, while the partial CF still existed near the cathode [35]. However, while oxygen ions migrate to the anode to annihilate oxygen vacancies during reset process, at the same time, oxygen vacancies are created, leading to formation of the CFs. Therefore, there may be a competition between the dissolution and formation of the CF during the reset process [43–46].

Also, there are different opinions about the growth and rupture direction/locations of CFs depending on the type of the oxides. The growth of CFs in the p-type oxides is from the anode to cathode, thus initiating switching process at the cathode and vice versa for n-type oxides [47–49].

4.2.1.2 ELECTROCHEMICAL METALLIZATION MEMORIES

ECM memories or devices can also be termed as solid electrolyte memory, programmable metallization cell, atomic switches, etc. Contrary to VC memories, the ECM memories are believed to possess metallic cations as the mobile species. The metallic cations usually come both from the electrode materials as well as from the doping metal ions. Therefore, one of the electrode materials (anode) is deliberately selected to be made up from an electrochemically active material, such as Cu [10,50], Ag [51], Ni [16], etc. However, the other electrode material (cathode) could be made from relatively inert materials such as Pt [10,17], W [52], Ir [53], or active materials such as Ag [54,55], etc. There are plenty of materials including chalcogenides

(Ag$_2$S [8], Cu$_2$S [56], Ge$_x$S$_x$ [57], Zn$_x$Cd$_{1-x}$S [58], GeTe [59]), many oxides (Ta$_2$O$_5$ [60], SiO$_2$ [11], HfO$_2$ [61], Al$_2$O$_3$ [14], WO$_3$ [62], MoO$_x$ [63], ZrO$_x$ [64], SrTiO$_3$ [65], TiO$_2$ [66], CuO$_x$ [67], ZnO [68], AlO$_x$ [14]), halides (AgI [69], RbAg$_4$I$_5$ [70]) as well as some organics or polymers [51,71] that were previously frequently reported as the insulating or RS materials sandwiched between cathode and anode. Generally, a forming process is required to trigger the switching process which involves the following steps [24]: first a positive voltage on the active electrode causes the oxidation of the atoms in the electrode, producing metallic cations, that is,

$$M \rightarrow M^{z+} + ze^- \qquad (4.3)$$

The electric field then drives the metallic cations across the electrolyte (insulating layer) to the inert cathode where they electrochemically transformed into metallic atoms and electro-crystallized on the inert cathode by the following relation:

$$M^{z+} + ze^- \rightarrow M \qquad (4.4)$$

As a consequence, the resistance of the device switches from high to low resistance (Off to ON) state. For a negative applied potential, the inert electrode can dissolve anodically the CFs consisting of active metallic atoms at the CF edge and switches the device from the LRS to HRS.

4.2.2 BARRIER-TYPE CONDUCTION

The barrier-type conduction processes are also exhibited in capacitor-like memristors with an insulating layer sandwiched between two electrodes. There have been many materials ranging from complex perovskite oxides [72–74], some binary oxides [18,75–78], sulfides, and selenides, such as CdSe [79], that are usually used as the insulating layer. In the Schottky barrier-type conduction, the Schottky barrier is assumed to be the most possible origin where switching process can take place [80]. Usually, an Ohmic contact is formed within the interface of metal and oxide material, and a Schottky barrier is formed at the other interface (i.e., Ohmic-metal/insulator/Schottky-metal). The contact resistance of the Schottky barrier is determined by the potential distribution of the barrier, i.e., the depletion layer. Furthermore, a Schottky barrier-type memristor usually expresses an Ohmic conduction at LRS, rectifying I–V behavior at the HRS [73,81,82].

The changes in the width or height of the Schottky barrier upon applied potential overall affect the device RS behavior. Different mechanisms have been proposed to explain the modulation in the Schottky barrier. The most plausible mechanism for the alteration of barrier height is distribution/redistribution of oxygen vacancies (or oxygen ions) under electric field [72,73,77,78,81–86] as illustrated in Figure 4.4.

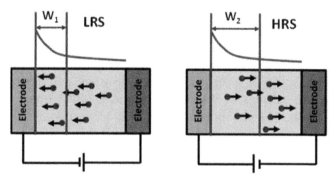

FIGURE 4.4   Modulation of Schottky barrier through redistribution of oxygen vacancies.

The oxygen vacancies (act as donors), being positively charged, is attracted toward left electrode and get localized at the interface between the left electrode and insulator on the implication of negative potential, thus increasing the donor density at the interface and reducing the depletion region, leading to a transition from the HRS to LRS (Fig. 4.6 in the left). While at applied relatively positive potential, the oxygen vacancies migrate away from the left electrode/insulator interface by decreasing the local donor density. Hence, the width of depletion region become wide resulting in the transition of HRS (Fig. 4.4 at right panel). There are some other opinions about the alternation of Schottky barrier that it is due to the trapping/detrapping of carriers in the interface regions [74,87–89]. Electrons, trapped/detrapped in/from the trapping centers, lead to the alternation of barrier height and thus the transition in conduction states is observed.

In summary, the technological status of RS memories, some basics of RS phenomena, candidate materials, and mechanisms are presented so far. A number of possible mechanisms were briefly reviewed and the first reason for the variety of the mechanisms is that the present understanding of the scientific and technological importance of the phenomena is not quite clear. The second reason is the lack of consistent experimental data reported in different publications. The same materials were found to exhibit different switching mechanisms, which may be attributed to different electroforming procedures or the stochastic nature of electroforming itself. At the initial-ization process of RS measurement, electroforming strongly affects the resulting RS behavior. From a technological point of view, this complicated initialization procedure should be avoided: electroforming-free RRAMs are preferable. Therefore, for better understanding, emphasis on systematic electroforming studies is important.

4.3 IMPORTANCE OF ELECTRODE MATERIALS

The "M" in the capacitor-like structure of MIM for RRAM not only refers to metals but it could be some low-resistive electronic conductor materials. The selection of electrode materials is very crucial to determine the overall device performances and their role will be discussed individually.

4.3.1 TOP ELECTRODES

The top-electrode material may comprise metal materials, thus, the mate-rials with different work function, electronegativity, inter-diffusion, inter-facial reactions, and oxygen affinity properties have strong influence on the device RS behavior. Many studies about the research on top-electrode effect of various RS matrixes on the RS behaviors have been proposed. In a study, Sawa et al. [90] proposed the formation of a Schottky barrier near metal p-type semiconductor interface and the origin of the nonlinearity in the current–voltage (I–V) curve was suggested by the modulation of Schottky barrier with electric stimuli. Afterward, the same conclusion was drawn by studying RS phenomenon in the $Pr_{1-x}Ca_xMnO_3$ [91] and $Nb:SrTiO_3$ [92] films in different studies. In contrast to former studies, Seo and coworkers [93] highlighted the formation of Ohmic contact between both interfaces by having a strong effective electric field within RS matrix.

The Ohmic contact can be distinguished from the Schottky contact in terms of symmetry. The Ohmic contact shows the symmetric I–V curve in both voltage polarities, whereas the Schottky contact exhibits the rectifying prop-erty causing the difference between the forward- and reverse-biased currents.

Recently, in the $Ag/La_{0.7}Ca_{0.3}MnO_3$ [94], top-electrode/$Cr:SrZrO_3$ [95], and Ti/ZrO_2 [96] systems, the existence of interface reaction and/or inter-diffusion between the top electrode and the RS film was identified. These interfacial reactions or interdiffusion caused an oxygen-depleted interface, thus resulting in space-charge-limited conduction (SCLC) process in these material systems.

In summary, when an electrode material contact with a RS-matrix, the contact resistance of the interface may be increased by introducing a modulated Schottky contact or by forming an Ohmic contact within the interface. The rectification of Schottky barrier could also be formed by interfacial reactions or interdiffusion that leads to the SCLC. Therefore, for more practical and reliable RS memories, it is very important to pick up the suitable combination of the electrodes and the RS matrices.

4.3.2 BOTTOM ELECTRODES

By having similar effects as of top electrode, such as work function, electronegativity, oxygen affinity, the interface reaction, and the interdiffusion, the crystallinity and crystal orientation of the RS film material are profoundly influenced by the bottom electrode that results in demonstrating considerably distinguishable RS properties.

Many studies revealed that the bottom electrode materials made of noble metals such as Au, Pt, Ru, etc. helped to suppress the interfacial reactions and the interdiffusion with the RS materials being possessing high-work function.

Very recently, Kim et al. [97] demonstrated significant improvement in the nickel nitride NiN-based RS device performances by using a nanopyramid patterned (NPP) Pt-bottom electrode as compared to flat Pt electrode. Their finding reveals that the sample fabricated on the NPP substrate and the conducting filaments composed of nitrogen vacancies are spontaneously formed between the electrodes during the fabrication process. Therefore, the device was initially less resistive and conducts the set and reset operations more easily by forming fewer conducting paths as compared with the use of flat-bottom electrodes. Almost similar type of behavior was reported by Liu et al. [98], where Cu nanocrystals (NCs) were introduced as the electric-field-concentrating initiator on a Pt-bottom electrode. The intensity of the electric field at the tip of the NC was assumed to be induced more than that in the planar region. As a result, conducting paths were uniformly formed and ruptured around the NC, thus improved the overall device performance.

In another work, Pham et al. [99] grew TiO_x thin films on different bottom electrodes such as FTO, Pt, and Ti. The reactivity of electrode materials with oxygen during TiO_x films' growth was suggested to be crucial in the observation of different switching behaviors. For the samples with inert Pt-bottom electrode, the threshold set voltage was lower than other films but had highest switching ratio, while the samples having active Ti electrode possess the largest set voltage and smallest memory window margin.

In summary, the bottom electrode materials are also key parameter for films growth which can strongly influence films RS performances.

4.4 SOLUTION PROCESSED TECHNIQUES TO FABRICATE RESISTIVE SWITCHING DEVICES

The RS processes in metal oxides seem to be limited to a small region in vertical directions, most likely near the interfaces between the electrodes and the metal oxide layer, which makes the use of nanometer scale (<50 nm) metal oxide films highly desirable. However, there is likely a barrier that the current optics-based lithography tools for making nanoscale features are prohibitively costly and complicated. As a result, sub-50-nanometer scale metal-oxide-based resistive devices were seldom reported. To fabricate RRAM devices on a sub-50-nanometer scale, it is necessary to form well-defined nanostructures such as nanodots, nanorods, and nanocubes by top-down or bottom-up approaches. However, the current hard-template-assisted process is not very convenient. Therefore, the design and fabrication of nanoscale building blocks that can interact anisotropically to form self-assembled structures without using hard templates remains challenging.

Current methods for the production of RS material thin films are based on the sequential deposition, patterning, and etching of selected semiconducting, conducting, and insulating materials. These sequential processes generally involve multiple photolithography and vacuum deposition processes at high temperature (e.g., 600°C), which contribute to their high manufacturing costs. Solution-processed methods offer the possibility of depositing RS layers using a bottom-up process that enables the fabrication of ultralow-cost sub-50-nm layers which would be able to exhibit high-performance RS characteristics.

In the following sections, we will focus on RS in metal oxide nanocrystalline thin films from solution-processed approaches, and the relationship between the phase, morphology, and composition of the nanocyrstalline thin films and their RS properties will be discussed in detail.

4.4.1 SOL–GEL

Sol–gel is one of the most exploited methods to produce thin film and powder catalysts. Its promise for the preparation of controlled and homogeneous compositions and for the fabrication of various forms such as powders, fibers, coatings, films, and monoliths has made it the potential answer for many of the problems which have been identified by people working with the processing of ceramics and glass. The technology has proved difficult to apply to some of the more classical sectors of ceramics such as powder and monolith preparation but notable successes have been achieved particularly where precise performance is required in components of limited dimensions (films, coatings, and fibers).

Many studies revealed that different variants and modifications of the process have been used to produce pure thin films or powders in large homogeneous concentration and under stoichiometry-control. In this process, oxides can be formed by network through polycondensation reactions of a molecular precursor in a liquid, which provides a good control over stoichiometry and reduced sintering temperature. The sol–gel technique has emerged as one of the most promising techniques to produce samples with good homogeneity at low cost.

In a study, Deshpande et al. [100] utilized sol–gel method to fabricate ZnO films for RS characteristics. Post-heat treatment was also carried out on the ZnO films to improve their crystallinity which ultimately results in their high R_{off}/R_{on}. Their findings also revealed the strong dependence of forming voltage on film thickness.

In another study, pure and Ag-doped ZrO_2 thin films on $Pt/Ti/SiO_2/Si$ substrates were also fabricated by sol–gel process for resistive random access memory (RRAM) application [101]. The Ag-doped films were found to exhibit better RS performance with high reproducibility and multi-bit storage capability over pure films. The improved RS behavior was attributed to Ag-doping effect, which promote the formation of the stable filamentary conducting paths during RS. Similar type of substrate material was used to fabricate sol–gel-derived pure and Mg-doped ZnO thin films by Cheng et al. [102] to study the RS characteristics.

In addition to binary metal oxides, sol–gel method was effectively utilized to fabricate perovskite-type oxides materials such as $SrTiO_3$ for RS characteristics [103].

In another study, Titanium isopropoxide (TIPP) film was spin coated on Pt-coated silicon substrate. Thermal annealing was carried out to transform TIPP film into TiO_2. Unipolar switching was observed. Set potential was recorded as 1.4 V, while reset potential was observed at 0.6V (as shown in Fig. 4.5).

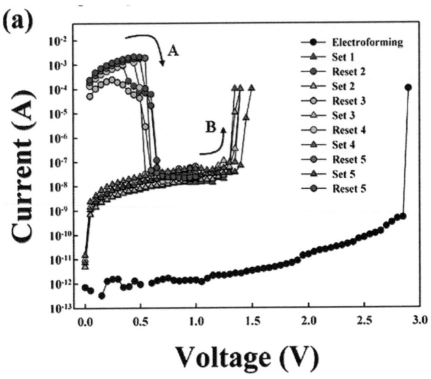

FIGURE 4.5 I-V curves of sol-gel TiO2 device showing repetitive switching cycles after initial electroforming process (reprinted with permission from Ref.[104]).

The reversible formation and annihilation of conducting filaments in sol–gel derived TiO_2 films were examined by conducting local current imaging through CS-AFM. The tip of AFM (Pt tip) was used instead of Ag electrode as top electrode. Randomly distributed conducting channels were observed during the electroforming, RESET and SET processes as shown in Figure 4.6. It was suggested that this observation may be attributed to the non-uniform current distribution between two electrodes. Moreover, the measured potential for reset and set process during CS-AFM experiments were higher than conventional IV measurements which may be due to the surface interfacial contamination on conducting AFM surfaces which cause additional energy barrier for the increasedvoltage thresholds.

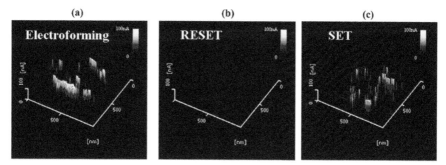

FIGURE 4.6 CS-AFM images of sol-gel-derived TiO$_2$ films after (a) electroforming at 10 V, (b) VRESET at 2.5 V and (c) VSET at 6 V (reprinted with permission from Ref. [104]).

The unipolar and bipolar RS behaviors were also reported in many sol–gel-derived materials, such as TiO$_2$ [104], Nb$_2$O$_5$ [105], ZnO [106], BiFeO$_3$ [107], which shows the great potential of this solution-based technique for diverse applications.

4.4.2 CHEMICAL BATH DEPOSITION

Among the many different methods available to deposit films of semiconductors, chemical bath deposition (CBD) must rank as conceptually the simplest. Being a soft solution route, alternative to conventional approaches for thin films deposition, CBD method has been widely used to deposit semiconducting and ceramic thin films in recent years [108]. CBD processes usually involve deposition of material on to an existing template via hydrolysis and condensation reactions of metal ions and complexes from aqueous solution [109]. CBD refers to depositions from solution (usually aqueous) where the required deposit is both chemically generated and deposited in the same bath. Control of size, shape, and crystallinity of the films is the fundamental topics for its practical applications, and the conditions of the solutions are controllable by adjusting the concentration of precursors, pH, and additives.

The method offers many advantages over other well-known vapor-phase synthetic routes. It may allow us to easily control the growth factors, such as film thickness, deposition rate, and quality of crystallites by varying the solution pH, temperature, and bath concentration. It does not require high-voltage equipment, works at room temperature, and hence it is inexpensive. The only requirement for this is an aqueous solution consisting of a few common chemicals and a substrate for the film to be deposited. It often suffers from a lack of reproducibility in comparison with other chemical processes; however, by the proper and careful optimization of the growth parameters, one can achieve reasonable reproducibility. There are few reports in which CBD method was deployed to fabricate thin films for the applications of RS memories.

Huang et al. [110] presented a homojunction diode and a memory device comprising vertically aligned ZnO_{1-x} nanorod layers (NRLs) coupled with ZnO thin films by using CBD and RF sputtering process on $Pt/Ti/SiO_2/Si$ substrate as shown in Figure 4.7.

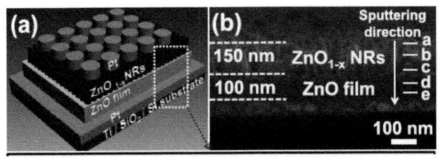

FIGURE 4.7 (a) Schematic of a Pt/ZnO_{1-x} NRs/ZnO TF/Pt resistive switching device. (b) Corresponding SEM image (reprinted with permission from Ref. [110]).

The bilayer structure expressed an excellent RS behavior, including a reduced operation power and high uniformity. The role of ZnO_{1-x} NR layer was suggested to a reservoir of oxygen vacancies which were responsible for enhanced device performances. Furthermore, the self-cleaning effect by having more hydrophobic surface was also observed in the bilayered structure demonstrating great potential of all ZnO-based system in diverse applications.

The major challenge encountered in these systems lies in doping metal ions especially in ZnO systems. It is well known that the addition of foreign ions into ZnO can be an important route for enhancing and controlling its optical, electrical, and magnetic performance. For example, ZnO doped with group III elements (e.g., Al, Ga, B) for improved electrical conductivity is regarded as the promising substitute of indium tin oxide [111]. Unfortunately, many previous studies have demonstrated the tendency for dopant ions to be excluded during crystal growth in solution. This is particularly true for ZnO films grown from aqueous solution, because it is difficult to control the kinetics in each individual reaction step which means simple addition of dopants to the solution hardly results in incorporation of dopants in ZnO crystal lattices. Therefore, an alternative way, electrodeposition has been proved to be successfully doping metal ions in ZnO matrix which shows that the solution processed technology still have great potential in the fabrication process.

4.4.3 ELECTROCHEMICAL DEPOSITION

Owing to its unique advantages of cost-effectiveness, reliability, and also the atom-by-atom replication on the given substrate surface profile; electrochemical process has been widely used for developing high-quality thin films. In addition, for diverse applications, electrochemical deposition allows us to easily alter both the bandgap and lattice constant by the modulation of composition by controlling growth parameters such as applied potential, pH, and bath temperature. By having such interesting features, the method of electrochemical deposition was widely deployed to fabricate various nanostructures for RRAM applications.

For example, a template-assisted electrodeposition synthesis of Au/Ni/Au nanowires was demonstrated as RS material by Lee et al. [112]. During synthesis process, an anodic aluminum oxide was used as a template followed by the thermal oxidation process to fabricate NiO. The thermal oxidation was process was found to cause Au diffusion into NiO nanowire matrix which ultimately trigger the unipolar RS phenomenon into NiO nanowire matrix. With the addition of Phosphorous into the interface between NiO and Au electrode, a transition from unipolar switching to bipolar switching was observed. Different conduction mechanisms including Ohmic conduction and FN-tunneling processes were proposed to explain the switching mechanisms in bundled and single nanowire architectures, respectively.

In another study, Koza and coworkers [113] also emphasized the versatility of electrochemical deposition process by fabricating VO_2 films for their applications as a RS material. The VO_2 film was thermally treated to observe the metal-to-insulator transition (MIT), which afterward found to exhibit RS behavior in the annealed films as shown in Figure 4.8. As the RS process was observed at very low voltages, therefore the local joule heating process that could drive MIT behavior in the presence of electric field was proposed to explain resistance transition behavior in these films.

Bipolar resistance switching behavior in electrochemically deposited TiO_x films was reported by Lee et al. [114]. The switching behavior of the prepared films was found to be strongly dependent on post-fabrication treatments. For example, the prepared films were post annealed in ambient O_2 and N_2 atmosphere to observe the chemical states of species in TiO_x films and their roles in switching performance. Their findings revealed that the reduction of non-lattice oxygen anions for annealing under O_2 was associated with the increase of Ti^{3+} in TiO_x films which strongly affect their RS performance.

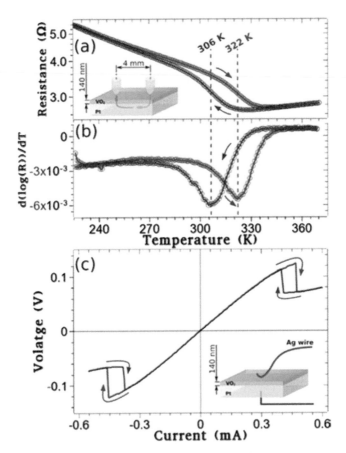

FIGURE 4.8   Electrical transport characterization of the annealed VO_2 film. (a) Temperature dependence of the normal to the film plane resistance showing the MIT temperatures upon heating (red) and cooling (blue) measured at 0.5 Kmin⁻¹. Transition temperatures were determined from (b) a plot of d(log(R))/dt vs. T. (c) I–V characteristic of the VO_2 film measured at room temperature showing reversible resistance switching between the low-temperature monoclinic (insulating) and high-temperature rutile (metallic) phases. The I–V curve was measured by scanning the current at a rate of 0.1 mAs⁻¹⁻¹ (reprinted with permission from Ref. [113]).

Recently, the enhanced RS performance of electrochemically deposited Co-doped ZnO thin films was demonstrated [115]. The Co doping in ZnO matrix was suggested to induce structural defects which might condense in the presence of electric field to form percolation paths between two electrodes; thus, a transition in device conduction states (from low to high conduction) was observed.

The process of electrochemical deposition have been effectively utilized in synthesizing 1D, 2D, and 3D-oriented metal oxide nanostructures for RS characterization. For instance 1D-oriented In_2O_3 nanorods were electrochemically deposited on FTO substrate for RS characteristics by Younis et al. [116]. To further tune the RS performances, Co doping was also carried out and a conducting filament-based model was proposed to explain the switching behavior in these devices. According to the proposed conducting filament model, the defects introduced by Co doping in In_2O_3 nanorods matrix may condense to form tiny conducting filaments. In the presence or absence of electric field, these tiny conducting filaments could accumulate and align themselves to generate long-conducting filaments within the electrodes (top and bottom). As a results, the device conductance state switch from high resistance to a LRS.

Recently, in another study, the inherent stochastic nature in 1D-oriented electrochemically deposited CeO_2 nanorods was explored by implementing weak programming conditions rather than to switch the device traditionally [117]. The traditional switching methods involve high fatigue testing that demand high programming voltages and long pulse durations to ensure the switching of the devices, which may cause reliability concerns. Also, there are few reports that state the random nature of filament formation and rupture process in RS memory devices. Therefore, to drag the randomness in filament formation process, weak programming tests were conducted in order to switch the device initial conduction state from high resistance to low resistance by imposing relatively much less set voltage than its actual value. The current was continuously monitored through the device to measure the time elapsed between the voltage application and the device to switch ON (called wait time). Interestingly, the wait time before switching shows plausible randomness as depicted in Figure 4.9.

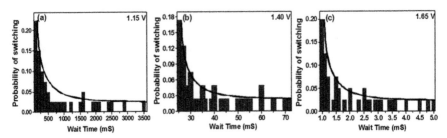

FIGURE 4.9   Random wait-time distribution for switching the device at (a) 1.15 V, (b) 1.40 V, and (c) 1.65 V (reprinted with permission from Ref. [117]).

2D nanosheets are considered to be excellent candidates for nanoelectronic applications as their electronic states play an important role in realizing the innovative electronic, optical, and magnetic functionalities. RS behavior of electrodeposited p-type Co_3O_4 nanosheets was recently reported by our group [118] (Fig. 4.10). The presence and distribution of oxygen vacancies through the film matrix was confirmed by chemical composition analysis. Furthermore, the nanosheet-based device was found to exhibit excellent and reliable RS behavior.

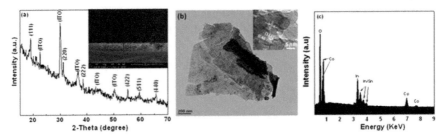

FIGURE 4.10   (a) X-ray diffraction pattern (inset, cross-sectional image), (b) TEM image of the mesoporous sheets (inset, HRTEM with lattice spacing), and (c) energy-dispersive X-ray spectroscopy (inset, surface morphology) (reprinted with permission from Ref. [118]).

A conducting filament-based model in which non-lattice oxygen and vacancies were drifted or diffused in the presence of electric field was proposed to explain the switching phenomenon in these devices.

In summary, the electrochemical approach is an effective and low-cost technique for producing high-quality thin films that could potentially be applied in diverse applications. Contrary to chemical and physical vapor-deposition methods, electrodeposition needs a substrate to be conducting. Although this can be a disadvantage for measuring the electrical properties of the films because of the substrate influence, it can be an advantage to selectively deposit film on an electrically poised conductor.

4.4.4 HYDROTHERMAL PROCESS

Hydrothermal process is defined as a heterogeneous reaction in the presence of aqueous solvents or mineralizers under high temperature and pressure conditions. In this process, the aqueous solution is heated up to a certain temperature to become supercritical fluids (SCFs), which are the moderate solvent and possess the characteristics of both gas and liquid solvents to provide a flexible condition for NC synthesis [119–121]. The dielectric constant of water decreases in high temperature that result the increase in solubility of organics. The function of organics at high temperature is crucial and the speed of hydrolysis can be slowed down in the SCFs, even in the alkaline environment, nanoparticles crystal growth can be also compressed. Besides, organic capping on the surface of nanoparticles can supply the repulsive force among each other when preparing dispersion, thus helped out to self-assemble the nanoparticles up to a distinct area.

Alike electrochemical deposition, the process of hydrothermal is also extensively utilized to synthesize well defined and morphology-controlled nanostructured films. Also, the use of this technique to fabricate RS materials ranging from 1D to 3D nanoarchitectures was proven from some reports.

For instance, Lai et al. [122] demonstrated hydrothermal synthesis of vertically aligned ZnO nanorods for RS characteristics. Furthermore, the presence of structural defects in ZnO nanorods matrix by copper doping was also observed which results to increase oxygen vacancies in the film. The bipolar RS characteristics were observed in both pure and Cu doped ZnO nanorods films. Due to excess of oxygen vacancies in doped films, the Cu-doped films expressed much more improved switching characteristics than undoped films in terms of distribution of set and reset voltages.

In another report, hydrothermally synthesized single VO_2 nanowire for nonvolatile memory operations was demonstrated [123]. Highly stable and multiple switchable resistance characteristics were controlled by imposing low biasing conditions, which was attributed to the inherent MIT nature of VO_2 material. Furthermore, local joule-heating effects were proposed to explain the resistance switchable effects in single VO_2 nanowire.

Recently, Sun et al. [124] demonstrated a white-light-assisted resistance switching and photovoltaic response in hydrothermally synthesized TiO_2/ZnO composite nanorods array as shown in Figure 4.11.

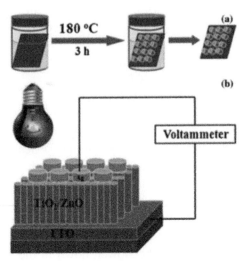

FIGURE 4.11   (a) The preparation process of TiO$_2$/ZnO composite nanorods array grown on FTO substrate. (b) The experimental test circuit (reprinted with permission from Ref. [124]).

The nanocomposite comprises TiO$_2$ nanorods array having average length and diameter as 3 μm and 200 nm, respectively, with ZnO nanoparticles embedded on TiO$_2$ nanorods matrix. The nanocomposite device was found to express enhanced bipolar RS characteristics and photovoltaic effects under white light operations.

In another report, a laterally bridged ZnO NR-based memory device with excellent RS characteristics by using a simple hydrothermal process was presented by Chuang et al. [125]. The ZnO NR bridge was formed by growing independently ZnO nanorods cluster across the gap between the Au electrodes as shown in Figure 4.12.

While testing the performance of the device, the memory device expressed a stable and rewritable memory performance with ultralow current level (less than pA range) at the HRS, and an on–off current ratio higher than 10^6. The region in between the boundaries between the opposite bridged ZnO NRs was proposed to be influential (high density of accumulated defects) in explaining switching characteristics.

In addition to the fabrication of 1D oriented nanoarchitectures, the hydrothermal process was also efficiently utilized to fabricate morphology-controlled 2D and 3D nanostructures for the applications for for RRAMs.

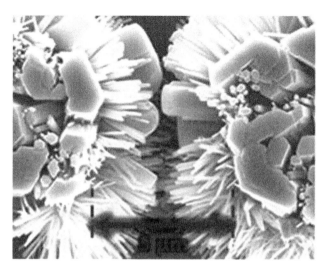

FIGURE 4.12   Typical FE-SEM plane view image of laterally bridged ZnO NRs (reprinted with permission from Ref. [125]).

Very recently, our group successfully presented hydrothermal synthesis of two-dimensionally designed $LaTiO_3$ nanosheets [126] for RRAM applications. In a typical synthesis process, solvothermally derived $LaTiO_3$ nanosheets dispersed into absolute ethanol was drop-coated into a gold-coated silicon substrate to form device structure. The device was found to exhibit excellent bipolar resistance switching characteristics with resistance ratio close to 100 with greater uniformity and stability at high temperatures.

The three dimensional nanostructures such as nanocubes can also be grown via solvothermal approach for RS characteristics. In a study, Chu et al. [127] presented facile synthesis of $BaTiO_3$ nanocubes for excellent RS characteristics in terms of ON/OFF ratio of 58–70, better reliability and stability over various polycrystalline $BaTiO_3$ nanostructures. In this study, the inter-cube junctions were considered responsible for such switching behavior and then a filament model was proposed to explain switching events in the $BaTiO_3$ nanocube-based RS device.

A similar type of detailed study was conducted for the hydrothermal synthesis of CeO_2 nanocubes and to identify them as a potential candidate for RS characteristics [128]. The interface engineered CeO_2 nanocubes architecture demonstrated excellent RS performances which are quite comparable to devices fabricated by complex and sophisticated fabrication techniques. Furthermore, the order of self-assembly in the CeO_2 nanocubes was also pointed as a critical factor to further tune the RS performances [129]. The presence of surfactant concentration in the suspension solution was high-lighted to play a key role in forming monolayer and self-assembled array composed of CeO_2 nanocubes as shown in Figure 4.13.

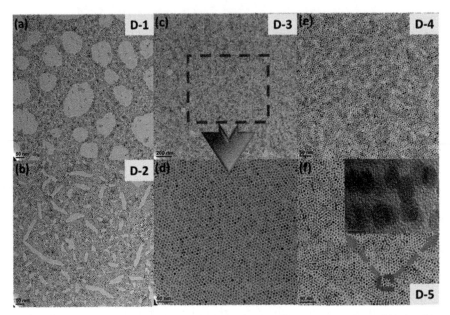

FIGURE 4.13   TEM images of CeO_2 nanocubes having OLA concentrations of (a) 0 vol.%, (b) 2 vol.%, (c) 3 vol.%, (e) 4 vol.%, and (f) 5 vol%. Image (d) shows the boxed area in panel (c) at high magnification. The inset in panel (f) is a high-magnification HRTEM image of the boxed region showing overlapping nanocubes (reprinted with permission from Ref. [129]).

Furthermore, the near perfect self-assembled architecture demonstrated great uniformity in switching parameters as compared to less self-assembled and/or overlapped nanostructures. The near perfect self-assembled nanocube-based RS device also expressed much faster writing (0.2 μs) and erasing (1 μs) response as compared to other devices as shown in Figure 4.14.

In summary, the fabrication of metal oxides or any other RS materials for RRAM applications are widely carried out by physical methods, such as radiofrequency magnetron sputtering, pulsed laser deposition, etc. These methods not only involve high fabrication cost but also limit the size and massive production. On the other hand, chemical methodologies, such as CBD and sol–gel, suffer from the problems of low crystallinity, disconnection of substrate and film, or high temperature calcinations. Compared with the aforementioned techniques, electrochemical deposition and hydrothermal methods provide an effective way to fabricate high-quality metal oxide thin films, at low temperatures and ambient atmosphere.

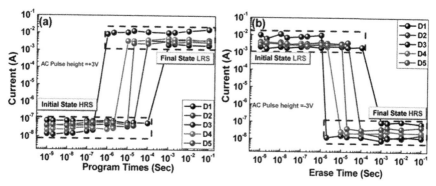

FIGURE 4.14   (a) Program and (b) erase characteristics for devices D-1–D-5 composed of CeO$_2$ nanocubes (reprinted with permission from Ref. [129]).

Moreover, in the process of electrodeposition, the deposition of metal oxide layers on the substrate is driven by the external electric field. Therefore, it is facile to precisely control the layer microstructure by this method and further to design heterostructures with novel functionalities. For instance, one can deposit constituent metal oxide layers with different compositions, defect concentrations, and crystal sizes. Also, a highly cost-effective method such as drop-coating could be efficiently be utilized to deposit films for hydrothermally prepared materials. Also by using these techniques, multilayered structures can also be formed which then effectively by utilized for diverse application areas, such as optoelectronics, spintronics, and nanoelectronics.

4.5   SUMMARY

In this book chapter, we have given an overview on the present understanding of the RS mechanisms in different types of RS devices presented so far. There are many more switching materials and mechanisms that surely are not covered in this chapter due to some reasons. The first of all is that a huge number of mechanisms which in terms of scientifically and technologically importance is not quite clear. The second reason is the lack of consistent experimental data reported in different publications. The result is various switching mechanisms for even the same switching materials. This may be due to different electroforming procedures or the stochastic nature of electroforming itself.

KEYWORDS

- **resistive random-access memories**
- **resistive switching**
- **organic compounds**
- **metal oxides**
- **electrochemical metallization process**

REFERENCES

1. Szot, K.; Dittmann, R.; Speier, W.; Waser, R. *Phys. Status Solidi* **2007,** *1,* R86.
2. Chua, L. *IEEE Trans. Circ. Theory* **1971,** 18 (5), 507–519.
3. Strukov, D. B.; Stewart, Snider, G. S.; D. R.; Williams, R. S. *Nature* **2008,** *80,* 453.
4. Liu, S. Q.; Wu, N. J.; Ignatiev, A. *Appl. Phys. Lett.* **2000,** *76,* 2749–2751.
5. Yang, J. J.; Pickett, M. D.; Li, X.; Ohlberg Douglas, A. A.; Stewart, D. R.; Williams, R. S. *Nat. Nano* **2008,** *3,* 429–433.
6. Wu, S. X.; Li, X. Y.; Xing, X. J.; Hu, P.; Yu, Y. P.; Li, S. W. *Appl. Phys. Lett.* **2009,** *94,* 253504.
7. Seo, J. W.; Park, J. W.; Lim, K. S.; Yang, J. H.; Kang, S. J. *Appl. Phys. Lett.* **2008,** *93,* 223505.
8. Wagenaar, J. J. T.; Morales-Masis, M.; van Ruitenbeek, J. M. *J. Appl. Phys.* **2012,** *111,* 014302.
9. Li, S.; Zeng, F.; Chen, C.; Liu, H.; Tang, G.; Gao, S.; Song, C.; Lin, Y.; Pan, F.; Guo, D. *J. Mater. Chem. C* **2013,** *1,* 5292.
10. Yang, X. Y.; Long, S. B.; Zhang, K. W.; Liu, X. Y.; Wang, G. M.; Lian, X. J.; Liu, Q.; Lv, H. B.; Wang, M.; Xie, H. W.; Sun, H. T.; Sun, P. X.; Sune, J.; Liu, M. *J. Phys. D: Appl. Phys.* **2013,** *46,* 245107.
11. Schindler, C.; Thermadam, S. C. P.; Waser, R.; Kozicki, M. N. *IEEE Trans. Electron Devices* **2007,** *54,* 2762.
12. Li, Y.; Zhong, Y. P.; Xu, L.; Zhang, J. J.; Xu, X. H.; Sun, H. J.; Miao, X. S. *Sci. Rep.* **2013,** *3,* 1619.
13. Sun, X. W.; Li, G. Q.; Zhang, X. A.; Ding, L. H.; Zhang, W. F. *J. Phys. D: Appl. Phys.* **2011,** *44,* 125404.
14. Sleiman, A.; Sayers, P. W.; Mabrook, M. F. *J. Appl. Phys.* **2013,** *113,* 164506.

15. Sun, X.; Sun, B.; Liu, L. F.; Xu, N.; Liu, X. Y.; Han, R. Q.; Kang, J. F.; Xiong, G. C.; Ma, T. P. *IEEE Electron Device Lett.* **2009,** *30,* 334.

16. Sun, J.; Liu, Q.; Xie, H. W.; Wu, X.; Xu, F.; Xu, T.; Long, S. B.; Lv, H. B.; Li, Y. T.; Sun, L. T.; Liu, M. *Appl. Phys. Lett.* **2013,** *102,* 053502.

17. Liu, D. Q.; Wang, N. N.; Wang, G.; Shao, Z. Z.; Zhu, X.; Zhang, C. Y.; Cheng, H. F. *J. Alloys Compd.* **2013,** *580,* 354.

18. Gao, X.; Xia, Y. D.; Ji, J. F.; Xu, H. N.; Su, Y.; Li, H. T.; Yang, C.; Guo, H. X.; Yin, J. A.; Liu, Z. G. *Appl. Phys. Lett.* **2010,** *97,* 193501.

19. Prakash, A.; Maikap, S.; Rahaman, S. Z.; Majumdar, S.; Manna, S.; Ray, S. K. *Nanoscale Res. Lett.* **2013,** *8,* 220.

20. Yoo, E. J.; Kim, J. H.; Song, J. H.; Yoon, T. S.; Choi, Y. J.; Kang, C. J. *J. Nanosci. Nanotechnol.* **2013,** *13,* 6395.

21. Kim, K. R.; Park, I. S.; Hong, J. P.; Lee, S. S.; Choi, B. L.; Ahn, J. *J. Korean Phys. Soc.* **2006,** *49,* S548.

22. Pantel, D.; Lu, H. D.; Goetze, S.; Werner, P.; Kim, D. J.; Gruverman, A.; Hesse, D.; Alexe, M. *Appl. Phys. Lett.* **2012,** *100,* 232902.

23. Yang, J. J. S.; Strukov, D. B.; Stewart, D. R. *Nat. Nanotechnol.* **2013,** *8,* 13.

24. Waser, R.; Dittmann, R.; Staikov, G.; Szot, K.; *Adv. Mater.* **2009,** *21,* 2632–663.

25. Almeida, S.; Aguirre, B.; Marquez, N.; McClure, J.; Zubia, D. *Int. Ferroelectr.* **2011,** *126,* 117.

26. Tsai, Y. T.; Chang, T. C.; Huang, W. L.; Huang, C. W.; Syu, Y. E.; Chen, S. C.; Sze, S. M.; Tsai, M. J.; Tseng, T. Y. *Appl. Phys. Lett.* **2011,** *99,* 092106.

27. Rozenberg, C. A.; Rozenberg, A. M. J. *J. Phys. Condens. Matter* **2009,** *21,* 045702.

28. Strukov, D. B.; Williams, R. S. *Appl. Phys., A: Mater. Sci. Proc.* **2009,** *94,* 515.

29. Yang, J. J.; Miao, F.; Pickett, M. D.; Ohlberg, D. A. A.; Stewart, D. R.; Lau, C. N.; Williams, R. S. *Nanotechnology* **2009,** *20,* 215201.

30. Menzel, S.; Waters, M.; Marchewka, A.; Böttger, U.; Dittmann, R.; Waser, R. *Adv. Funct. Mater.* **2011,** *21,* 4487.

31. Medeiros-Ribeiro, G.; Perner, F.; Carter, R.; Abdalla, H.; Pickett, M. D.; Williams, R. S. *Nanotechnology* **2011,** *22,* 095702.

32. Omura, Y.; Kondo, Y. J. *J. Appl. Phys.* **2013,** *114,* 043712.

33. Waser, R.; Aono, M. *Nat. Mater.* **1007,** *6,* 833.

34. Kwon, D. H.; Kim, K. M.; Jang, J. H.; Jeon, J. M.; Lee, M. H.; Kim, G. H.; Li, X. S.; Park, G. S.; Lee, B.; Han, S.; Kim, M.; Hwang, C. S. *Nat. Nanotechnol.* **2010,** *5,* 148.

35. Chen, J. Y.; Hsin, C. L.; Huang, C. W.; Chiu, C. H.; Huang, Y. T.; Lin, S. J.; Wu, W. W.; Chen, L. J. *Nano Lett.* **2013,** *13,* 3671.

36. Jang, W. L.; Lu, Y. M.; Hwang, W. S.; Dong, C. L.; Hsieh, P. H.; Chen, C. L.; Chan, T. S.; Lee, J. F. *EPL* **2011,** *96,* 37009.

37. Lee, D. Y.; Tseng, T. Y. *IEEE Electron Device Lett.* **2012,** *33,* 803.

38. Chen, Y. S.; Kang, J. F.; Chen, B.; Gao, B.; Liu, L. F.; Liu, X. Y.; Wang, Y. Y.; Wu, L.; Yu, H. Y.; Wang, J. Y.; Chen, Q.; Wang, E. G. *J. Phys. D: Appl. Phys.* **2012,** *45,* 065303.

39. Xu, Q. Y.; Wen, Z.; Shuai, Y.; Wu, D.; Zhou, S. Q.; Schmidt, H. *J. Superconduct. Novel Magn.* **2012,** *25,* 1679.

40. Lim, H.; Jang, H. W.; Lee, D.-K.; Kim, I.; Hwang, C. S.; Jeong, D. S. *Nanoscale* **2013,** *5,* 6363.

41. Russo, U.; Lelmini, D.; Cagli, C.; Lacaita, A. L. *IEEE Trans. Electron Devices* **2009,** *56,* 193.

42. Lelmini, D.; Nardi, F.; Cagli, C. *Nanotechnology* **2011,** *22,* 254022.

43. Yu, Q.; Liu, Y.; Chen, T. P.; Liu, Z.; Yu, Y. F.; Fung, S. *Electrochem. Solid State Lett.* **2011,** *14,* H400.

44. Wang, S. Y.; Lee, D. Y.; Tseng, T. Y.; Lin, C. Y. *Appl. Phys. Lett.* **2009,** *95,* 112904.

45. Hu, S. G.; Liu, Y.; Chen, T. P.; Liu, Z.; Yang, M.; Yu, Q.; Fung, S. *IEEE Trans. Electron Devices* **2012,** *59,* 1558.

46. Chen, A. *Appl. Phys. Lett.* **2010,** *97,* 263505.

47. Kim, K. M.; Choi, B. J.; Song, S. J.; Kim, G. H.; Hwang, C. S. *J. Electrochem. Soc.* **2009,** *156,* G213.

48. Kim, K. M.; Jeong, D. S.; Hwang, C. S. *Nanotechnology* **2011,** *22,* 254002.

49. Nagashima, K.; Yanagida, T.; Oka, K.; Kanai, M.; Klamchuen, A.; Kim, J. S.; Park, B. H.; Kawai, T. *Nano Lett.* **2011,** *11,* 2114.

50. Liu, T.; Verma, M.; Kang, Y. H.; Orlowski, M. K. *IEEE Electron Device Lett.* **2013,** *34,* 108.

51. Wu, S. M.; Tsuruoka, T.; Terabe, K.; Hasegawa, T.; Hill, J. P.; Ariga, K.; Aono, M. *Adv. Funct. Mater.* **2011,** *21,* 93.

52. Maikap, S.; Rahaman, S. Z. *ECS Trans.* **2012,** *45,* 257.

53. Schindler, C.; Weides, M.; Kozicki, M. N.; Waser, R. *Appl. Phys. Lett.* **2008,** *92,* 122910.

54. Liu, D. Q.; Zhang, C. Y.; Wang, G.; Shao, Z. Z.; Zhu, X.; Wang, N. N.; Cheng, H. F. *J. Phys. D: Appl. Phys.* **2014,** *47,* 085108.

55. Zhang, J. J.; Sun, H. J.; Li, Y.; Wang, Q.; Xu, X. H.; Miao, X. S. *Appl. Phys. Lett.* **2013,** *102,* 183513.

56. Nayak, A.; Ohno, T.; Tsuruoka, T.; Terabe, K.; Hasegawa, T.; Gimzewski, J. K.; Aono, M. *Adv. Funct. Mater.* **2012,** *22,* 3606.

57. van den Hurk, J.; Havel, V.; Linn, E.; Waser, R.; Valov, I. *Sci. Rep.* **2013,** *3,* 2856.

58. Wang, Z.; Griffin, P. B.; McVittie, J.; Wong, S.; McIntyre, P. C.; Nishi, Y. *IEEE Electron Device Lett.* **2007,** *28,* 14.

59. Choi, S. J.; Kim, K. H.; Park, G. S.; Bae, H. J.; Yang, W. Y.; Cho, S. *IEEE Electron Device Lett.* **2011,** *32,* 375.

60. Tsuruoka, T.; Hasegawa, T.; Terabe, K.; Aono, M. *Nanotechnology* **2012,** *23,* 435705.

61. Wang, Y.; Liu, Q.; Long, S. B.; Wang, W.; Wang, Q.; Zhang, M. H.; Zhang, S.; Li, Y. T.; Zuo, Q. Y.; Yang, J. H.; Liu, M. *Nanotechnology* **2010,** *21,* 045202.

62. Li, Y. T.; Long, S. B.; Lu, H. B.; Liu, Q.; Wang, Q.; Wang, Y.; Zhang, S.; Lian, W. T.; Liu, S.; Liu, M. *Chin. Phys. B* **2011,** *20,* 017305.

63. Lee, D.; Seong, D. J.; Jo, I.; Xiang, F.; Dong, R.; Oh, S.; Hwang, H. *Appl. Phys. Lett.* **2012,** *90,* 122104.

64. Long, S. B.; Liu, Q.; Lv, H. B.; Li, Y. T.; Wang, Y.; Zhang, S.; Lian, W. T.; Zhang, K. W.; Wang, M.; Xie, H. W.; Liu, M. *Appl. Phys. A* **2011,** *102,* 915.

65. Yan, X. B.; Li, K.; Yin, J.; Xia, Y. D.; Guo, H. X.; Chen, L.; Liu, Z. G. *Electrochem. Solid State Lett.* **2010,** *13,* H87.

66. Busani, T.; Devine, R. A. B. *J. Vac. Sci. Technol. B* **2008,** *26,* 1817.

67. Li, Y.; Zhao, G. Y.; Su, J.; Shen, E. F.; Ren, Y. *Appl. Phys. A* **2011,** *104,* 1069.

68. Zhuge, F.; Peng, S. S.; He, C. L.; Zhu, X. J.; Chen, X. X.; Liu, Y. W.; Li, R. W. *Nanotechnology* **2011,** *22,* 275204.

69. Tappertzhofen, S.; Valov, I.; Waser, R. *Nanotechnology* **2012**, *23*, 145703.

70. Valov, I.; Staikov, G. *Solid State Electrochem.* **2013**, *17*, 365.

71. Gao, S.; Zeng, F.; Chen, C.; Tang, G. S.; Lin, Y. S.; Zheng, Z. F.; Song, C.; Pan, F. *Nanotechnology* **2013**, *24*, 335201.

72. Chen, X. M.; Zhang, H.; Ruan, K. B.; Shi, W. Z. *J. Alloys Compd.* **2012**, *529*, 108.

73. Tang, X. W.; Zhu, X. B.; Dai, J. M.; Yang, J.; Chen, L.; Sun, Y. P. *J. Appl. Phys.* **2013**, *113*, 043706.

74. Shuai, Y.; Zhou, S.; Bürger, D.; Helm, M.; Schmidt, H. *J. Appl. Phys.* **2011**, *109*, 124117.

75. Yang, J. B.; Chang, T. C.; Huang, J. J.; Chen, S. C.; Yang, P. C.; Chen, Y. T.; Tseng, H. C.; Sze, S. M.; Chu, A. K.; Tsai, M. J. *Thin Solid Films* **2013**, *529*, 200.

76. Gao, S. Z. F.; Chen, C.; Tang, G. S.; Lin, Y. S.; Zheng, Z. F.; Song, C.; Pan, F. *Nanotechnology* **2013**, *24*, 335201.

77. Park, J.; Biju, K. P.; Jung, S.; Lee, W.; Lee, J.; Kim, S.; Park, S.; Shin, J.; Hwang, H. *IEEE Electron Device Lett.* **2011**, *32*, 476.

78. Yang, J. J.; Pickett, M. D.; Li, X.; Ohlberg, D. A. A.; Stewart, D. R.; Williams, R. S. *Nat. Nanotechnol.* **2008**, *3*, 429.

79. Wu, D.; Jiang, Y.; Yu, Y. Q.; Zhang, Y. G.; Li, G. H.; Zhu, Z. F.; Wu, C. Y.; Wang, L.; Luo, L. B.; Jie, J. S. *Nanotechnology* **2012**, *23*, 485203.

80. Sawa, A. *Mater. Today* **2008**, *11*, 28.

81. Xu, Z. T.; Jin, K. J.; Gu, L.; Jin, Y. L.; Ge, C.; Wang, C.; Guo, H. Z.; Lu, H. B.; Zhao, R. Q.; Yang, G. Z. *Small* **2012**, *8*, 1279.

82. Zhou, Q. G.; Zhai, J. W. In *2012 International Workshop on Information Storage and Ninth International Symposium on Optical Storage*, Shanghai, China, 2013, Vol. 8782; p 87820E.

83. Wang, W. H.; Dong, R. X.; Yan, X. L.; Yang, B.; An, X. L. *IEEE Trans. Nanotechnol.* **2012**, *11*, 1135.

84. Rubi, D.; Gomez-Marlasca, F.; Bonville, P.; Colson, D.; Levy, P. *Physica B* **2012**, *407*, 3144.

85. Lee, H. S.; Choi, S. G.; Choi, H. J.; Chung, S. W.; Park, H. H. *Thin Solid Films* **2013**, *529*, 347.

86. Choi, S. J.; Park, G. S.; Kim, K. H.; Yang, W. Y.; Bae, H. J.; Lee, K. J.; Lee, H. I.; Park, S. Y.; Heo, S.; Shin, H. J.; Lee, S. Cho, S. *J. Appl. Phys.* **2011**, *110*, 056106.

87. Zou, X.; Ong, H. G.; You, L.; Chen, W. G.; Ding, H.; Funakubo, H.; Chen, L.; Wang, J. L. *AIP Adv.* **2012**, *2*, 032166.

88. Sun, J.; Jia, C. H.; Li, G. Q.; Zhang, W. F. *Appl. Phys. Lett.* **2012**, *101*, 133506.

89. Huang, H.-H.; Shih, W.-C.; Lai, C.-H. *Appl. Phys. Lett.* **2010**, *96*, 193505.

90. Sawa, A.; Kawasaki, M. T. F.; Tokura, Y. *SPIE 2005*, 59322C.

91. Kim, C. J.; Chen, I. W. *Thin Solid Films* **2006**, *515*, 2726.

92. Sim, H.; Choi, H.; Lee, D.; Chang, M.; Choi, D.; Son, Y.; Lee, E. H.; Kim, W.; Park, Y.; Yoo, I. K.; Hwang, H. Excellent Resistance Switching Characteristics of $Pt/SrTiO_3$ Schottky Junction for Multi-bit Nonvolatile Memory Application. In: *IEDM Tech. Dig.* 2005; p 777.

93. Seo, S.; Lee, M. J.; Kim, D. C.; Ahn, S. E.; Park, B.-H.; Kim, Y. S.; Yoo, I. K.; Byun, I. S.; Hwang, I. R.; Kim, S. H.; Kim, J.-S.; Choi, J. S.; Lee, J. H.; Jeon, S. H.; Hong, S. H.; Park, B. H. *Appl. Phys. Lett.* **2005**, *87*, 263507.

94. Shang, D. S.; Chen, L. D.; Wang, Q.; Zhang, W. Q.; Wu, X. M.; Li, X. M. *Appl. Phys. Lett.* **2006,** *89,* 172102.

95. Lee, H. S.; Bain, J. A.; Choi, S.; Salvador, P. A. *Appl. Phys. Lett.* **2007,** *90,* 202107.

96. Lin, C. Y.; Wu, C. Y.; Wu, C. Y.; Tseng, T. Y.; Hu, C. *J. Appl. Phys.* **2007,** *102,* 094101.

97. Hee-Dong, K.; Min Ju, Y.; Seok, H. M.; Tae, G. K. *Nanotechnology* **2014,** *25,* 125201.

98. Liu, Q.; Long, S.; Lv, H.; Wang, W.; Niu, J.; Huo, Z.; Chen, J.; Liu, M. *ACS Nano* **2010,** *4,* 6162.

99. Pham, K. N.; Nguyen, T. D.; Ta, T. K. H.; Thuy, K. L. D.; Le, V. H.; Pham, D. P.; Tran, C. V.; Mott, D.; Maenosono, S.; Kim, S. S.; Lee, J.; Pham, D. T.; Phan, B. T. *Eur. Phys. J.—Appl. Phys.* **2013,** *64,* 30102.

100. Deshpande, S.; Nair, V. V. Resistive Switching of Al/Sol–Gel ZnO/Al Devices for Resistive Random Access Memory Applications, Advances in Computing, Control, & Telecommunication Technologies, 2009. In: ACT '09. International Conference on, 28–29 Dec. 2009, 2009; pp 471–473.

101. Sun, B.; Liu, L.-F.; Han, D. D.; Wang, Y.; Liu, X.-Y.; Han, R.-Q.; Kang, J.-F. *Chin. Phys. Lett.* **2008,** *25,* 2187.

102. Cheng, H.-C.; Chen, S.-W.; Wu, J.-M. *Thin Solid Films* **2011,** *519,* 6155.

103. Tang, M. H.; Wang, Z. P.; Li, J. C.; Zeng, Z. Q.; Xu, X. L.; Wang, G. Y.; Zhang, L. B.; Xiao, Y. G.; Yang, S. B.; Jiang, B.; He, J. *Semin. Sci. Technol.* **2011,** *26,* 075019.

104. Chanwoo Lee, Inpyo Kim, Wonsup Choi, Hyunjung Shin, and Jinhan Cho. *Langmuir.* **2009,** *25,* 4274.

105. Lee, C.; Kim, I.; Choi, W.; Shin, H.; Cho, J. *Langmuir* **2009,** *25,* 4274.

106. Baek, H.; Lee, C.; Choi, J.; Cho, J. *Langmuir* **2012,** *29,* 380.

107. Kim, S.; Moon, H.; Gupta, D.; Yoo, S.; Choi, Y. K. *IEEE Tran. Electron. Dev.* **2009,** *56,* 696.

108. Yin, K.; Li, M.; Liu, Y.; He, C.; Zhuge, F.; Chen, B.; Lu, W.; Pan, X.; Li, R.-W. *Appl. Phys. Lett.* **2010,** *97,* 043102.

109. Govender, K.; Boyle, D. S.; Kenway, P. B. O'Brien, P. *J. Mater. Chem.* **2004,** *14,* 2575.

110. Huang, C.-H.; Huang, J.-S.; Lin, S.-M.; Chang, W.-Y.; He, J.-H.; Chueh, Y.-L. *ACS Nano* **2012,** *6,* 8407.

111. Suzuki, A.; Nakamura, M.; Michihata, R.; Aoki, T.; Matsushita, T.; Okuda, M. *Thin Solid Films* **2008,** *517,* 1478.

112. Saeeun, L.; Donguk, K.; Hyeonjin, E.; Woo-byoung, K.; Bongyoung, Y. *Jpn. J Appl. Phys.* **2014,** *53,* 024202.

113. Koza, J. A.; He, Z.; Miller, A. S.; Switzer, J. A. *Chem. Mater.* **2011,** *23,* 4105.

114. Lee, S.; Na, H.; Kim, J.; Moon, J.; Sohn, H. *J. Electrochem. Soc.* **2011,** *158,* H88.

115. Chu, D.; Younis, A.; Li, S. *ISRN Nanotechnol.* **2012,** *2012,* 1.

116. Younis, A.; Chu, D.; Li, S. *RSC Adv.* **2013,** *3,* 13422.

117. Younis, A.; Chu, D.; Li, S. *Appl. Phys. Lett.* **2013,** *103,* 253504.

118. Younis, A.; Chu, D.; Lin, X.; Lee, J.; Li, S. *Nanoscale Res. Lett.* **2013,** *8,* 1.

119. Shah, P. S.; Hanrath, T.; Johnston, K. P.; Korgel, B. A. *J. Phys. Chem. B* **2004,** *108,* 9574.

120. Ziegler, K. J.; Doty, R. C.; Johnston, K. P.; Korgel, B. A. *J. Am. Chem. Soc.* **2001,** *123,* 7797.

121. Weingärtner, H.; Franck, E. U. *Angew. Chem., Int. Ed.* **2005,** *44,* 2672.

122. Lai, Y.; Wang, Y.; Cheng, S.; Yu, J. L. *J. Electron. Mater.* **2014,** *43,* 2676.

123. Bae, S.-H.; Lee, S.; Koo, H.; Lin, L.; Jo, B. H.; Park, C.; Wang, Z. L. *Adv. Mater.* **2013,** *25,* 5098.

124. Sun, B.; Zhao, W.; Liu, Y.; Chen, P. *J. Mater. Sci.: Mater. Electron.* **2014,** *25,* 4306.

125. Chuang, M.-Y.; Chen, Y.-C.; Su, Y.-K.; Hsiao, C.-H.; Huang, C.-S.; Tsai, J.-J.; Yu, H.-C. *ACS Appl. Mater. Interfaces* **2014,** *6,* 5432.

126. Lin, X.; Younis, A.; Xiong, X.; Dong, K.; Chu, D.; Li, S. *RSC Adv.* **2014,** *4,* 18127.

127. Chu, D.; Lin, X.; Younis, A.; Li, C. M.; Dang, F.; Li, S. *J. Solid State Chem.* **2014,** *214,* 38.

128. Younis, A.; Chu, D.; Mihail, I.; Li, S. *ACS Appl. Mater. Interfaces* **2013,** *5,* 9429.

129. Younis, A.; Chu, D.; Li, C. M.; Das, T.; Sehar, S.; Manefield, M.; Li, S. *Langmuir* **2014,** *30,* 1183.

CHAPTER 5

ZnO THIN FILMS AND NANOSTRUCTURES FOR ACOUSTIC WAVE-BASED MICROFLUIDIC AND SENSING APPLICATIONS

HUA-FENG PANG[1,2], J. K. LUO[3], and Y. Q. FU[2*]

[1]*Department of Applied Physics, School of Science, Xi'an University of Science and Technology, Xi'an, PR China*

[2]*Faculty of Engineering and Environment, Northumbria University, Newcastle upon Tyne NE1 8ST, United Kingdom*

[3]*Centre for Material Research and Innovation, University of Bolton, Deane Road, Bolton BL3 5AB, United Kingdom*

[*]*Corresponding author. E-mail: Richard.fu@northumbria.ac.uk*

CONTENTS

ABSTRACT

Progress in ZnO thin films and nanostructures for the acoustic wave micro-fluidic and sensing applications are reviewed in this chapter. ZnO thin films with good piezoelectric properties possess large electromechanical coupling coefficients and can be fabricated for the surface acoustic wave (SAW) and film-bulk acoustic resonator (FBAR) devices with a good acoustic perfor-mance. The SAWs can be excited to mix, stream, pump, eject, and atomize the liquid, and precision sensing can be performed using SAWs and FBARs. Therefore, the ZnO SAW devices are attractive to be integrated into a lab-on-chip system where the SAWs can transport bio-fluids to the desired area, mix the extracted DNA or proteins, and detect the changes of the signals using SAWs or FBARs. The ZnO SAW and FBAR devices in combination with different sensing layers could also be used to successfully detect gas, UV light, and biochemicals with remarkable sensitivities.

5.1 INTRODUCTION

Zinc oxide (ZnO) is a binary compound via a covalent bonding between the transition-metal zinc atom and oxygen atom. ZnO thin films and nanostruc-tures are multifunctional materials, which have attracted much attention from as early as 1930s until today due to various fundamental electronic, chemical, physical, and optical properties and applications [1–3]. In recent years, the development of new growth technologies of ZnO thin films and nanostructures and their new applications has renewed lots of interest on obtaining the high-quality thin films and single crystals [4–7], and further investigation of their growth mechanism, band structures, excitons, and deep centers in luminescence, nonlinear optics, and UV lasing [8,9]. With in-depth understanding of the semiconducting, optical, electronic, piezoelectric, and pyroelectric properties, ZnO has now been widely applied for microfluidics, optoelectronics, piezotronics, sensors, solar cell, and actuators [10–12]. These immense applications also have boosted the extensive researches on the fundamentals and growth techniques of the ZnO-based materials.

Microfluidics is focused as one of the important applications, which is an interdisciplinary science of controlling and manipulating the flow of liquids typically at micron and submicron dimensions in a miniaturized system and the corresponding technologies for such systems. This multidisciplinary technology is comprehensively based on physics, nanotechnology, biotech-nology, chemistry, and electronic engineering. Microfluidics is crucial to the developments of the inkjet print-heads, DNA chips, and lab-on-a-chip (LOC) technologies [13]. The typical flow of liquid with very small volume is conveniently handled through generating, transporting, separating, mixing, nebulizing, and heating in the microfluidic system [14]. This endows the microfluidics with distinctive advantages such as low fluidic volume consumption, high-throughput, compactness, high sensitivity, fast response, high-speed processing, and low energy consumption [12,14–17]. Therefore, a rapid increase in the development of new methods to modulate fluid flow at microscale has been supported with the urgent needs from healthcare, medical research, life science, drug-development sectors. However, the microscale fluids obviously differ with the flow features of the conventional large fluid with macro scale due to differences in their surface tension, energy dissipation, and fluidic resistance, etc. Some physical and chemical effects and phenomena become dominant at a microscale level, such as capillary forces, surface roughness, and undesired chemical interactions. These may be difficult to accurately predict and design, which results that the fabrication processes of these microsystems is more complex and difficult than those for the conventional ones. Various technologies, including eletrokinetics, electrowetting, acoustic wave technology, etc. have been integrated into the LOC microfluidic systems [15], which are used to overcome above problems. The planar chip which combines the surface acoustic wave (SAW) technique into the droplet microfluidics is more efficient and attractive than the conventional channel-based chip [16]. Therefore, the SAW microfluidics become an important and fundamental field of the enormous researches on the programmable microfluidic chip [17,18].

Acoustic wave is generated from the piezoelectric materials using electric fields applied onto the electrodes of the acoustic wave device. The features of the acoustic waves are determined by the propagation ways because of the piezoelectricity and the boundaries in the materials. Different acoustic waves can be categorized into SAW, bulk acoustic wave, shear-horizontal wave, acoustic plate mode wave, Love wave and Lamb wave, etc. [19]. The SAW techniques offer simpler and more compact devices without moving parts when the acoustic wave is designed to interact with the fluids, and SAW microfluidics has been extensively studied in recent years because of considerable interest and developments on SAW technology in the past decades. The microfluidic coupling among the fluid, acoustic wave, and microstructure is modulated to be more efficient, reliable, and controllable for the integrated LOC microfluidic device [14]. Compared with bulk piezoelectric materials (e.g., quartz, lithium tantalite, and lithium niobate), thin films (e.g., ZnO, aluminum nitride, and lead zirconium titanate or PZT) have advantages of simple, inexpensive and large-scale production, and flexibile integration with the microfluidic structures and controller circuits. Therefore, ZnO-based microfluidics is recently presented as one of the key the main applications for ZnO thin films and nanostructures.

A sensor is a device or instrument that converts the physical/chemical/ electrical/mechanical quantities into visible or readable signals. It normally consists of three units, including the input port, sensing unit, and output port, with a functional relationship between the input and output quantities in a form of electrical or optical signals. The sensor always appears as a probe device that widely exists around our world, covering natural sensors in living organisms such as eye, nose, and ear, and the artificial sensors including biochemical sensors, gas sensors, physical sensors such as humidity and temperature sensors, pressure sensor, and viscometers [20,21]. The sensor technologies have made a remarkable leap in the last a few decades owing to the development of micro-electromechanical systems (MEMS) in micro-electronic engineering [20]. This allows multiple sensors to be manufactured at micro or nanoscale as microsensors or nanosensor, which can reach a significantly higher selectivity and sensitivity compared with the macroscopic sensors. Take thin-film bulk acoustic resonator (TFBAR) for instance, it operates with a frequency in the range of GHz and offers a high sensitivity to the variations of mass load [22]. Tremendous advances and latest technologies of the sensor structure, manufacturing technology, and signal-processing algorithms have been incorporated into micro and nano-sensors and wireless sensor networks [23]. The sensors have now been broadly applied as an integral part in medical diagnostics, chemical, and biological recognition systems, health care, automobile and industrial manufacturing, and environmental monitoring. Among the various sensing materials, ZnO thin films and nanostructures have been widely used for designing and developing of the sensors due to their high sensitivity to the physical, chemical, and biological environment [21].

This chapter will provide an overview of ZnO thin films and nanostructures and their in acoustic wave devices used for microfluidics and sensors applications.

5.2   ZnO THIN FILMS AND NANOSTRUCTURES

ZnO is an "old" semiconductor material, which have been studied for 80 years. The renewed interest is fueled by availability of the new findings on the fundamentals of ZnO and novel applications. ZnO can stably be crystallized in cubic zinc blende or hexagonal wurtzite structure. The rochsalt structure of ZnO only exists in relatively high pressures. Different crystallized states of ZnO include thin films, nanostructured crystals, and single crystals. However, much attention has been paid for ZnO materials with low dimensions due to the increasing demand of the miniaturized devices, which leads to thorough and extensive investigations on the ZnO thin films and nanostructures. In this section, fundamentals, growth, and deposition techniques of ZnO thin films and nanostructures will be briefly presented; and high-quality ZnO thin films and nanostructures are discussed for acoustic wave devices and sensing applications.

5.2.1 FUNDAMENTALS OF ZnO

The electronic band structure of ZnO is one of the basic properties and has been investigated through the theoretical calculations and experimental determination. The band structure calculation was first proposed using a Greens function method in 1969 [24]. Later, density functional theory method was used. A band gap of 3.77 eV was obtained between the valence band maxima and the conduction band minima at Γ point of the Brillouin zone using a local density approximation (LDA) and atomic self-interaction corrected pseudopotentials [25]. However, simply using the LDA could underestimate the band gap for 0.2 eV [26]. Experimentally, the band gap of the ZnO is 3.37 eV at room temperature and 3.44 eV at low temperatures, which is significantly dependent on the temperature and pressure due to the change of the lattice constants. The Varshni's empirical relation between the band gap and temperature can be written as follows [27]:

$$E_g = E_{g0} - \frac{\alpha T^2}{T + \beta} \tag{5.1}$$

where E_{g0}, α, and β correspond to the transition energy at 0 K, a temperature fitting coefficients of -5.5×10^{-4} eV·K^{-1} and -900 K, respectively. Such a direct and wide band gap of ZnO is beneficial to the optoelectronic applications in the blue and UV regions, including light-emitting diodes, UV laser, and photodetectors that will be discussed later in this chapter. In order to get a lager band gap of ZnO, band-structure engineering has been adopted by doping or alloying with MgO and CdO, which has been considered as an alternative to the wurtzite gallium nitride (GaN)-based optical devices [28,29]. In addition, the high exciton binding energy of 60 MeV ensures efficient excitonic emission of optoelectronics based on excitonic effects [30], and the laser worked with exciton transition is expected.

Piezoelectric effect is a reversible process that exhibits a linear electromechanical interaction between the mechanical and the electrical state in crystalline materials. Piezoelectricity was discovered by French physicists Jacques and Pierre Curie [31]. As an important characteristic of hexagonal wurtzite ZnO, piezoelectricity originates from the polarity that is composed of tetrahedral coordination. The direction of the polarity is along the c-axis from cation to anion, which results in the primary polar plane (0 0 0 1) with the lowest energy. When the external mechanical stress is applied and induces lattice distortion of the wurtzite ZnO materials, the centers of Zn cation and O anion are displaced in the noncentrosymmetric structure and local dipole moments are formed. Accordingly, piezoelectricity along the [0 0 0 1]-direction appears due to the macroscopic polarization in the ZnO crystal. A large electromechanical coupling of k^2 ranging from 1% to 5.2% can be obtained due to the highest piezoelectric tensor of the tetrahedrally bonded ZnO in the II–VI compounds with wurtzite structure [32,33].

In the practical applications, good piezoelectricity requires that the ZnO single crystal or thin film possesses a strong texture, low defects, an accurate stoichiometric ratio of Zn atoms to O atoms, a smooth surface with a low roughness, and an appropriate thickness. Various technologies and methods have been developed to obtain high-quality piezoelectric ZnO materials [34–37]. For instance, in order to increase the piezoelectric constant, different transition-metal atoms (e.g., Fe, V, Cr) doped in ZnO have been reported, and the piezoelectric properties have been significantly improved [34,35]. The piezoelectric properties can also be tailored by tuning the Mg composition in Mg-doped ZnO [36,37]. The size reduction of the ZnO materials to nanoscale can enhance the piezoelectricity. Recent advance on the theoretical computations using the first-principles method has showed that the effective piezoelectric constant of ZnO nanowire is much larger than that of bulk ZnO material due to their free boundary [38]. Giant piezoelectric size effects in the ZnO nanowire were also reported, and the piezoelectric coefficient of 50.4 C m^{-2} can be obtained when the diameter of the nanowires was reduced to 0.6 nm calculated using density functional theory [39].

The optical properties of the ZnO include luminescence and photoconductivity, as well as the refractive index and absorption index. The luminescence of ideal ZnO crystals only refers to the intrinsic near-band-edge emission at the UV region. The free exciton emission can be observed at a low temperature (e.g., 2–10 K) [40]. A number of studies on the free or bound excitons were performed to observe the low-temperature luminescence [41–43].

However, the defects in ZnO materials are inevitable because of the growth conditions during the process of the synthesis and preparation. These result in diverse characteristics of the luminescence at room temperature such as the blue emission, green emission, red emission as well as UV emission. The different forms of the photoluminescences are possibly originated from the oxygen vacancies, zinc interstitials, zinc vacancies, or doubly ionized oxygen vacancies, as well as the free excitons and intrinsic transition [41–43]. Taking the blue emission in ZnO for instance, it was assigned to the transitions of electrons from the shallow donor of oxygen vacancy to the valence band and from the conduction band to the acceptor of zinc vacancy [44]. The later transition was argued that it might not exist due to fewer creations of zinc vacancies in the sputtered ZnO thin film [45]. The interstitial-zinc-related defects were also considered to contribute to the blue emission [46]. The luminescence is heavily dependent on the defects that are sensitively varied in the different growth conditions. The intrinsic transition near band-edge for the high-quality ZnO materials can be used to demonstrate the UV lasing under optical pump conditions [47–49]. Experimental and theoretical investigations have provided an insight into the formation mechanism of those different defects, by combining the luminescence spectroscopy with electron paramagnetic resonance measurements and the theoretically accurate computations [9,50]. The formation of the different defects depends on the growth conditions. Therefore, control of the defect formations in the ZnO becomes an important field in engineering and development of ZnO materials.

Photoconductivity and surface conductivity of ZnO thin film and nanostructures are sensitive to the exposure of the surface to the light and absorbed molecules or atoms, respectively. Photoconductivity is a phenomenon in which the electrical conductivity depends on the absorption of the electromagnetic radiation. Photoconductivity of ZnO, which was first observed by Mollow and Miller, is now extensively investigated [51,52]. Different ZnO materials including thin film, nanostructures, and single crystals have been involved in the evaluation of the performance of photoconductivity. The photoresponse of ZnO changes from tens of microseconds to a few minutes because of the different transitions governed by the intrinsic inter-band or excitonic transition, the surface-related oxygen adsorption/desorption process, or the recombination process [53–56]. Normally, the photoresponse induced by the intrinsic inter-band transition is much faster than those dominant by the oxygen adsorption and the recombination. The different mechanisms of the photoconductivity have been proposed such as the hole-capture model and exciton transition in ZnO, as well as the oxygen adsorption/

desorption for ZnO materials with low dimensions [57,58]. Therefore, high-quality ZnO thin films and nanocrystals are the basic requirement for the high-speed photosensor with a fast response.

The surface conductivity of ZnO is changed with the charge accumulation near the surface due to the band bending when the charge transfer occurs after absorbing various gaseous molecules or atoms. The surface conductivity of ZnO significantly depends on the exposure of the gas and prone to be changed. Surface conductivity of ZnO was investigated by annealing and adsorption of atomic hydrogen or of oxygen in an ultrahigh vacuum chamber [59]. A layer of the surface electrons accumulated on the annealed single crystals was confirmed that could exist in vacuum and disappear in ambient air [60]. The surface defects formed by chemical bonding or physical treatment also play important roles on the surface conductivity, which can increase or hinder the charge transfer [61–63]. The appearance of surface states normally reduces the mobility of the carriers and lead to the decrease in surface conductivity [64]. Furthermore, the research on the surface conductivity is helpful to understand the mechanism of ZnO-based sensors discussed in the later sections.

Owing to the main focus of this review on microfluidic and sensing applications, some related properties of ZnO such as nonlinear optics, pyroelectricity, and thermal conductivity are not discussed here, which can be obtained from the other references [65,66].

5.2.2 ZnO THIN FILMS

ZnO thin film is a layered material with thicknesses ranging from a few nanometers (monolayer) to several micrometers. Lots of the studies on ZnO-thin films mainly focus on five topics, including the properties, growth mechanism, methods of the preparation, preparation technologies, and device applications. The crystalline properties, including the texture, orientation, microstructure, morphology, stress, adhesion, substrate, and defects, are intrinsic to piezoelectric and sensing performance. However, the growth conditions significantly influence on the growth dynamics; therefore, in order to acquire high-quality ZnO thin films, the growth parameters are needed to be optimized. This preliminary work is thoroughly performed before the application of ZnO thin films to fabricate high-performance ZnO-based devices [12,33]. The improvements of the deposition techniques, using physical vapor deposition (PVD), chemical vapor deposition (CVD), and wet chemical method, offer a better control of the crystallinity, piezoelectric, and electrical properties of ZnO thin films. A reproducible, stable, and robust process is expected for the large-scale production of ZnO thin films with a high quality. Therefore, the crystallinity of ZnO thin films and how to grow ZnO thin films using different deposition techniques are presented and discussed in the following sections.

5.2.2.1   CRYSTALLINE CHARACTERISTICS

Texture in ZnO thin films normally refers to the distribution of the crystallographic orientations. It is changed from nontexture in the polycrystalline thin film to perfect texture in single crystals due to the different growth processes. ZnO thin films with a strong texture can be prepared using various growth techniques including the magnetron sputtering, CVD, pulsed laser deposition (PLD), thermal evaporation, atomic layer deposition, molecular beam epitaxy (MBE), wet chemical method, etc. In a highly textured ZnO thin film, the preferred orientation is commonly along the [0 0 0 2] direction in a hexagonal wurtzite structure, and the cross-sectional microstructure is generally columnar or rod-like as shown in Figure 5.1a, whereas nontextured polycrystalline ZnO thin films consist of particle-like fine grains in microstructure. Furthermore, the film texture and microstructure determine the hardness, stress, elastic, and piezoelectric properties. Therefore, control and tailoring the crystalline texture of ZnO thin films on demand is of critical importance toward the device-based applications.

Inclined ZnO thin films with a tilted columnar microstructure, as shown in Figure 5.1b, can be deposited to allow the excitation of a novel mode wave (e.g., shear wave or dual mode waves) in the acoustic devices [67–70]. Common methods of varying the substrate tilt angle or the angle between the substrate and target have been proposed to acquire above ZnO thin films [71–73]. The c-axis zig-zag ZnO thin films have been grown in a multilayered structure using SiO$_2$ buffer layer, which was used to control the generation of the shear waves [74,75]. Shear waves with suppressed longitude wave can be excited in such thin films when the inclined angle of the ZnO thin film approaches a special value.

Substrates and interlayer (or buffer layer) are considered as one important factor to prepare high-quality ZnO thin films. Various substrates such as silicon, glass, fused silica, sapphire, diamond, metal, MgO, and flexible polymer have been applied in the deposition of ZnO thin films [76,77]. The substrate can directly determine the lattice mismatch and thermal expansion mismatch, which results in different microstructures, morphologies and strains in the deposited ZnO thin films. The c-axis-oriented ZnO thin films generally grow faster on the special crystallographic surfaces than those on amorphous substrates [77]. In order to improve the film quality, a buffer layer of ZnO has been used as a seed layer to direct the growth orientation [78,79]. Interlayers of silicon carbide (SiC), GaN, diamond, nanocrystalline diamond (NCD) or diamond-like carbon (DLC) on Si substrates has been used to enhance the texture and crystallinity of ZnO thin films [80–83].

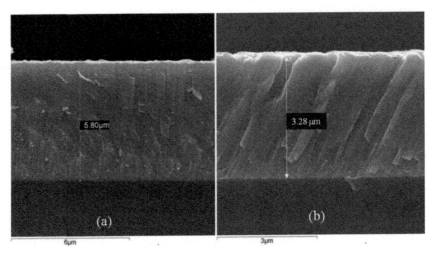

FIGURE 5.1   Textures of the ZnO films on the silicon substrates (a) with columnar structure and (b) with tilted columnar structure measured using scanning electron microscopy (SEM).

Film stress and adhesion to substrate for the ZnO thin films are considered as the key factors to successfully fabricate ZnO-based device in the process of MEMS. The stresses in the ZnO thin films usually originate from the mismatch of the lattice and thermal expansion, which is affected by different growth conditions using various growth techniques. The quantity of the stress can be evaluated using the Stoney formula [84]. Take magnetron sputtering for instance, the compressive stresses often arise through bombardment of the growing film with energetic ions and atoms controlled by the radio-frequency (RF) power, chamber pressure, and deposition rate. Large stress in ZnO thin films could lead to a poor adhesion, resulting in early adhesion failure or delamination. This is one of the major obstacles in manufacturing ZnO-based devices with a high performance. Therefore, reduction in the film stress is critical to improve the adhesion and obtain high-performance ZnO-based devices. Recently, some strategies were adopted, such as introducing the buffer layer or interlayer, using free-standing substrate or self-standing without substrates, annealing treatment, as well as the optimization of the growth conditions [85,86]. New deposition technique such as high-target utilization sputtering (HiTUS) has also been developed to significantly decrease the stress as shown in Figure 5.2 [84]. Helicon-wave-excited-plasma sputtering was also reported to prepare the ZnO thin film that could exhibit a smooth surface morphology with 0.26-nm-high monolayer atomic steps [87]. In addition, good stoichiometric ratio, low defects, and surface roughness in the ZnO thin films also contribute to crystalline properties. Thereafter, control and improvement of the crystallinity is fundamental. However, it is complicated to enhance the piezoelectric, optical, and electrical properties of ZnO thin films.

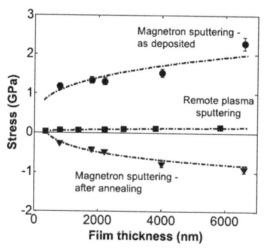

FIGURE 5.2 Comparison of the stresses in ZnO films deposited using the HiTUS technique and normal magnetron sputtering (reprinted with permission from Ref. [84]). (Reprinted with permission from Garcia-Gancedo, L.; Pedros, J.; Zhu, Z.; Flewitt, A. J.; Milne, W. I.; Luo, J. K.; Ford, C. J. B. Room-temperature remote-plasma sputtering of c-axis oriented zinc oxide thin films. *J. Appl. Phys.* 2012, *112*, 014907. © 2012, AIP Publishing LLC.)

5.2.2.2 GROWTH OF ZnO THIN FILMS

The growths of ZnO thin films have different characteristics due to the various external conditions, which can obviously change the growth dynamics and modulate the crystallizing process. Therefore, the properties of ZnO thin films are various and complicated due to the different growth mechanisms using various techniques such as PVD, CVD, and wetting chemical methods. The balance among the surface free energies of the substrate, film materials, and their interfaces plays a remarkable role on the growth kinetics that determines the growth mode [88–92]. The interfacial energy is mainly contributed by the strains in ZnO thin films that were mainly formed by the lattice mismatch. When the total free energy of the interface and the film surface is equal to the free energy of the substrate surface, a layer-by-layer (or Frank–van der Merwe) growth occurs [88,89]. For this two-dimensional (2D) growth mode, nucleation of each new layer will only appear after the previous layer is completed. In the opposite case, if the interactions among adatoms are stronger than those of the adatoms within the surface, the three-dimensional (3D) islands or clusters are formed to minimize the interfaces between the thin film and substrate. It is described as the Volmer–Weber growth [90]. However, a common process is an intermediary of the 2D layer and 3D island growth named as Stranski–Krastanov growth [91,92]. A few monolayers usually form first, and subsequently the layer-by-layer growth is transit to islands growth when the thickness of the layer reaches a critical value that is varied with the chemical and physical properties of the substrate and film. Furthermore, the three primary growth mechanism can be transformed due to the variations of the free energies and lattice parameters, which lead to different morphologies and microstructures in ZnO thin films [90]. Appropriate modification of the growth condition could well control the growth mode to prepare the ZnO thin films with special properties. For instance, ZnO quantum dots (QDs) were reported grown on SiO_2/ Si substrates using metalorganic CVD (MOCVD) based on the Stranski–Krastanov growth [93]. ZnO QDs were self-assembled by a vapor-phase transport process, and ZnO nanodots with tunable optical properties were achieved on solid substrates in the islands growth mode [94].

PVD technologies are vacuum deposition methods that produce vaporized film materials, following with the condensation at the wafer surface. They consist of sputtering, PLD, thermal evaporation, electron-beam evaporation, and MBE. As a common feature of PVD, ZnO thin-film growth involves a process beginning with the random nucleation and following with stages of nucleation and growth. Nucleation and growth is significantly dependent upon the deposition parameters including the temperature and pressure, growth rate, and substrate surface conditions. This leads to different crystalline phases, orientations, microstructures, film stresses, and associated defects in the ZnO thin films. The unique properties (e.g., size effect) are exhibited from the atomic growth, which cannot be observed in bulk materials.

Sputtering, including direct current sputtering, RF magnetron sputtering and reactive sputtering, is a preferred and popular technique with advantages of simplicity, good reproducibility, low cost, better adhesion, low operating temperature, and compatibility with microelectronics and MEMS processing. It is a leading choice for the deposition of ZnO thin films in device fabrication. In the sputtering process, the energetic particles bombard on or beneath the surface of the target and generate a lot of atoms or ions that are transported and impinged onto the substrate. The chemical or physical adsorption of atoms or ions leads to the nucleation. Subsequently, the sputtered atoms collide and diffuse among the ZnO crystal grains, and they grow up and form the ZnO thin film through further condensation and recrystallization. The correlations among the deposition parameters (e.g., chamber pressure, substrate temperature, sputter voltage, bias voltage, and deposition rate), microstructure and the film growth have been discussed thoroughly using a modified Thornton model [95].

Although magnetron sputtering is the most scalable method, plasma bombardment at the surface of the growing film is unavoidable during a typical sputtering deposition. Significant physical changes in the crystallinity could be resulted due to the increase in the plasma density, which leads to more defects and large intrinsic stress in ZnO thin films. High temperature deposition and/or post-deposition annealing can partially improve the film quality; however, they could have problems of incompatibility with the MEMS process. In order to resolve the above problem, a side arm has been developed to generate the plasma that can be launched into the chamber and further steered onto the target using an electromagnet in HiTUS [96,97]. Thus, the low Ar$^+$ bombardment on the growing film leads to a low ion-induced damage and better control of the roughness and stress for the ZnO thin film (e.g., a roughness of 1.7 nm for 360-nm-thick films) [86].

PLD is a versatile technique that uses the high-power pulse laser beam to ablate and evaporate the surface of the target material. The vaporized materials subsequently form the plasma plume and are impinged onto the substrate to be deposited as a thin film. PLD has advantages of relatively lower substrate temperatures and relative higher oxygen-partial pressure with a wide range than other PVD techniques [98–100]. High-quality ZnO thin films is significantly related to the process parameters of PLD such as the incident laser fluence, oxygen pressure, substrate temperature and substrate to target distance [98,99]. High oxygen background pressure normally results in compressive strain on sapphire and Si substrates, whereas the honeycomb-like morphology is formed at relative low pressure [100]. Thus, the deposition rate and the kinetic energy of ejected species are influenced by oxygen pressure because of the strong collisions between background gas molecules and the ablated species. Owing to the different optimum regimes of oxygen pressure and substrate temperature, a two-step method was proposed to epitaxially grow ZnO thin films, where the homo-buffer layer was deposited at a low pressure or low temperature, and then subsequently deposited the ZnO thin films at a relatively high pressure or high temperature [101,102]. This method will reduces the surface roughness of the ZnO thin film on glass and sapphire up to 1–2 nm. The substrates (e.g., ScMgAlO$_4$) with a close lattice match to ZnO can also improve the crystal-linity and reduce the defect densities and increase the Hall mobility up to 440 cm$^2 \cdot$V$^{-1} \cdot$s^{-1} [103,104]. Recent study showed that the in-plane misfit of epitaxial ZnO thin film on r-plane sapphire were controlled in the range from −1.5% for the [0 0 0 1]ZnO//[1 $\bar{1}$ 0 $\bar{1}$] sapphire to −18.3% for the [$\bar{1}$ 1 0 0] ZnO//[$\bar{1}$ $\bar{1}$ 2 0] sapphire direction [105]. The strains in the in-plane direc-tions are considered to be generated from the anisotropic lattice matching and thermal contraction of the sapphire substrate. In order to prepare the high-quality ZnO thin films, many buffer layers were also used including platinum, GaN, MgO, SiC, etc. [106–109]. The improvement and opti-mization of the target–substrate distance and geometry has been done to deposit the ZnO thin films with a large area up to a few inches on sapphire substrate [110]. Novel Aurora PLD method was also developed by applying a magnetic field to the plasma plume, which could significantly reduce the substrate temperature and enhance the photoluminescence [111,112].

MBE is a vacuum atomic layer by atomic layer growth technique invented at the late 1960s [113]. The epitaxial growth is attributed to the crystallographic relation between the film and substrate. The precise control of the reactions of the molecular or atomic beams on the heated crystalline substrate can deposit extremely high purity and highly crystalline thin films. In situ characterization using reflection high-energy electron diffraction offers real-time growth information to monitor, optimize, and control of the surface structures, lateral uniformity, and growth process [114]. The fraction of nanometer range can be reached for the film thickness using MBE. The evaporated zinc metal and oxygen (or dihydrogen dioxide, H$_2$O$_2$) generally are used as the source materials. The growth process of ZnO thin films using MBE is complicated which involves the adsorption, desorption, surface diffusion, incorporation, and decomposition [114]. Furthermore, the growth rate, composition, and doping concentration are mainly dependent on the arrival rates of different species in the collimated beams.

The substrates are crucial to the epitaxial growth of ZnO thin films. The preferred choice is to use the lattice-matched substrates that efficiently reduce the mismatch at the interface, resulting in the decrease in the strain and dislocation between the ZnO film and substrate. The sapphire with a-, c-, or R-plane, GaN, MgO, and ScAlMgO$_4$ are most frequently used in the various substrates [115–117]. An epitaxial relationships were found to be (1 1 $\bar{2}$ 0) ZnO/(0 1 $\bar{1}$ 2) Al$_2$O$_3$ and [1 $\bar{1}$ 0 0] ZnO/[1 1 $\bar{2}$ 0] Al$_2$O$_3$ with R-plane, and a low density of threading dislocations was observed in the [1 $\bar{1}$ 0 0] direction, leading to a high piezoelectric coupling coefficient of 6% [118]. However, the strain is generally large in epitaxial ZnO thin films due to the difference of the thermal expansion coefficients. The buffer layer (e.g., GaN, MgO, ZnS, or SiC) were often used to solve the above problem [119]. An MgO buffer layer has been introduced and promoted the epitaxial growth of ZnO thin films on c-sapphire, resulting in the improvement of the crystallinity and reduction of the screw dislocation density of ZnO layers from 6.1×10^8 cm^{-2} to 8.1×10^5 cm^{-2} [120]. Recent advance further showed that the thermal annealed MgO buffer layer can effectively control stress accumulation and produce a high-quality ZnO thin film on sapphire [121].

The MBE growth allows accurate donor or acceptor in the desired thin films. Undoped ZnO is typical n-type semiconductor because acceptors are compensated by native defects. The p-type ZnO thin films can be deposited with good reproducibility using MBE. The acceptor concentration larger than the unintentional donor concentration can be obtained via doping the elements such as N, P, As, and Sb, which is important to develop the ZnO-based light-emitting devices. [122,123]. However, the p-type ZnO thin films are controversial due to their poor reproducibility and instability of electric properties in the acceptor-doped samples. Take ZnO:N for instance, it changed from p-type conductivity to n-type one after a few days and the lattice constants were relaxed to its undoped value [124]. The sputtered ZnO:N thin film with p-type conductivity on glass converted to n-type after repeated measurements in the dark and recovered p-type with exposure to sunlight [125]. The slow transition from p- to n-type conductivity was tentatively assigned to the acceptor migration from the substitutional to the interstitial position [126,127].

In order to deposit ZnO thin films for a large-scale production, CVD technique is developed at the expense of more complicate setup [128–130]. CVD growth is a chemical process that involves the chemical reactions of the gaseous precursors that are delivered into the growth zone using the carrier gas, and the condensation and crystallization of the compounds onto the substrates. Various CVD techniques are categorized according to their operating pressure, such as atmospheric pressure CVD, Low-pressure CVD, and ultrahigh vacuum CVD. Plasma processing was also used to enhance chemical reaction rates of the precursors, leading to the CVD variants including the microwave plasma-assisted CVD, plasma-enhanced CVD, and remote plasma-enhanced CVD [128]. When the metalorganic species is used as the precursors, this method is known as the MOCVD. It is a preferred and standard technique to grow epitaxial ZnO thin films that exhibits the advantages of good reproducibility, uniform distribution over large area, availability of a wide variety of source materials, excellent control of composition, flexibility of low or high operating pressure, and ability to coat complex shapes [128].

Recently, advances of ZnO thin films have been successfully grown using CVD on different substrates such as glass, sapphire, Si, Ge, GaAs, GaP, InP, GaN, and ZnO [129,130]. The (1 1 $\overline{2}$ 0) nonpolar a-plane ZnO films on (0 1 $\overline{1}$ 2) r-sapphire substrates can be epitaxially grown, which was expected to fabricate the multimode SAW devices [131]. The prereactions in the gas phase using pure oxygen usually cause a significant reduction of the reactor pressure. Thus, alternative oxygen precursors have been proposed including *iso*-propanole, butanole, ozone, ethanol, or N_2O [132,133]. The resolutions for above problem can also be performed by modifying the MOCVD system. The methods such as separating of metal-organics and oxygen flow, controlling of flow patterns and gas residence time, and using horizontal or vertical reactors and high-speed rotation reactors, are efficient to eliminate the prereaction [134,135]. Different metal–organic sources (e.g., zinc acetylacetonate, bis(acetylacetonato) zinc(II), and diethylzinc) are used to produce the high-quality ZnO thin films with better uniformity and reproducibility [136–139].

Although the above growth technologies of ZnO thin films have their advantages and limitations, they are popular and compatible with microelectronics in the device-based application because they could meet the requirements of the MEMS processing. The sol–gel deposited ZnO thin films could not approach the requirements of the fabrication for the SAW devices in spite of the facts that the sol–gel technique is easily operated in low temperature and widely used to prepare the polycrystalline ZnO thin films [140].

5.2.3 ZnO NANOSTRUCTURES

ZnO nanostructures are defined as the ZnO materials consisting of the structural elements with at least one dimension at nanoscale that ranges from 0.1

to 100 nm. The individual ZnO nanostructure refers to the QD, nanoparticle, nanowire, nanorod, nanotube, nanobelt, and nanoplatelet, whereas the collections of the ZnO nanostructures normally are shown in forms of arrays, assembly, and hierarchical nanoarchitectures based on the above individual ZnO nanostructure. Considering the dimensions at nanoscale, the ZnO nanostructures also are categorized as follows [5,66]:

- zero dimension (0D) nanomaterials, for example, nanoparticle, nanocluster, nanocolloids, and nanocrystals;
- one-dimension (1D) ones, for example, nanowire, nanorods, nanobelt, and nanotube;
- two-dimension (2D) ones, for example, nanoplatelet and nanodisk; and
- hierarchical three-dimension (3D) nanoarchitectures, for example, nanowire arrays, ordered porous structure, core–shell structure, nanoflower, and brushed shape.

The physical and chemical properties of the ZnO nanostructures are significantly dependent on the size, growth direction, specific shape and microstructure, and this is different with the bulk ZnO materials. Therefore, controllable synthesis of the ZnO nanostructures with multiple functionalities offers the possibility of engineering and developing novel ZnO-based devices.

The unique effects of the ZnO nanostructures include size effect, surface effect, and quantum confinement. They can modulate and tailor the optical, electrical, and piezoelectric properties up to extremely wide regions for specific applications [141]. The size effect is a typical phenomenon linking with quantum effect with the decrease in the size of the ZnO materials to nanoscale, mainly due to the confinement of electrons to very small regions of space in one, two, or three dimensions [141,142]. When the size of the ZnO materials reaches nanoscale the ratio of the surface to volume becomes very large. Thus, the catalytic properties are significantly enhanced, which is usually considered as the surface effect. It is crucial to the sensing applications. The feature of the ZnO nanomaterials in the quantum region is the quantum confinement which was first analyzed by Kubo to interpret the quantum size effect of metals at nanoscale [142]. The continuous energy levels of the electronic structures are split into discrete energy levels. This effect can enhance the optical properties of ZnO nanostructures, for instance, it led to the increase in the band-gap energy of ZnO nanocrystals with the sizes ranging from 3 to 5.4 nm [141].

5.2.3.1   0D ZnO NANOSTRUCTURES

ZnO QDs are typically thought as the 0D ZnO nanostructure. Their discrete energy levels are generated from the excitons confined by the 3D potential well when the dimensions of the ZnO QDs approach or become less than the exciton Bohr radius [142,143]. Many investigations were conducted to understand the excitonic properties of the ZnO QDs, as well as their influences on the optical properties. The exciton Bohr radius for the bulk ZnO is evaluated as 0.9 nm, and the size of the QDs is considered to be two to three times larger than the size of the bulk exciton [143]. The electron–hole interaction and quantum confinement effects have comparable strengths for the size ranging from 2 to 6 nm [144]. The quantum confinement of the ZnO QDs leads to the blue shift for the free exciton transition and size dependence of the coupling strength between electron and longitudinal optical phonon in the Raman analysis [145].

Recent advances of the growth of ZnO QDs have mainly been made using the wet chemical methods such as sol–gel technique, solvothermal technique, and hydrothermal technique [146–148]. These methods have changed the environment of the reaction solution including the temperature, the precursor concentration, pH value, and the surfactant forms the external factors that could modulate the size of the ZnO QDs, and subsequently resulting in their tunable optical properties. The obtained ZnO QDs have been used to label the cells with antibacterial activity, low cytotoxicity, and proper labeling efficiency [149]. In order to obtain the emission at special region of the light wavelength, the ZnO QDs were reported to be embedded in graphene, SiO_2 matrix, and PMMA matrix [150–152]. These techniques allowed the applications of the ZnO QDs for the light-emitting devices, for example, the ZnO QDs wrapped in a shell of single-layer graphene could be made a white-light-emitting diode with a brightness of 798 cd·m^{-2}. The blue-light emitting ZnO QDs combined with biodegradable chitosan (N-acetyl-glucosamine) were used for tumor-targeted drug delivery [153]. Owing to the surface effect, the ZnO QDs were also utilized as sensing materials that exhibited good response to the gas and UV light detections [154,155]. In addition, high-quality self-assemble ZnO QDs can be also grown using the techniques of MOCVD and vapor phase transport [156–158].

The ZnO QDs are special nanoparticles with the size ranging from 1 to 20 nm. When the size is larger than 20 nm and less than 100 nm, the normal ZnO nanoparticles could be stably synthesized using the wet chemical method. The wet chemical method is a low-cost and environmentally friendly technique, which allows control of the shape, size and crystal phase of ZnO nanoparticles in the solution with mild reaction [159–163]. Since the sol–gel technique is one of the preferred wet chemical growth techniques, the ZnO nanoparticles could be synthesized by the preparation of ZnO sols in the liquid phase from homogeneous ethanolic solutions with precursors of sodium hydroxide and zinc acetate [159,160]. Thus, numerous studies were carried out to explore the growth mechanisms of ZnO nanoparticles with the variations of different solution parameters, which are expected to reproducibly grow ZnO nanoparticles with stable properties for the specific applications such as gas sensing, biosensing, bioimaging, and photoelectronic devices [161–163]. For instance, the cytotoxic properties of the biocompatible ZnO nanoparticles against cancerous cells are significantly dependent on the size, and the toxicity becomes greater when the size is reduced. Therefore, an optimum design of the size range could improve the cancer cell electivity and minimize the toxicity against the normal body cells [164].

5.2.3.2   1D AND 2D ZnO NANOSTRUCTURES

1D ZnO nanostructures are the most attractive nanomaterials because of the easy control of the nucleation sites and the diverse device applications. As a typical 1D ZnO nanostructure, ZnO nanowires (or nanorods) has a large aspect ratio of the length to width. The physical properties are remarkable changed with the reduction of the diameter. The Young's modulus and effective piezoelectric constant of the ZnO nanowires are much large than those of the bulk ZnO based on the first-principle computation and experimental measurements [165,166]. However, the structural defects in the ZnO nanowires could delay the response of the electromechanical coupling [167]. In order to grow high-quality ZnO nanowires, various techniques such as vapor phase transportation growth [168] and hydrothermal growth at low temperature [169,170] were developed to control the orientation, diameter, morphology, and position.

Vapor phase transportation growth normally consists of catalyst assisted vapor–liquid–solid (VLS) process and catalyst free vapor–solid (VS) process, which is one of the most popular techniques to synthesize the ZnO nanowires. The metal-catalyzed VLS technique proposed in 1964 is very flexible because the growth parameters are tunable and controllable during the growth process [171]. The typical VLS growth begins with introducing liquid droplets of a catalyst metal with nanosize on the substrate, where the vapor of the ZnO source is condensated, and the nucleated seeds at the liquid–solid interface direct the crystal growth to form ZnO nanowires [172]. The key factor of growing ZnO nanowires is the size and physical properties of the metal catalyst which determines the size and positions of the ZnO nanowire. The minimization of the total free energy dominates the growth direction which is mainly contributed from the free energy of the interface between ZnO and metal catalyst. When the size of the metal catalysts is reduced, it is a challenging issue to grow ZnO nanowires with such a small diameter because the growth dynamics becomes difficult with the increase in the chemical potential of the liquid alloy droplet. Different metal catalysts (e.g., Au, Ag, Cu, Ni, and Sn) were used to modify the growth directions and morphologies of the ZnO nanowires [172–174]. The ZnO nanowires with a (0 0 0 1) orientation usually can be grown on the a-plane sapphire because of the nearly lattice match epitaxial growth that results in the vertical alignment of the nanowires [175,176]. Some other substrates such as GaN, $Al_{0.5}Ga_{0.5}N$, and AlN also allow the vertically aligned growth of the ZnO nanowire beside the sapphire substrate [177]. Novel crawling growth of ZnO nanowires can be driven on (0 0 0 1) GaN with CVD due to the diffusion of Au from the primary Au-catalyzed particle with large size [178]. Furthermore, the ZnO nanowires can only grow in the activated area with metal catalysts, which allows patterning of the metal catalysts arrays using various lithographical technologies including soft lithography, e-beam lithography, and photolithography [179–181]. Thus, ZnO nanowire arrays with different patterns are controllably grown, which could serve as the highly appropriate device-based materials for the applications of the solar cell, field emission, nanogenerator, and nanolaser [182,183]. The nonlithographical technique using anodic aluminum oxide membranes as a mask can also pattern the Au catalyst on GaN substrate to prepare the ZnO nanowire arrays [184]. It is also noted that the introduction of metal catalysts leads to the impurity level in band gap that could degrade the optical properties, and the metal catalysts may be not compatible with the COMS process in the device fabrication [185]. Naturally considering of the catalyst free VS process, the ZnO thin film on Si substrate was used as a seed layer and the ZnO nanowires were grown using a simple solid–vapor phase thermal sublimation technique [186]. Another method without metal catalysts is MOCVD, which could also obtain the ZnO nanowires on sapphire with high optical quality [187].

Hydrothermal synthesis of the ZnO nanowires at low temperature is a popular growth technique in a solution phase, which has potential for wafer-scale production to fabricate ZnO nanowire-based devices [188]. The ultra-long ZnO nanowires are grown up to several micrometers with the growth time increasing as shown in Figure 5.3a–d. The aqueous solutions of zinc nitrate and hexamethylenetetramine (HMTA) are often used to prepare the

ZnO nanowires on the ZnO-seeded substrates, with a typical growth process shown in Figure 5.3e. The high water-soluble HMTA is decomposed to release the hydroxyl ions reacting with Zn^{2+} ions to form ZnO, and the ZnO seeds are nucleated and grown to be the nanowires. The mild reaction conditions are flexibly modified including different substrates, temperatures, reaction times, pH values, surfactants, and substituting zinc acetate and ammonia for the precursors [6,189]. The morphologies of the resulting ZnO nanomaterials with various density and size could be changed from the normal nanowires to patterned nanowires, nanotubes, nanoplate, and

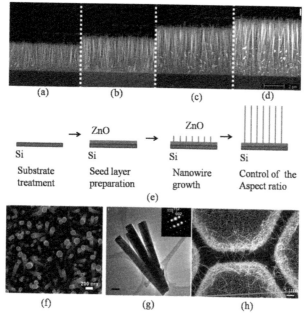

FIGURE 5.3   (a–d) The lengths of the ZnO nanowires increasing with growth times [169]; (e) typical hydrothermal growth process in the aqueous solutions; (f–h) images of SEM and transmission electron microscope for the ZnO nanowires (reprinted with permission from Refs. [168,188,238]). (Reprinted with permission from a–d: Tak, Y.; Yong, K. J.; Controlled Growth of Well-Aligned ZnONanorod Array Using a Novel Solution Method *J. Phys. Chem. B* 2005, *109*, 19263.© 2005, American Chemical Society.f: McPeak, K. M.; Le, T. P.; Britton, N. G.; Nickolov, Z. S.; Elabd, Y. A.; Baxter, J. B. Chemical Bath Deposition of ZnO Nanowires at Near-Neutral pH Conditions without Hexamethylenetetramine (HMTA): Understanding the Role of HMTA. *Langmuir* 2011, *27*, 3672.© 2011 American Chemical Society.g: Greene, L.; Law, M.; Goldberger, J.; Kim, F.; Johnson, J. C.; Zhang, Y. F.; Saykally, R. J.; Yang, P. D. Low-Temperature Wafer-Scale Production of ZnO Nanowire Arrays. Angew. *Chem., Int.* Ed. 2003, *42*, 3031.Copyright © 2003 WILEY-VCH Verlag GmbH & Co. KGaA, Weinheim.h: Yang, P.; Yan, H.; Mao, S.; Russo, R.; Johnson, J.; Saykally, R.; Morris, N.; Pham, J.; He, R.; Choi, H.-J.; Controlled Growth of ZnO Nanowires and Their Optical Properties. *Adv. Funct. Mater.* 2002, *12*, 323.© 2002 WILEY-VCH Verlag GmbH, Weinheim, Fed. Rep. of Germany.)

flower-like shape [190]. The hydrothermal synthesis of ZnO nanowires can also be assisted by the technologies of applying the electric field, microwave and laser that allow the rapid growth of long nanowires [191–194]. In addition, the ZnO nanowires can also be prepared using the other wet chemical techniques such as ultrasonic sol–gel technique, spray pyrolysis, and electrochemical deposition assisted by the templates of nanoporous membranes and polystyrene (PS) sphere monolayer [195–198].

ZnO nanobelts normally are long ribbons with a rectangular cross section and well-defined crystallographic surfaces. Wurtzite ZnO possesses two important characteristics: noncentral symmetry and polar surfaces. They dominate the growth of novel structures of ZnO nanobelts by controlling of the growth rates along different directions. The nonpolar facets of $\{2\,\bar{1}\,\bar{1}\,0\}$ and $\{0\,1\,\bar{1}\,0\}$ have lower energies than the $\{0\,0\,0\,1\}$ facets, thus, the ZnO nanobelts dominated by the $(0\,0\,0\,1)$ polar surface will grow along the a-axis [199]. They bend and fold into shapes of nanoring, nanohelixes, and nanospring to achieve the minimized energy and neutralize the local polar charges as shown in Figure 5.4a–c [4,200]. The polarization of the ZnO nanobelts is perpendicular to the spiral axis for the structures of nanohelixes and nanospirals, and a seamless nanoring can be formed with a rotation of 90° for the polarization [201]. The temperature is an important factor to determine the growth direction during the process of the thermal evaporation growth. The ZnO nanobelts began to growing along $[0\,1\,\bar{1}\,3]$ before the temperature reach 1475°C and subsequently switched to growing along $[0\,1\,\bar{1}\,0]$ with the temperature stabilizing at 1475°C [202]. The effective piezo-coefficient (d_{33}) of the $(0\,0\,0\,1)$ surface of the ZnO nanobelt measured using piezoresponse force microscopy was found to be frequency-dependent, which is much larger than that of the $(0\,0\,0\,1)$ surface of the bulk ZnO [203]. Recent investigations showed that the field-emission properties could be enhanced for the ZnO nanobelts [204]. The lasing emission in PL spectra was also found for the ZnO nanobelts, and the sensing application of the ZnO nanobelts for detecting H_2 and NO_2 exhibited fast response and good repeatability [205,206].

ZnO disks and platelets at nanoscale are widely reported as the 2D ZnO nanostructures with example presented in Figure 5.4d–f. The typical method to synthesize the ZnO disks and platelet at nanoscale is solution phase growth including the microemulsion and hydrothermal growth. When the surfactant of sodium bis(2-ethylhexyl) sulfosuccinate was used during the preparation of an oil-in-water microemulsion, large-scale uniform ZnO disks with diameters of 2–3 μm and thicknesses of 50–200 nm were produced. The morphology of the disk would be modified as hexagonal rings with the increase in temperature from 70 to 90°C as shown in Figure 5.4d,e [207]. If the growth along [0 0 0 1] direction of the wurtzite ZnO is strongly suppressed by the surfactant or capping reagent (e.g., polyacrylamide and ethylenediaminetetraacetic acid), the growth directions along six symmetric directions of ±[1 0 $\bar{1}$ 0], ±[1 $\bar{1}$ 0 0], and ±[0 1 $\bar{1}$ 0] are favored to form the hexagonal ZnO disk during the hydrothermal process. For instance, the ZnO disks and plates can be hydrothermally grown from the aqueous reaction of sodium nitrate and HMTA [208]. Hydrothermal growth of ZnO nanodisks with an average diameter of 150 nm and thickness of 40 nm were reported using ethylenediaminetetraacetic acid as an organic ligand [209]. Double surfactants of hexadecyltrimethylammonium bromide and sodium dodecyl sulfate recently are used to controllably grown the regular hexagonal twinned

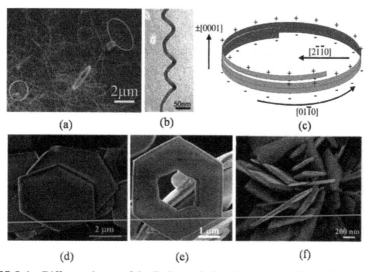

FIGURE 5.4 Different shapes of the ZnO nanobelts: (a) nanoring [4] and (b) nanohelixes [200]; (c) the crystallographic orientation of a ZnO nanoring [4]; 2D ZnO nanostructures: (d) hexagonal disk [208], (e) hexagonal ring [208], and (f) platelet [211] (reprinted with permission from Refs. [200,208,211]). (Reprinted with permission from a and c: Ding, Y.; Wang, Z. L. Doping and planar defects in the formation of single-crystal ZnOnanorings. *J. Phys. Chem. B* 2004, *108*, 12280.© 2004, American Physical Society. b: Li, F.; Ding, Y.; Gao, P.; Xin, X.; Wang, Z. L. Structure Analysis of Nanowires and Nanobelts by Transmission Electron Microscopy Angew. *Chem.* 2004, *116*, 5350.© 2004, American Chemical Society. d–e: Li, F.; Ding, Y.; Gao, P.; Xin, X.; Wang, Z. L. Single-Crystal Hexagonal Disks and Rings of ZnO: Low-Temperature, Large-Scale Synthesis and Growth Mechanism. Angew. *Chem.* 2004, *116*, 5350. © 2004, John Wiley and Sons. f: Chen, H.; Wu, X.; Gong, L.; Ye, C.; Qu, F.; Shen, G. Hydrothermally Grown ZnO Micro/Nanotube Arrays and Their Properties. Nanoscale Res. *Lett.* 2009, 5, 570. https://creativecommons.org/licenses/by/4.0/)

ZnO nanodisks [210]. The growth along [0 0 0 1] direction of the wurtzite ZnO was recovered without surfactants, then ZnO nanoplates were formed as shown in Figure 5.4f [211]. The mixtures of zinc oxide and graphite powders were used as source materials in vapor-phase transport method, resulting in hexagonal nanodisks with about 3 μm in diagonal and 300 nm in thickness [212].

5.2.3.3 HIERARCHICAL ARCHITECTURES OF ZnO NANOSTRUCTURES

Hierarchical architectures of the ZnO nanostructures refer to the 3D structures with multiple levels that are uniformly arranged or constructed in order by the relatively low-dimension nanostructures. Figure 5.5 shows the commonly observed hierarchical ZnO nanostructures, which are categorized as ordered porous nanostructure, core–shell, flower-like shape, and branched nanostructures [213–216]. The unique functionality of the hierarchical ZnO nanostructure can be enhanced from the cooperative coupling of the multiple-level, multiple-dimension, and different compositions. They are stimulating lots of interest on the control of growing various hierarchical architectures, which are expected to be applied for the design and development of highly efficient ZnO-based devices [213–216].

Ordered porous ZnO nanostructures are normally fabricated using the template method. The templates can be formed using different materials including PS sphere, porous alumina membranes, colloidal crystals, Schiff-base amine, liquid crystals, and ordered mesoporous carbon [213,217,218]. The ZnO materials are filled into the space of the template that is removed later, thus porous ZnO nanostructures are obtained. The pore size, pore density, and periodicity are determined by the physical parameters of the template, for instance, the smaller PS sphere leads to the rapid reduction of the pore size. Some unique properties of the ordered porous ZnO nanostructures are exhibited, such as superhydrophobicity and high surface area [219,220]. When the PS spheres with a size of 193 nm had been used as the template the products also were considered as inverted 3D photonic crystal structure which helps enhancing the optical property of ordered ZnO nanostructures [221].

FIGURE 5.5 Hierarchical structured ZnO nanomaterials including (a) ordered porous structure [213], (b and c) core–shell structure [222,223], (d and e) flower-like structure [215,235], (f and h) branched structure [243–245] (reprinted with permission from Refs. [213,215,222,223,235,243–245]). (Reprinted with permission from a: Ding, G. Q.; Shen, W. Z.; Zheng, M. J.; Fan, D. H. Synthesis of ordered large-scale ZnOnanopore arrays. *Appl. Phys. Lett.* 2006, *88*, 103106.© 2006, AIP Publishing LLC.b: Liu, H. L.; Wu, J. H.; Min, J. H.; Zhang, X. Y.; Kim, Y. K. Tunable synthesis and multifunctionalities of Fe3O4–ZnO hybrid core-shell nanocrystals. *Mater. Res. Bull.* 2013, *48*, 551. © 2013, Elsevier.c: Wang, K.; Chen, J. J.; Zeng, Z. M.; Tarr, J.; Zhou, W. L.; Jiang, C. S.; Pern, J.; Mascarenhas, A. Synthesis and photovoltaic effect of vertically aligned ZnO/ZnS core/shell nanowire arrays. *Appl. Phys. Lett.* 2010, *96*, 123105. © 2010, AIP Publishing LLC. d: Elias, J.; Levy-Clement, C.; Bechelany, M.; Michler, J.; Wang, G.-Y.; Wang, Z.; Philippe, L. Hollow Urchin-like ZnO thin Films by Electrochemical Deposition. *Adv. Mater.* 2010, 22, 1607. © 2010, John Wiley and Sons.e: Pugliese, D.; Bella, F.; Cauda, V.; Lamberti, A.; Sacco, A.; Tresso, E.; Bianco, S. A Chemometric Approach for the Sensitization Procedure of ZnO Flowerlike Microstructures for Dye-Sensitized Solar Cells. ACS Appl. Mater. Interfaces 2013, 5, 11288. © 2013, American Chemical Society. f: He, F.-Q.; Zhao, Y.-P. Growth of ZnOnanotetrapods with hexagonal crown. *Appl. Phys. Lett.* 2006, 88, 193113.© 2006, AIP Publishing LLC.g: Fan, H. J.; Scholz, R.; Kolb, F. M.; Zacharias, M. Two-dimensional dendritic ZnO nanowires from oxidation of Zn microcrystals. *Appl. Phys. Lett.* 2004, 85, 4142. © 2004, AIP Publishing LLC.h: Lao, J. Y.; Wen, J. G.; Ren, Z. F. Hierarchical ZnO Nanostructures. *Nano Lett.* 2002, *2* (11), 1287. © 2002, American Chemical Society.)

Core–shell ZnO-based nanostructures consist of the core–shell nanoparticles and core–shell nanowires, where the ZnO nanocrystals can be grown as either the shell or the core [214,222–225]. Owing to the bandgaps and the relative position of electronic energy levels, two types of the core–shell ZnO-based nanostructures are mainly focused for investigation. The first one is that the band gap of shell nanocrystals is larger than that of the core and the carriers are confined in the core. The second one is that the carriers are partially or completely confined in the shell which is dependent on the thickness of the shell [214]. When the ZnO nanocrystals were selected as the shell materials, they provide energy barriers with the different core nanomaterials such as FePt, Mn:ZnS, Zn, and oxides (e.g., Co_3O_4, Fe_3O_4, and TiO_2) at the interface. This allows the properties of the core–shell nanostructures to be tuned via precisely controlling of the size, shape, and composition of the core. For instance, the strong violet PL at 425 nm from the ZnO shell of core–shell Zn/ZnO nanoparticles can be well controlled by adjusting the ZnO shell thickness or annealing during the preparation process using laser ablation in liquid media [224]. FePt/ZnO core/shell nanoparticles exhibited a wide range of semiconducting, magnetic, and piezoelectric properties that were expected to modulate the material's response to magnetic, electrical, optical, and mechanical stimuli [225]. Considering that the nanoparticles with ZnO shell often are applied as QDs system for the bioprobe, the core–shell Mn:ZnS/ZnO QDs have been demonstrated that can be glutathione-functionalized as the time-resolved Förster resonance energy transfer (TR-FRET) bioprobes, which are sensitive to detect a trace amount of biomolecules [226]. When the ZnO nanocrystals were used as the core, selenide (e.g., CdSe), sulfides (e.g., ZnS, Ag_2S, and CuS), and oxides (e.g., ZnMgO and MgO) are widely used as the shell materials [227–229]. Recent advances show that ZnO/CdSe core/shell nanowire arrays can be used as efficient photoanodes for photoelectrochemical water splitting [230], and ZnO/CdSe core/shell nanoneedle arrays significantly help improving the photo-to-current conversion efficiency for solar cell [231]. ZnO/TiO_2 core–shell nanowires were designed as the electrode film to reduce the recombination rate in dye-sensitized solar cells. Owing to the relation of the conduction band potentials, the energy barrier hinders the reaction of electrons in core with the oxidized dye in electrode, resulting in the enhancement of the solar-cell performance [232]. Homogeneous core–shell ZnO/ZnMgO nanowires could be grown as quantum well heterostructures, which showed a high optical efficiency at room temperature [233]. Hydrogenated ZnO/amorphous ZnO-doped MnO_2 core–shell nanocables on carbon cloth were reported to be used as supercapacitor electrodes, resulting in excellent electrical performance (e.g., an area capacitance of 138.7 mF cm^{-2} and specific capacitance of 1260.9 F g^{-1}) [234].

Figure 5.5d,e presents the typical flower ZnO nanostructures. The nanowire, nanorod, nanoplatelet, and nanodisk often serve as the petal, which self-assembly forms into the flower nanostructures with different morphologies. ZnO flowers were assembled with prism-shaped platelets as shown in Figure 5.5e, which could be deposited on the FTO-covered glass acting as a photoanode of the dye-sensitized solar cell [215]. The nanowire bundles with a radial shape, snow flake, star-like, and urchin-like morphologies may be thought as the variants of flower-like structures [235,236]. The solution-phase growth with low temperature is traditionally performed as a simple route to get the flower ZnO nanostructures. Hydrothermally synthesized ZnO flower-like nanostructures using the aqueous reaction of zinc nitrite (or chloride) with sodium hydrate (or ammonia) are normally made up of nanowires and nanorods at low temperature with a range of 60–90°C, whereas the temperature increases up to 120–200°C the flower structure is assembled by nanoplatelets or nanodisks at a relatively high pressure in the Teflon-lined autoclave [237–240]. Although the studies of ZnO flower-like nanostructures are partially driven from the interesting and aesthetic points of view, the functionality of these structures are one of the most attractive factors in the diverse ZnO device-based applications [6–11]. The ZnO nanoflowers with a diameter of 200 nm provide a large surface area and direct channels of electron transport from petals to the stem, which exhibited a good overall conversion efficiency that are higher than that of the films of nanorod arrays with comparable diameters and array densities in solar-cell applications [241,242].

Considering the common feature of the collections of the nanowires, namely, the ZnO nanowires are arranged in order around the core wire; brush-like, tetrapod, dendritic, tree-like, and bunched nanowires are assigned to the branched ZnO nanostructures as shown in Figure 5.5f–h [243–245]. The 3D branched ZnO nanostructures possess advantages of large surface area and direct transport pathway for charge carriers, which are applied for the energy conversion and storage devices [216]. The strategies of growing the branched ZnO nanostructures are usually based on the techniques of sequential catalyst-assisted VLS and solution phase growth. The technique of catalyst-assisted VLS as described in the previous section of ZnO nanowire growth could be adopted to prepare the heterobranched structures. For instance, the hexagonal ZnO nanorods with diameters of about 20–200 nm were reported to grow either perpendicular to or slanted to all the facets of the core In_2O_3 nanowires with diameters of about 50–500 nm; and the vapor phase transport resulted in the typical branched ZnO/In_2O_3 nanostructures [244]. When the self-catalytic liquid–solid method was combined with the vapor–solid process, ZnO dendritic nanowires were observed in form of 2D web-like structure with homoepitaxial interconnections at the branch-to-arm and branch-to-branch regions [245]. As for the solution phase growth, when the Au catalysts were used to assist the growth of ZnO nanowires, comb-like ZnO nanostructures were produced using the hydrothermal technique [246]. Brush-like ZnO–TiO$_2$ hierarchical heterojunctions nanofibers were prepared via hydrothermally growing highly dispersed ZnO nanorods on electrospun TiO$_2$ nanofibers [247]. Nanoforest of hierarchical ZnO nanowires can be observed via modulating the growth rate and controlling of the length-wise growth of the core nanowire and the branched growth during the hydrothermal synthesis [248]. The overall light-conversion efficiency based on the branched ZnO nanowire was almost five times higher than the efficiency using upstanding ZnO nanowires for the dye-sensitized solar cell [248].

The growth methods of the hierarchical ZnO nanostructures are various, and there are many controllable factors during the synthesis process. The hierarchical nanostructure is sensitively changed due to the variations of the growth conditions, resulting in the different mechanism of the assembly process [213–216]. Furthermore, the relation of the difference of the hierarchical structures to the properties of the ZnO materials still deserves study with in-depth; and good theory of the accurately controllable growth has not be elucidated. All the above issues are crucial for hierarchical ZnO nanostructures in the various applications.

5.3   ENGINEERING AND DEVELOPMENT OF ZnO-BASED DEVICES

Engineering and development of ZnO-based devices is complicated as it involves accurately designing and modeling the desired device with a special configuration, preparing the ZnO materials with high quality, pretreating the ZnO materials, fabricating the device, characterizing the devices, and optimizing the design and remodeling. Since the discovery of ZnO materials, each segment of the above process has been devoted with lots of efforts of science and technology.

Various novel ideas for designing and developing ZnO-based devices have been proposed based on the fundamental properties of ZnO thin films and nanostructures [6,10,12,21,100]. The ZnO associated devices are categorized as follows:

- traditional microelectronic devices, for example, diode, transistor, and supercapacitors;
- piezoelectric devices, for example, SAW device, TFBAR, thin film ultrasonic transducer, and self-power device (e.g., nanogenerator);

- sensors, for example, gas and chemical sensor, biosensor, UV sensor, humidity sensor, wireless sensor, and strain sensor;
- mirofluidic devices, for example, pump, mixer, ejector, and atomizer;
- photoelectric devices, for example, solar cells, photoelectrochemical cells, and UV photodetector; and
- optical devices, for example, optical resonator, UV laser and light-emitting device.

The main goal of the engineering and development is to improve the performance of the ZnO-based device. The reproducibility, stability, sensitivity, and efficiency are critical for assessment of a device. Reduction of the size and manufacturing cost also deserves attention to obtain the highly integrated device. In order to realize these, multiple functionalities have been integrated as one unit in some applications [6–12].

Design of the microfluidic attenuation and manipulation using SAWs is one of the most interesting and attractive project, and the SAW microfluidic devices could be comprehensively applied for biomolecular and cellular manipulation, biosensor, drug delivery, biomaterials synthesis, and point-of-care diagnostics [249]. ZnO thin films possess advantages over bulk materials such as device design flexibility, low cost and convenience of fabrication, and compatibility with other integrated microfluidics and sensing technologies. Thus, ZnO SAW microfluidic devices are expected to be highly integrated, more efficient, and stable. Multifunctionality of ZnO SAW microfluidic devices offers a better choice to improve the performance of the microfluidic devices as well as the enhancement of the microfluidic performance and optimization of the acoustic properties [249]. The standard processing in technological details also ensures more reproducible for ZnO SAW microfluidic devices with the same structures.

When ZnO thin films, nanostructures or the mixture of them are exploited for tracing different minor samples in air, bioliquid, and optical environment, the engineering and development of the ZnO-based sensors would be a complicated issue. When various ZnO-based sensors are designed, the high sensitivity, good selectivity, and strong stability are important factors to be pursued. The fundamental properties of the ZnO materials are needed to be optimized using different technologies including growth techniques, fabrication methods, and analytical techniques.

The following sections focus on the advances of the researches on the applications of ZnO SAW device, ZnO SAW microfluidics, and ZnO SAW sensors.

5.4   ZnO ACOUSTIC WAVE DEVICES

The fundamental parameters such as the insertion loss, phase velocity of the acoustic wave, electromechanical coupling coefficient, and quality factor determine the performance of ZnO acoustic wave devices. In order to obtain high-performance ZnO acoustic wave devices, the following factors are necessary to be considered or improved:

- High-quality piezoelectric materials normally contribute large electromechanical coupling coefficient. The stoichiometric ratio, cut direction, dielectric constant, and effective piezoelectric constant all have played important roles on the coupling coefficient of the bulk materials, whereas the piezoelectricity of ZnO thin films are also dependent on the film crystallinity including orientation, surface roughness, thickness, and substrates.
- Pattern design of the electrodes is crucial to the excitation of the acoustic waves. Good structure of the electrode patterns significantly improves the characteristic of the relation between the resistance and frequency as well as the quality of the acoustic signal. For instance, the number of finger pairs, spatial periodicity, aperture, and distance of the two ports for the interdigital transducers (IDTs) could influence the intensity, frequency, and propagation direction of the acoustic waves.
- As for fabricating the device, the physical properties of the electrode materials and the parameters of the configuration for the electrodes provide a complex impedance of the equivalent circuit. The compatibility of the fabrication with silicon technologies is necessary. Advanced lithography technique such as e-beam lithography could manufacture the high-frequency acoustic device with frequencies at GHz level.
- A standard probe station could help to get accurate information of the device. The connection of the cables among equipment, bonding using different metal wires and the measurement skills are vital to the characterization of the acoustic wave devices.

Herein, ZnO acoustic wave devices mainly refer to the ZnO SAW device, ZnO TFBAR, and ZnO ultrasonic transducer.

5.4.1   SURFACE ACOUSTIC WAVE DEVICES

The ZnO SAW devices are mainly composed of filters, resonators, delay lines, and convolver. An IDT is two sets of connected metallic fingers that are interspaced between each other. A typical ZnO SAW filter has two ports of IDTs that are patterned on the ZnO substrates. The acoustic waves are excited from the input IDT by the applied electric signals, and subsequently the received acoustic waves at the output port are reconverted into electric signals. The acoustic properties of the device could be measured from the signals between the input port and output port, which allow the evaluation of the performance for the ZnO SAW device. Phase velocity (v) of SAW is related to the wavelength (λ) and resonant frequency (f). It is calculated based on the following equation:

$$v = \lambda f \tag{5.2}$$

where the wavelength, λ, corresponds to the spatial period of the fingers in IDT. The phase velocity is dependent on the thickness of the piezoelectric substrate. If the UV-based nanoimprint lithography is combined with lift-off processes, ultrahigh-frequency ZnO SAW device on silicon substrates could be fabricated that have operated resonance frequency at a range of about 4–23.5 GHz [250].

The electromechanical coupling coefficient (k^2) is used to estimate the efficiency of the energy conversion between the electric signal and the SAW, which is generally calculated using the following equation [251]:

$$k^2 = \frac{\pi}{4N}\left(\frac{G}{B}\right)_{f=f_o} \tag{5.3}$$

where N is the finger pairs, and G and B are the radiation conductance and susceptance at the central frequency, respectively. G and B can be measured from the Smith charts of the reflection coefficients at the central resonant frequency of the SAW signals. The strong electromechanical coupling normally leads to a large coupling coefficient, which is influenced by piezo-electricity of the ZnO thin films and the fabrication processing.

In order to obtain a high-performance ZnO SAW device, some hurdles need to be overcome, including the thickness effect, substrate effect, and the second-order effects including the triple-transit effect, metallization ratio effect, SAW reflections and diffractions, and bulk wave generations, etc.

The acoustic properties of the ZnO SAW device are dependent on the thickness of the ZnO thin films. When the thickness of ZnO thin film increases the acoustic velocity in thin film is varied to be close to the bulk value of about 2700 m s^{-1} [252], whereas when the thickness is rapidly reduced, it leads to a decrease in the phase velocity of the SAW that could approach the Rayleigh velocity of the substrate material. This is caused by the stronger penetration of the SAW into the substrate and more energy localization in the substrate. The acoustic wave mode is also changed with the thickness of the ZnO thin film. The Rayleigh SAW could be observed in the thinner ZnO film; however, higher harmonic modes (e.g., second mode known as Sezawa mode) appear with the increase in the thickness of the ZnO thin film as shown in Figure 5.6.

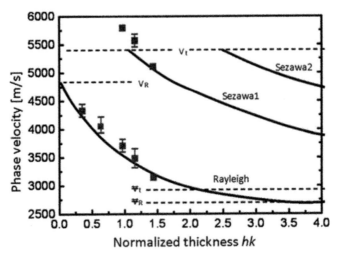

FIGURE 5.6 Phase velocities as a function of normalized thickness (reprinted with permission from Ref. [252]). (Reprinted with permission from Du, X. Y.; Fu, Y. Q.; Tan, S. C.; Luo, J. K.; Flewitt, A. J.; Milne, W. I.; Lee, D. S.; Park, N. M.; Park, J.; Choi, Y. J.; Kim, S. H.; Maeng, S. ZnO film thickness effect on surface acoustic wave modes and acoustic streaming. *Appl. Phys. Lett.* 2008, *93*, 094105. © 2008, AIP Publishing LLC.)

Substrate effect means that the difference of the acoustic velocities between the piezoelectric thin film and the substrate affects the acoustic properties of the ZnO SAW devices. The phase velocity in a ZnO thin film is dependent on the acoustic properties of the substrate. The phase velocity of the SAW becomes larger when the ZnO thin film with a fixed thickness is deposited on the substrate with high acoustic velocity such as sapphire, diamond, NCD, or DLC [253–255]. As for diamond materials, including

NCD and ultra-nanocrystalline diamond (UNCD), they have the highest SAW propagation velocity, largest elastic modulus, and lowest thermal expansion coefficients. Theoretical simulation and experimental analysis showed that the acoustic energy can be limited within the ZnO and diamond layers, and the propagation loss of the SAW can be significantly reduced using the diamond films [254,256]. For instance, theoretical modeling using the Campbell and Jones method showed that ZnO/diamond SAW device could approach a phase velocity in the range of 7180–10, 568 m s^{-1}, k^2 of 1.56–7.01%, and a temperature coefficient of frequency (TCF) of 22–30 ppm °C^{-1} [257].

Second-order effects often hinder the formation of good acoustic signals. Tripe-transit effect is generated from the interference between acoustic waves that propagate different paths. The useful signals are reflected from the output IDT and sent back to the input IDT, where it is reflected back to the output IDT. The multipeaks in the form of spurious ripples are often observed for the frequency response of the higher order mode of the SAW. In order to minimize this effect, acoustic absorbers are generally used, and mismatching the finite source and load impedances of the input and output IDTs enable the reduction of this effect at the expense of the increased insertion loss [258]. The effect of metallization ratio is also necessary to be considered in the design and fabrication of the ZnO SAW device. The ratio of the finger width to spatial period has significant contributions to the harmonic signals (e.g., frequency and insertion loss). The thickness of the finger electrodes also affects the central frequency, and variation of the electrode thickness could lead to a distortion of the frequency response. SAW reflections and diffractions are not circumvented due to the acoustic wave properties, thus, appropriately designed structures of the IDTs could minimize this effect.

Recent advances of the ZnO SAW devices have shown that their performances can be enhanced by using high-quality ZnO thin films with multilayer structures, designing of novel structures of the electrodes (e.g., curved IDTs and circle IDT), and flexible substrates (e.g., polyimide film) [12,259]. Considering the bulk piezoelectric materials which have large coupling coefficients (e.g., LiTaO$_3$ and LiNbO$_3$), ZnO thin films have often been deposited on top of them to fabricate the ZnO-layered SAW devices or called Love mode SAW [260,261]. When Fe-doped ZnO films were used to fabricate high-frequency SAW filters on Si, a better performance was found compared with the SAW filters on undoped ZnO films, for example, the electromechanical coupling coefficient and bandwidth increases up to 75.7% and 14.8%, respectively, while the insertion loss decreases 20.3% [34]. Low-cost deposition of the smooth and continuous ZnO films with densely packed vertical ZnO nanorods on silicon substrates was achieved using a chemical solution method, which was used to successfully fabricate the ZnO SAW device with a well-defined resonant signal [262].

5.4.2 FILM BULK ACOUSTIC RESONATOR

ZnO film bulk acoustic resonators (FBARs) are the devices that consist of a ZnO film sandwiched between two thin electrodes, which were first demonstrated in 1980 [263]. The mechanical resonance is generated from the applied electric field on the input electrode. In order to enhance the resonant signals, the working piezoelectric unit of FBARs is isolated acoustically from the supporting substrates. Three types of the structures can be designed for the back electrodes of FBARs such as Bragg reflector type, air-gap type, and back trench. Considering the ZnO film with a definite acoustic velocity of V, the relation between the thickness (d), and resonant frequency (f_n) is determined as follows [264]:

$$f_n = \frac{\pi(n+1)V}{2d} \tag{5.4}$$

where the natural number, n, corresponds to different resonant modes [261]. This relation clearly shows a thickness effect for FBARs, for example, the thinner ZnO film results in a higher resonant frequency.

The design of FBARs has less degrees of freedom than that of SAW devices because the frequency is only determined by the layer stack rather than by the lithography technique. The typical electrode materials are gold or aluminum and later Mo was proved to be good success in depositing low-stress electrodes. The performance of FBARs is simply evaluated from the electromechanical coupling coefficient (k^2), quality factor, TCF, area efficiency, and environmental robustness. Here, the area efficiency refers to the utilization of the piezoelectric layer apart from the dead area between resonators in topology and area consumed by interconnects and packaging. FBARs usually work in the range of a high frequency above 1 GHz, whereas the ZnO SAW devices have low resonant frequency, much lower than 2 GHz. Therefore, FBARs exhibit many advantages of small base mass, high sensitivity, efficient parallel detection by array of FBARs. They can be also integrated with LOC system and microfluidics.

Comparing with those using a longitudinal wave, the damping of the shear wave in a liquid is significantly reduced; therefore, the shear mode in FBARs is a good choice in its applications in liquid. Dual-mode ZnO FBARs with tilted c-axis orientation were investigated, which showed that material properties and bulk wave properties were are strongly dependent on the tilt angle. Pure thickness longitudinal modes was found at 0° and 65.4° for the ZnO FBARs, and pure thickness shear modes occur at 43° and 90° [265]. c-Axis inclined ZnO thin films were used in FBAR with Bragg reflector that operated at 850 MHz in the shear vibration mode, with a coupling coefficient of 1.7% and quality factors of 312 in air and 192 in water [266]. Recent study showed that a lateral field excited ZnO FBAR operated in pure-shear mode has a resonant frequency near 1.44 GHz and a Q-factor up to 360 in air and 310 in water, which reached the mass sensitivity of 670 $Hz \cdot cm^2 \cdot ng^{-1}$ and the high mass resolution of 0.06 $ng \cdot cm^{-2}$ [267]. Direct comparison of the gravimetric responsivities of ZnO-based FBARs and solidly mounted resonators (SMRs) indicated that the FBARs' mass responsivity was about 20% greater than that of the SMRs' for the specific device design and resonant frequency at about 2 GHz, which was mainly due to the acoustic load at one of the facets of the piezoelectric films in the Bragg reflector [268].

5.4.3 ULTRASONIC TRANSDUCERS

ZnO ultrasonic transducers operated in a large range of 20 kHz–20 MHz are typically applied for nondestructive testing and biomedical imaging [269–271]. Typical ZnO ultrasonic transducers consist of the top electrode, ZnO piezoelectric film, back electrode, and the matching layer on backing materials. The simplest single-element structure is the piston transducer. In the various designs of the ZnO ultrasonic transducers, the piezoelectric properties of ZnO film, acoustic wave modes, electrode size, effects of backing and matching, and array configuration are considered to enhance the performance of ZnO ultrasonic transducers. The resonant frequency (f) in the thickness mode in influenced by the thickness of the ZnO film and their relation is described by the following equation [272]:

$$f = \frac{nV_p}{d} \tag{5.5}$$

where n is an odd integer; the lowest resonant frequency corresponds to $n = 1$; V_p and d corresponds to the acoustic wave velocity and the thickness of the ZnO film. This relation indicates that the resonance only occurs when the thickness is equal to odd multiples of one-half wavelength of the acoustic wave.

Different substrates are widely used to deposit the ZnO films such as metal, silicon, pyrex-glass, and aluminum foil. Thin metal sheets (e.g., stainless steel, aluminum, and carbon steel) are used in the ZnO ultrasonic transducers which are widely used in the commercial ultrasonic transducers [272–275]. The Al foil supported ZnO film was proposed for the underwater ultrasonic transducers in the very high frequency and ultrahigh frequency ranges, and theoretical analysis showed that with spurious-response-free, wide-bandwidth, high-efficiency transducers could be fabricated based on the properly optimizing the thicknesses of the Al foil and the ZnO film [270,271]. The direction of the ultrasonic wave could be focused by depositing the ZnO film on the curved backing substrate as shown in Figure 5.7a [272]. The dome-shaped diaphragm has also been fabricated in the ZnO transducer for 200 MHz cellular microstructure imaging [273]. High frequency ZnO transducers with single element have been extensively reported that could be operated in the region of 100–300 MHz for ultrasonic imaging [274,275]. An acoustic lens-based ZnO transducer with a focus distance of 50–80 μm and lens aperture size of 100–400 μm has been fabricated that possesses a very large resonant frequency of 1 GHz as presented in Figure 5.7b [275].

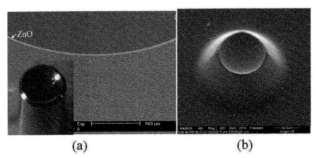

(a)   (b)

FIGURE 5.7 Deposited ZnO films on the curved backing substrates: (a) sputtering ZnO film on the Al rods [272], (b) the acoustic lens-based ZnO transducer [275] (reprinted with permission from Refs. [272,275]). (Reprinted with permission from a: Cannata, J. M.; Williams, J. A.; Zhou, Q. F.; Sun, L.; Shung, K. K.; Yu, H.; Kim, E. S. Self-focused ZnO transducers for ultrasonic biomicroscopyJ. Appl. Phys. 2008, 103 (8), 084109. © 2008, AIP Publishing LLC. b: Zhou, Q.; Lau, S.; Wu, D.; Shung, K. K. Piezoelectric films for high frequency ultrasonic transducers in biomedical applications. *Prog. Mater. Sci.* 2011, *56*, 139. © 2011, Elsevier.)

However, some issues need to be overcome such as poor resolution out of focal zone and limited frame rate. The scheme of linear arrays is a better way combining the technique of electronically sweeping a beam to solve above problem [276]. The shear mode ZnO ultrasonic transducers were also reported to be operated at a high frequency using the inclined ZnO films [277,278]. The maximum amplitude of the shear mode wave could be accessed when the inclined angle changes at 41°. The c-axis zig zag ZnO film ultrasonic transducers were also designed for longitudinal and shear wave resonant frequencies and modes using the multilayered c-axis 23° tilted ZnO films [279]. Therefore, high-frequency ZnO ultrasonic transducer array is a promising trend in this field for the biomedical detection and imaging.

5.5 ZnO-BASED ACOUSTIC WAVE MICROFLUIDICS

SAW microfluidics provides a popular and attractive method for LOC system. Various fluidic phenomena can be observed that are applied to enrich the functionality of the microfluidic device. When the SAWs are refracted into the liquid, the acoustic energy is transferred mainly as fluidic kinetic energy in the liquid. The microfluidic actuation and manipulation depends on the coupling of the fluid with SAW, which provides the SAW-based microfluidic devices with advantages of simple, compactness, no moving parts, low-cost, highly biocompatibility, versatility, and easy integration [280].

Comparing with the bulk piezoelectric substrates (e.g., $LiNbO_3$, $LiTaO_3$, and quartz), ZnO thin films on silicon substrates allow integration of silicon processing technology, microfluidic, and sensing functions into one device, which exhibits advantages of low cost, flexibility, and large-scale production [12,252,254]. This offers new options for the SAW microfluidics using ZnO SAW devices, and tremendous efforts have been devoted to investigate the microfluidic manipulation and control.

5.5.1 THEORY OF SAW-BASED MICROFLUIDICS

The Rayleigh SAW and its harmonics (e.g., Sezawa wave) are often excited on the ZnO SAW devices, which consist of a longitudinal and a vertically polarized shear component. When the Rayleigh SAW encounter a liquid (e.g., a droplet) on the propagation path it refracts into the liquid at a Rayleigh angle that is determined by the Snell law of diffraction, as shown in Figure 5.8 [281]. The Rayleigh angle is calculated using the following formula:

$$\theta_R = \arcsin\left(\frac{c_1}{c_s}\right) \tag{5.6}$$

where c_l and c_s correspond to the velocity of the Rayleigh SAW wave in the liquid and substrate, respectively. For instance, when the Rayleigh wave on a ZnO SAW device is coupled into water liquid, the SAW velocity in water is 1490 m s^{-1}, and that on the ZnO thin film is about 4200 m s^{-1}, and the Rayleigh angle, θ_R, is found to be 20.95°. Of course, the Rayleigh angle will be changed with the wetting property of the divers liquid.

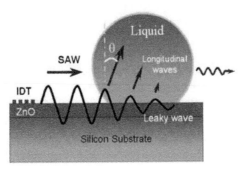

FIGURE 5.8 Schematic illustration of the coupling between the Rayleigh SAW and a liquid droplet (reprinted with permission from Ref. [281]). (Reprinted with permission from Du, X. Y.; Fu, Y. Q.; Luo, J. K.; Flewitt, A. J.; Milne, W. I. ZnO film thickness effect on surface acoustic wave modes and acoustic streaming. *J. Appl. Phys.* 2009, *105*, 024508. © 2009, AIP Publishing LLC.)

When the SAW propagates into the liquid, a net pressure gradient, P, is generated along the direction, which is described as following:

$$P = \rho_0 v_s^2 \left(\frac{\Delta \rho}{\rho_0} \right)^2 \tag{5.7}$$

where ρ_0 and $\Delta \rho$ correspond to the liquid density and the slight density change due to the acoustic pressure. An effective force in the liquid is formed from the net pressure gradient. When the force is not large enough to deform the liquid shape, SAW-induced acoustic streaming occurs in liquid. The streaming pattern is significantly changed with the shape of the confined liquid (e.g., droplet), as well as the incident position, angle, and the operating frequency of the SAW [282]. The induced flow streaming in the liquid droplet is governed by the incompressible 3D Navier–Stokes equations, which are driven by an external acoustic force, F [283]:

$$\frac{\partial U}{\partial t} + (U \cdot \nabla)U = \frac{1}{\rho}F - \nabla P + \eta \nabla^2 U, \tag{5.8}$$

$$\nabla \cdot U = 0 \qquad (5.9)$$

where U is the streaming velocity, P is the kinematic pressure, v is the kinematic viscosity, and F is the acoustic body force due to the presence of the sound source [283]. When the applied RF powers are less than a special value, the droplet can keep in its original shape without inducing significant distortion. Consequently, stress free boundary conditions can be applied at the droplet–air interface. The nonlinear term $(U \cdot \nabla)U$ in Eq. (5.9) was ignored, which gives a transient version of linearized hydrodynamic model. Eq. (5.9) is simplified as following:

$$0 = \frac{1}{\rho} F - \nabla P + \eta \nabla^2 U. \qquad (5.10)$$

Different flow patterns of the SAW induced acoustic streaming in liquid droplet can be simulated combining with the parameters of the SAW and liquid. The careful comparisons of streaming features induced by the SAWs from ZnO and LiNbO$_3$ SAW devices are expected, which is thought to thoroughly expand SAW induced microfluidics to a more integrated platform for ZnO SAW microfluidic device.

When the applied RF powers are large enough to deform the shape of the liquid, the acoustic force confined in the droplet can be calculated from the asymmetry in these contact angles and the droplet size using the following formula [281]:

$$F_s = 2R\gamma_{LG} \sin\left(\frac{\theta_t + \theta_l}{2}\right)(\cos\theta_t - \cos\theta_l) \qquad (5.11)$$

where R is the radius of the droplet and γ_{LG} is the liquid–gas surface energy; θ_t and θ_l correspond to the trailing edge and the leading edge of the droplet on the substrate [281]. This force can generate pumping, ejection, and atomization at extremely large driven powers. These SAW-induced microfluidic phenomena are very complicated due to various boundary conditions, which results in only a few simulations on their nonlinear fluidic dynamics. Although many experimental investigations of the ZnO SAW microfluidics have been reported, the accurately theoretical analysis is still urgently needed to quantitatively reveal the influencing factors in the SAW induced microfluidics.

5.5.2 ZnO-BASED MICROFLUIDIC MANIPULATION AND CONTROL

Using the Rayleigh SAWs excited from the ZnO SAW devices, various studies have demonstrated that the ZnO-based SAW could successfully manipulate the microfluid and particles in liquid and control the fluidic behaviors. Liquid droplets are the main targets that are mixed, pumped, ejected, and heated in the ZnO SAW microfluidics. This offers great potential for the increased throughput and scalability of the microfluidics systems. For instance, the droplet can be used as microreactors, which can be individually streamed, mixed, pumped, ejected, and atomized.

5.5.2.1 MICROSTREAMING AND PARTICLE CONCENTRATION

The microstreaming induced by the SAWs is dependent on the coupling intensity of the SAWs with the liquid. The signal power, droplet position, aperture of the IDTs, and surface roughness could remarkably change the flow patterns in a droplet. When the droplet is placed on the propagation path of the SAW, a butterfly flow pattern with a double vortex pattern is resulted due to the symmetrical distribution of the acoustic energy in the droplet. The quadrupolar streaming patterns could be obtained when the single IDT is immersed in liquid and the bidirectional SAWs are excited.

The streaming velocity in the droplet was found to be linearly changed with the applied voltage of the RF signal, which ranges from 0.2 cm s^{-1} to 1 cm s^{-1} as shown in Figure 5.9a [281]. The Sezawa wave may be more efficient than the Rayleigh wave for the acoustic streaming. A smaller size of the droplet could obtain a larger velocity when the applied voltage was constant. This has been interpreted from the force–voltage relation that is expressed as the following equation [281]

$$F_s = \alpha(V - V_{th})^2 \left(\frac{\sin\left[\beta(V - V_{th})\right]}{\beta(V - V_{th})} \right)^2 \tag{5.12}$$

where α is a force–voltage coupling coefficient and β corresponds to the constant of proportionality relating the input voltage to the temperature-induced frequency shift [281]. The power fed into the acoustic wave has two effects including (1) increasing the amplitude of the acoustic wave and the force acting on the droplet; and (2) a shift in the resonant frequency due to heating. When a flexible ZnO/polyimide SAW device was used in the acoustic induced streaming, Figure 5.9b indicates that a better efficiency of streaming could be reached with a speed of 3.4 cm s^{-1} at a signal voltage of 9.5 V for a 10-mL droplet, as well as the linear relation between the streaming velocity and applied voltage [259].

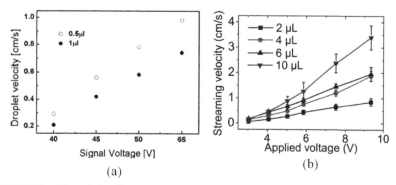

(a)   (b)

FIGURE 5.9 The relation between the applied voltage and streaming velocity based on the ZnO film (a) on silicon substrate [281] and (b) on the flexible polyimide substrate [259] (reprinted with permission from Refs. [281,259]). (Reprinted with permission from a: Du, X. Y.; Fu, Y. Q.; Luo, J. K.; Flewitt, A. J.; Milne, W. I. Microfluidic pumps employing surface acoustic waves generated in ZnO thin films. *J. Appl. Phys.* 2009, *105*, 024508.© 2009, AIP Publishing LLC.b: Jin, H.; Zhou, J.; He, X.; Wang, W.; Guo, H.; Dong, S.; Wang, D.; Xu, Y.; Geng, J.; Luo, J. K.; Milne, W. I. Flexible surface acoustic wave resonators built on disposable plastic film for electronics and lab-on-a-chip applications. *Sci. Rep.* 2013, 3, 2140. © 2013, Nature Publishing Group.)

An asymmetric distribution of SAW radiation in the droplet results in the concentration of particles when a droplet with half contact area is placed on the propagation path. Figure 5.10a shows the concentration of the

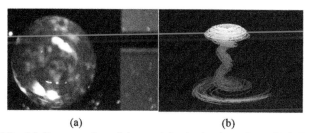

(a)   (b)

FIGURE 5.10 (a) Concentration of the particles in droplet using a ZnO SAW device and (b) a mechanism of the particle concentration (reprinted with permission from Ref. [283]). (b: Reprinted with permission from Alghane, M.; Chen, B. X.; Fu, Y. Q.; Li, Y.; Luo, J. K.; Walton, A. J. Experimental and numerical investigation of acoustic streaming excited by using a surface acoustic wave device on a 128° YX-LiNbO3 substrate. *J. Micromech. Microeng.* 2011, 21, 015005.© IOP Publishing. http://iopscience.iop.org/article/10.1088/0960-1317/21/1/015005/meta)

particles in droplet using a ZnO SAW device with resonance frequency of 61.2 MHz. This phenomenon can be explained using the similar mechanism proposed from the particle concentration on LiNbO$_3$ SAW device as shown in Figure 5.10b [283]. The surface friction hinders the streaming velocity at the bottom of the droplet to be reduced to zero. Thus, a secondary bulk circulation flow is generated from the primary azimuthal rotation flow around the droplet periphery, which produces the swirl-like flow at the bottom of the droplet [14]. The fluid close to the bottom of the droplet moves toward the center stagnation point and takes the particles along with it; thus, the particles can be concentrated near the stagnation area.

5.5.2.2   MICROPUMPING AND LIQUID JETTING

When the RF signal with a high power is applied on the ZnO SAW device, an acoustic force in the droplet is large enough to move the droplet as a micropump. Significant liquid jetting from the original droplet has also been realized at relatively high powers. The surface wetting property of the ZnO film is hydrophilic, that is, the droplet could be stretched along the direction of the SAW propagation. Therefore, the surface treatments are often performed using the hydrophobic polymer layer (e.g., Teflon, CYTOP, or silanization).

The effects of the droplet volume on the pumping velocity of the droplet are shown in Figure 5.11 using a ZnO/Si SAW device [281]. The pumping velocity increases exponentially with the input power. Based on the comparison of the pumping velocities on the ZnO/Si SAW device with those on the ZnO/UNCD/Si SAW device, the pumping efficiency is enhanced for the same droplet size when the acoustic driving force increases as shown in Figure 5.11b. The pumping velocities are less than 1 cm s^{-1} as shown in Figure 5.11a when the UNCD has not been used. The enhancement of the pumping velocity using UNCD interlayer is a combination effect from (1) the increase in the coupling efficiency between the SAWs and the fluids, and (2) the retention of the SAWs near the surface of the devices. The high-quality ZnO films deposited using the advanced HiTUS technique allow the efficient excitation of the SAW due to the high piezoelectricity. The UNCD interlayer with a very low surface roughness can further reduce the dissipation of the acoustic energy into the Si substrate and reduce penetration depth of the Rayleigh wave into the Si substrate, resulting in a strong Rayleigh mode wave and the reduction of the propagation loss of the wave [254,256]. The electromechanical coefficient is also increased by the UNCD-modified

ZnO/Si structure [82,254]. However, a quantitative separation of the two effects is difficult. Therefore, a larger driving force is formed by the increased amplitude of the SAW in the acoustic–liquid interaction.

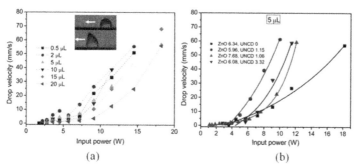

(a)   (b)

FIGURE 5.11 (a) Pumping velocities of the droplets on ZnO/UNCD SAW devices and as a function of the input power and (b) pumping velocities of the droplets measured from different layered structures (reprinted with permission from Ref. [284]). (Reprinted with permission from Pang, H. F.; Fu, Y. Q.; Garcia-Gancedo, L.; Porro, S.; Luo, J. K.; Placido, F.; Wilson, B.; Flewitt, A. J.; Milne, W. I.; Zu, X. T. Enhancement of microfluidic efficiency with nanocrystalline diamond interlayer in the ZnO-based surface acoustic wave device. *Microfluids Nanofluids* 2013, *15*, 377. With kind permission from Springer Science and Business Media.)

An empirical relation between the pumping velocity, V_{pumping}, and input power, P, was found to be the following relation [284]:

$$V_{\text{pumping}} = ae^{P/\beta} + \beta \tag{5.13}$$

where a, b, and β are constants that are mainly dependent on the droplet volume, wetting properties of the surface, the liquid viscosity, and the acoustic property of the ZnO-based SAW devices.

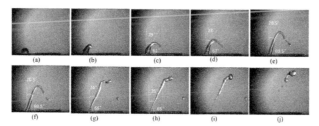

FIGURE 5.12 High-speed images of the liquid ejection for a 5-μl water droplet on the surface of the ZnO/UNCD SAW device (reprinted with permission from Ref. [284]). (Reprinted with permission from Pang, H. F.; Fu, Y. Q.; Garcia-Gancedo, L.; Porro, S.; Luo, J. K.; Placido, F.; Wilson, B.; Flewitt, A. J.; Milne, W. I.; Zu, X. T. Enhancement of microfluidic efficiency with nanocrystalline diamond interlayer in the ZnO-based surface acoustic wave device. *Microfluids Nanofluids* 2013, *15*, 377. With kind permission from Springer Science and Business Media.)

If the streaming inside the liquid droplet possesses sufficient inertia to overcome the capillary forces acting on the interface of the fluid and surrounding media, the drop interface will be deformed, and a liquid jet in the form of an elongated beam was emanated from the free surface of the drop. Early reports of the liquid jetting using the ZnO acoustic wave devices are aiming to the inkjet printing, fuel and oil ejection, and biotechnology [12].

High-speed images provide the detailed process of the liquid ejection for a 5-μL water droplet on the ZnO/UNCD SAW device as shown in Figure 5.12. The maximum jetting angle is also dependent on the droplet volume. Figure 5.13 shows that the shape of the liquid beam changes with the time and droplet size due to the strong propagating loss of the acoustic radiation at the end of the extremely bent liquid beam.

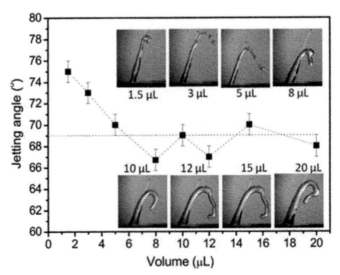

FIGURE 5.13 Maximum jetting angle and beam shape change with the droplet size (reprinted with permission from Ref. [284]). (Reprinted with permission from Pang, H. F.; Fu, Y. Q.; Garcia-Gancedo, L.; Porro, S.; Luo, J. K.; Placido, F.; Wilson, B.; Flewitt, A. J.; Milne, W. I.; Zu, X. T. Enhancement of microfluidic efficiency with nanocrystalline diamond interlayer in the ZnO-based surface acoustic wave device. *Microfluids Nanofluids* 2013, *15*, 377. With kind permission from Springer Science and Business Media.)

5.5.2.3 NEBULIZATION AND LIQUID HEATING

Nebulization is the generation of an aerosol of small solid particles or liquid droplets at micron and submicron scale. It is important to a wide range of the numerous applications such as internal combustion engines, drug-delivery device, mass spectrometry, nanoparticle synthesis, agriculture, and cosmetics [280]. The nebulization or atomization has been reported using the ZnO-based SAW device [285]. Some characteristics of the nebulization on the ZnO SAW device are presented here. First, the substrate surface is hydrophilic and the frequencies of the Rayleigh SAWs are not large (e.g., 11–37 MHz). Second, the nebulization in macroscopic view is like a micro-volcano. Third, the droplet volume is very small (e.g., a few microliter or below) and the ejected satellite tiny droplets are not uniform in size. Figure 5.14 shows the details of the nebulization for a 0.3-μL water droplet. The acoustic pressure induces significant capillary waves on the surface of the liquid droplet that quickly becomes unstable. Some satellite droplets are ejected from the surface following with the generation of a significant mist in the vicinity of the top of the droplet.

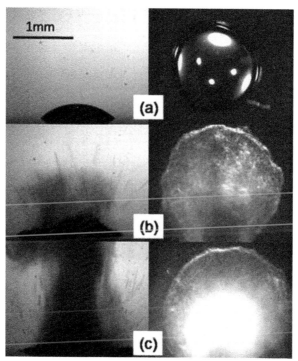

FIGURE 5.14 Typical process of the nebulization on the surface of the ZnO-based SAW device (reprinted with permission from Ref. [285]). (Reprinted with permission from Fu, Y. Q.; Li, Y.; Zhao, C.; Placido, F.; Walton, A. Surface acoustic wave nebulization on nanocrystallineZnO film. *J. Appl. Phys. Lett.* 2012, *101*, 194101. © 2012, AIP Publishing LLC.)

Liquid heating is a thermal effect of the SAWs and widely exists in the SAW induced microfluidics [281]. The increased temperature results in shifts of the resonant frequency of a SAW device, which has an important influence on the accuracy of microfluidic control and the sensing using the SAW devices. The surface temperature of the SAW devices was changed with the applied voltage of the RF signal that is measured using a thermocouple. The temperature range was varied from room temperature to about 120 in 1 min [281]. The thermal image in Figure 5.15b clearly shows that the temperature near the IDT is higher than those in the other area of the SAW device due to the acoustic heating effect from SAW. The temperature variations are proportional to the applied power for the heating rate of the thermal effect in the ZnO SAW device. It is expected to improve the temperature stability by effectively reducing the thermal effect of the ZnO/UNCD SAW device using a diamond layer [284]. The heat could damage the integrity of biological substances and decrease the bioactivity; however, the acoustic thermal effect of the SAW is a good option to assist the polymerase chain reaction (PCR) [286]. Well controlling the increase of the temperature in liquid may be an attractive application in SAW-assisted PCR for the ZnO SAW devices.

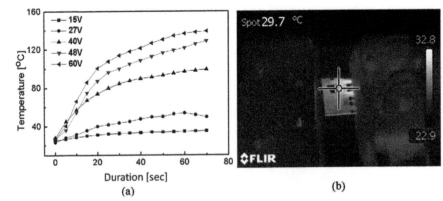

FIGURE 5.15 (a) Thermal effects of the ZnO SAW device and (b) IR images of the ZnO SAW device (reprinted with permission from Ref. [281]). (a: Reprinted with permission from Du, X. Y.; Fu, Y. Q.; Luo, J. K.; Flewitt, A. J.; Milne, W. I. Microfluidic pumps employing surface acoustic waves generated in ZnO thin films *J. Appl. Phys.* 2009, 105, 024508. © 2009, AIP Publishing LLC.)

5.6 ZnO THIN FILMS AND NANOSTRUCTURES FOR SENSING APPLICATION

5.6.1 ZnO SAW SENSORS

ZnO thin films and nanostructures are extensively used for the SAW sensors as well as the other types of the sensors based on the resistance, field-effect transistor, thin-film transistor, and diode. For the principle of the SAW sensors, the SAW will be modulated on the path of the SAW propagation by the sensitive area because the physical properties of the sensing materials are changed by the adsorbed gas molecules, UV light, biological molecules in liquid, or temperature. The responses of the ZnO SAW sensors are resulted in forms of the phase changes, reduction of the insertion loss, or frequency variations. Furthermore, the perturbations in the velocity of the SAW may be caused from the variations of many parameters including mass load (Δm), conductivity ($\Delta\sigma$), mechanical constant (Δc), dielectric constant ($\Delta\varepsilon$), temperature (ΔT), and pressure (ΔP), which is provided as following formula [287]:

$$\frac{\Delta V}{V} = \frac{1}{V}\left(\frac{\partial V}{\partial m}\Delta m + \frac{\partial V}{\partial\sigma}\Delta\sigma + \frac{\partial V}{\partial c}\Delta c + \frac{\partial V}{\partial\varepsilon}\Delta\varepsilon + \frac{\partial V}{\partial T}\Delta T + \frac{\partial V}{\partial P}\right) \quad (5.14)$$

where the dominated factors for the SAW sensors are varied due to the changes in the sensing target. The performance of the sensors could be estimated from the parameters such as the sensitivity, detect limit, selectivity, response time, recovery time, stability, and reproducibility.

   In this section, the SAW sensors based on ZnO thin films and nanostructures are focused that mainly involve detecting gases, chemicals, UV lights, and biological molecules.

5.6.1.1 ZnO SAW GAS SENSORS

SAW gas sensors are the techniques that deposit a sensing layer on the SAW devices. The adsorbed gas molecules on the sensing layer could change the acoustic properties of the SAWs on the propagating path and subsequently the variations of the attenuation and velocity of the SAW occur. The perturbation of the SAWs is remarkably dependent on the number of the adsorbed molecules and a large concentration of the gas results in strong modulation of the SAWs [284]. In order to improve the sensitivity, oscillation circuits are designed and a dual-device configuration is often adopted to reduce the interference of the environment (e.g., humidity and temperature). Some other factors also need to be considered in the sensor design and fabrication, such as acoustic vibration mode, substrates, geometry of the IDTs, and the sensing film depositions. Therefore, a good SAW gas sensor can be resulted from the depositing highly sensitive materials, a good compatibility, and integration of the sensor element with the detection system and appropriate analysis methods of the sensing data.

As for ZnO SAW gas sensors, they have two structural types: ZnO sensing layer on the SAW device and the sensing layer with different materials on the ZnO SAW device. For the first type, the ZnO sensing layer with nanostructures is deposited on the surface of the piezoelectric substrates using various growth techniques. The ZnO nanorods were grown on a layered ZnO/64°-YX LiNbO$_3$ SAW device using a liquid solution method for H$_2$ sensing; and sensor response was measured as 274 kHz toward 0.15% of H$_2$ at 265°C [288]. The electrospun ZnO nanostructured thin film was prepared on a commercial quartz Rayleigh SAW device operated at 433 MHz using an electrospray technique. This sensor exhibited responses to different volatile organic compounds such as acetone, trichloroethylene, chloroform, ethanol, n-propanol, and methanol vapor. It is expected to be applied for the sensor array for detecting different volatile organic compounds in a special environment [289].

The different materials are used as the sensing layer on the ZnO SAW device, forming the second-type sensor design. A uniform nanostructured InO$_x$ layer was sputtered on the ZnO/LiNbO$_3$, which served as a sensor that demonstrated a high sensitivity such as positive frequency shifts of 91 kHz for 2.125 ppm of NO$_2$ and negative frequency shifts of 319 kHz for 1% of H$_2$ in synthetic air [260]. Platinum (Pt) and gold (Au) catalyst activated tungsten trioxide (WO$_3$) selective layers were deposited on a 36° Y-cut LiTaO$_3$ SAW devices for sensing hydrogen (H$_2$) concentrations; and frequency shifts of 705 and 118 kHz toward 1% of H$_2$ in air could be approached, respectively [290]. A 150-nm WO$_3$ sensing layer was introduced on the 36° YX LiTaO$_3$-layered SAW device with a 1.2-μm ZnO intermediate layer for sensing ethanol in dry and humid air. It detected 500 ppm of ethanol in synthetic air with frequency shifts of 119, 90, and 86 kHz, corresponding to the relative humidity of 0%, 25%, and 50%, respectively [291]. Palladium (Pt)-coated ZnO nanorods were used as the selective layer that was prepared on the 128°YX-LiNbO$_3$ SAW device for hydrogen detection. The SAW sensor operated at a frequency of 145 MHz showed a frequency shift of 26 kHz for a hydrogen concentration variation of 6000 ppm at room temperature [292]. A bilayer SAW sensor with a layered structure of WO$_3$ (about 50 nm) and

Pd (about 18 nm) thin films on the Y-cut LiNbO$_3$ substrate for hydrogen gas-sensing application; the sensing results indicated that the frequency changes correlated very well with decreases of the bilayer structure resistance and the sensitivity was at a level of 2.4 kHz for 4% of hydrogen concentration in dry air [293]. The SAW device with new structures was also reported to be used as a gas sensor, for example, a ball-type SAW sensor was fabricated by sputtering Pt-coated ZnO film as a sensitive film. The sensing measurements showed that the sensor could detect H$_2$ at a concentration of 20 ppm and the amplitude change of −0.79 dB. When the sensor was wetted by water, it could detect 20 ppm of H$_2$ with an amplitude change of 0.4 dB, which indicated that the sensitive film was not deteriorated by wetting [294].

In order to improve the sensitivity of the ZnO SAW gas sensors, some assisted techniques are good options, for instance, exploring new sensing materials, designing a sensor array, etc. The selectivity of the ZnO SAW gas sensors are enhanced by adopting appropriate analytical methods including cluster analysis, factorial analysis and regression analysis [295]. Although the ZnO SAW gas sensors have good sensitivity, the selectivity still needs to be improved for the mixture of different gases because the metal or tran-sitional-metal oxides have high affinities to other gases and humidity. If the sensing layer is not active at room temperature, these sensors are normally operated at a high temperature, which limits the wide applications of the ZnO SAW devices in various extreme environments.

5.6.1.2   ZnO SAW UV SENSORS

The SAW UV sensor is to detect the UV light with a wavelength ranging from 400 to 10 nm. The UV-sensitive layer is grown on the SAW propagation path atop of the piezoelectric substrate. When the sensitive layer is exposed for the UV light, electron–hole pairs will be generated and interact with the propagating SAW, thus, the insertion loss of the SAW device increases, and the acoustic velocity will be reduced.

ZnO films serve as a photoconducting layer that can be deposited on the SAW propagating path. A good photoresponse to the slight variation of the acoustic–electric interaction could be detected, which arise from the changes of the sheet resistivity and carrier concentration in the ZnO film [296]. The variations of the velocity and amplitude for the SAW signal can be exploited to monitor the physical variation of the sensing layer, surface, and interface of the multilayers among sensing layers and piezoelectric substrates. They are also influenced by the wavelength and optical power density of the UV light.

Typical ZnO-sensing layers have been deposited on the SAW UV sensors. If the thickness of the ZnO film is smaller than the acoustic wavelength, the velocity shift (ΔV) and the attenuation ($\Delta\Gamma$) induced by the acoustic–electric interaction could be estimated by the following equations [297]:

$$\frac{\Delta V}{\Delta V_0} = -\frac{k^2}{2}\frac{\sigma_{sh}^2}{\sigma_{sh}^2 + V_0^2 C_s^2} \tag{5.15}$$

$$\Delta\Gamma = \frac{k^2}{2}\frac{V_0^2 C_s \sigma_{sh}}{\sigma_{sh}^2 + V_0^2 C_s^2} \tag{5.16}$$

where V_0, ΔV, k^2, C_s, and σ_{sh} corresponds to the SAW velocity on free surface, SAW velocity difference, coupling coefficient, capacitance per unit length of the surface, and the sheet conductivity of the ZnO film, respectively. This indicates that a small downshift of the frequency results in a decrease of the phase velocity, and the amplitude change correspond to SAW attenuation under the UV irradiation. It also indicates that the coupling coefficient k^2 is the key factor to influence the change of velocity and attenuation. A higher k^2 is desirable for a larger acoustoelectric effect and hence a higher sensitivity.

For ZnO SAW UV sensing, the Rayleigh-mode SAW devices can be used to detect the acoustic–electric interaction, and the sensing mechanism have been interpreted [298,299]. For instance, a 200-nm-thick ZnO film was sputter-deposited onto a LiNbO$_3$ SAW device operated at 37 MHz, which exhibited a large frequency shift of 170 kHz when illuminated using 365-nm UV light with an intensity of 10 mW·cm^{-2} [300]. When the 71-nm-thick ZnO thin film was deposited on a LiNbO$_3$ SAW filter with an operating frequency of 36.3 MHz, its sensitivity could reach a low-level intensity of 450 nW·cm^{-2} for the 365-nm UV light [301]. A polycrystalline (poly)3C–SiC buffer layer had been introduced in ZnO/Si structures, which could enhance the sensitivity of the UV SAW sensors from 25 Hz·(mW·cm^{-2})$^{-1}$ to 85 Hz·(mW·cm^{-2})$^{-1}$ [302]. Furthermore, nanocystalline ZnO films with various morphologies of nanoparticles, nanorods, and nanowires have been grown on the Rayleigh mode SAW devices to enhance their photoconductivity [303–306]. Besides the fundamental Rayleigh mode, the Sezawa mode and third-mode harmonics has also been adopted to detect the low-intensity UV light [297,304]. A Mg:ZnO layer was introduced to isolate the semiconducting ZnO layer from the piezoelectric ZnO layer in the ZnO/Mg:ZnO/ZnO/Si structure for the ZnO SAW filter, which exhibited a good sensitivity to a power density of 810 µW·cm^{-2} when operated using the Sezawa mode

TABLE 5.1 Parameters and Sensitivity of the ZnO SAW UV Sensors

Layered structure	Resonance mode	Operating frequency (MHz)	Optical power density ($\mu W\ cm^{-2}$)	Frequency shift (kHz)	References
ZnO/LiNbO$_3$	Rayleigh wave	37	10	170	[300]
ZnO/LiNbO$_3$	Rayleigh wave	36.3	0.45	28	[301]
ZnO/LiNbO$_3$	Rayleigh wave	145	500	40	[305]
ZnO particles/LiNbO$_3$	Rayleigh wave	64	691	–	[306]
ZnO/quartz	Rayleigh wave		19,000	45	[303]
ZnO/3C-SiC/Si	Rayleigh wave	122.25	600	150	[302]
ZnO/Mg:ZnO/ZnO/r-Al$_2$O$_3$	Sezawa wave	711.3	810	1360	[297]
ZnO/Si	Sezawa wave	842.8	551	1017	[298]
ZnO/Si	Third mode		3000	400	[304]
ZnO nanorods/quartz	Love wave	117	–	–	[308]
ZnO/LiTaO$_3$	Love wave	41.5	350	150	[261]

SAW of 711.3 MHz [297]. A large frequency shift of 1.017 MHz could be observed for the 385-nm UV light at a power density of 551 $\mu W \cdot cm^{-2}$ when the Sezawa mode SAW (with a frequency of 842.8 MHz) was used [298]. The response of the ZnO SAW UV sensors were often measured using a network analyzer. However, the ZnO SAW UV oscillator could be designed using an amplifier oscillator circuit, which significantly enhanced the sensitivity [301].

Love mode SAW is a guided wave that is generated from a shear-horizontal acoustic wave propagating in a guiding layer on top of the piezoelectric substrate [261]. SAW device with a Love mode has a good sensitivity to the mass change and photoconductivity variation for applications in liquid biosensing, gas sensing, and photodetection. A 1-μm-thick sputtered ZnO film on a 90° rotated ST-cut (42°45′) quartz SAW device had been reported that did not generate any apparent response texposed using a 365 nm UV light; however, ZnO nanorods grown on the ZnO/90° rotated ST cut (42°45′) quartz SAW device significantly increased the UV sensitivity [307]. Therefore, the photoconductivity of ZnO films is related to their crystalline structure and defect properties, which can be varied by controlling the film growth conditions. This enables the sensitivity of the Love mode SAW UV sensor to be optimized through sputtering the ZnO film on a shear-horizontal SAW device under appropriate growth conditions. It is known that rotated 90° ST-cut quartz has a low electromechanical coupling coefficient (k^2) of 0.11% and low dielectric constant of 4.5 [308,309]. However, the 36° Y-cut LiTaO$_3$ possesses a large k^2 of 4.7% and high dielectric constant of 47, which is beneficial to enhance the sensitivity of the Love mode SAW UV sensor. Love mode SAW UV sensor based on the ZnO/36° Y-cut LiTaO$_3$ structure was operated at ~41.5 MHz; the amplitude response is −6.4 dB and the frequency shift approached ~150 kHz under a 254-nm illumination at the power density of 350 $\mu W \cdot cm^{-2}$ [261]. The summarized performances of the ZnO SAW UV sensors are listed in Table 5.1.

5.6.1.3   ZnO SAW BIOSENSORS

SAW biosensors combine the SAW technique with the biological detector. They form the analytical devices for the analyte sensing based on the biochemical reaction, which could show selective and quantitative responses to the trace amounts of biological samples. SAW biosensors normally consist of the biochemical recognition component, SAW device coated with biospecific layer, and electronic component that process and amplify the output signal [310–312]. The biochemical recognition refers to the biore-ceptor that is designed from the interactions of the biomolecules with the target analyte, which is crucial to determining the functionality of the SAW biosensor. The SAW biosensor offers label-free detection of biomolecules and analysis of binding reactions. High selectivity of the SAW biosensor is based on the bioreceptor selectively interacting with the specific analyte against the various chemical and biological components [310–312]. Further-more, the biochemical reactions in the SAW biosensors are categorized as enzymes, antibody/antigen, nucleic acids/DNA, and cellular structures/cells [312–314]. The ZnO SAW biosensor is utilized based on the advantages of the ZnO piezoelectric films. The ZnO film deposited on the SAW device allows the SAW propagating in this guiding layer and forms the Love mode SAW device, which could reduce the acoustic loss into the liquid due to the confinement of the shear-horizontal wave in the guiding layer. Thus, the Love mode SAW is sensitively modulated by the changes of the sensor surface on the propagation path. The guiding layers are deposited using different materials, such as ZnO, polymer (e.g., polydimethylsiloxane and polymethylmethacrylate), SiO_2, and Au [260]. The ZnO and SiO_2 thin films were prepared on the 90° rotated ST-cut quartz SAW devices for immune sensing [314]. The measured mass sensitivity of the Love mode SAW device with ZnO layer was larger than that of with SiO_2 layers, and a high sensi-tivity of 950 cm^2 g^{-1} was detected for the adsorption of rat immunoglob-ulin G [314]. A highly sensitive Love mode SAW biosensor with a layered structure of $ZnO/SiO_2/Si$ was reported to be designed for the detection of interleukin-6 (IL-6) [315]. Preliminary sensing measurements showed a successful detection of IL-6 protein with a low level when it was operated at 747.7 MHz or 1.586 GHz [315]. The mass sensitivity reached 4.456 $\mu m^2/$pg when the active areas were functionalized by immobilizing the mono-clonal IL-6 antibody onto the ZnO biosensor surface [316]. Mn-doped ZnO multilayer structures were recently used to fabricate Love mode SAW biosensor for detecting blood sugar. The results showed that the reactions of the glucose oxidase with glucose are sensitive with 7.184 MHz mM^{-1} and stable for about 1 month [317].

Owing to the good integration of the ZnO SAW device with MEMS and COMS processes, novel designs of the ZnO SAW bionsensor were recently developed, for instance, COMS-SAW biosensor [318]. A SAW delay line biosensor was fabricated in standard CMOS technology for detecting cancer biomarkers, and the results based on the streptavidin/biotin reaction indi-cated that the sensitivity reached 8.704 pg Hz^{-1} and a mass sensitivity of 2810.25 m^2 kg^{-1} was detected [318]. Different microfluidic techniques, such as electrowetting on dielectrics, phononic crystal structure, and surface plasmon resonance, were also expected that would be combined with the ZnO SAW biosensors [12,15,286].

5.6.2 ZnO FBAR SENSORS

As a new, effective, and important acoustic sensor technique, the ZnO FBAR sensors could sensitively detect the chemical, biological, and optical variations from the adsorbed species because of their high performance such as the high quality factor and large operating frequency in the GHz regimes. The sensing principle of the ZnO FBAR sensors is similar to the sensor using the commercial quartz crystal microbalances (QCMs) which operates in thickness shear mode [319]. The resonant frequency of the ZnO FBAR sensors is highly sensitive to the changes of mass, UV light, and biochemical reaction in the working area. This results that the ZnO FBAR sensors are superior to the QCM sensors due to the relatively low resonant frequencies (i.e., less than 100 MHz) for the latter, which provides the advantages with the ZnO and FBAR sensors such as high sensitivity, low hysteresis, label free, good selectivity, and excellent compatibility with the integrated circuit [320]. Therefore, the ZnO FBAR sensors have been widely applied for the mass sensor, biosensor, gas sensor, and UV sensor.

Owing to the high mass sensitivity, the ZnO FBAR sensor with a high operated frequency will be an alternative to the traditional QCM sensors. Typical mass sensing using the ZnO FBAR sensors is that mass loading materials are deposited layer by layer on the top electrode, and the sensitivity is calculated based on the frequency shift, the film area, and film mass. Early work showed that the SMR-type FBAR sensors possessed a high sensitivity of 500 Hz·cm^2 ng^{-1} that was five orders of magnitude higher than that of 0.057 Hz·cm^2 ng^{-1} for the commercial 5 MHz QCM [321]. Combining with the CMOS technique, the ZnO FBAR oscillator array with a resonant frequency of 905 MHz could be fabricated for mass sensing, which exhibited a mass sensitivity of 328 Hz·cm^2 ng^{-1} [322]. A ZnO FBAR sensor with a resonant frequency of 2.44 GHz was measured using a network analyzer and a probe station, which could achieve an excellent sensitivity of 3654 Hz·cm^2 ng^{-1} for detecting the mass loading layers of titanium [319]. The further comparison of FBAR and QCM-D sensitivity showed that the larger resonant frequency of the ZnO FBAR could increase the sensitivity to the variations in the viscoelasticity of the adsorbent [323]. Based on the mass-loading effect, the ZnO FBAR sensor operated at 830 MHz in shear mode could detect the viscosity of the glycerol solutions up to 10 mPa·s [324]. The ZnO FBARs with dual longitudinal modes were designed and fabricated to parallel monitor the mass and temperature changes [325].

Based on the low damping of the bulk wave in liquid, biosensing is one of the important applications for the ZnO FBAR sensors. The backside of the silicon wafer was also etched using deep reactive ion etch to release the free-standing membrane. In the typical biosensing process, the antibodies were adsorbed or coated on the top electrode or the back membrane; the analytes were detected in the liquid through a specific reaction with the antibody on the working area. In order to keep the Q-factor from significantly damping, the channel or cavity structures are generally used to direct or confine the liquid flowing over the functionalized surface of the ZnO FBAR sensor [325].

There are three trends of the ZnO FBAR biosensors that are rapidly developed including the improvement of the sensitivity, enhancement of the integration with other functions, and expansions of the fields of the biological detections. For instance, when a 750-nm-thick ZnO thin film was sandwiched between the Au and Pt electrodes with a Bragg reflector, a high resonant frequency of 3.94 GHz was obtained [22]. It was used to detect the bio-immobilization mass and the mass of coupling protein up to 380 and 630 pg, respectively, which corresponds to a sensitivity of 8970 Hz·cm^2 ng^{-1} [22]. Recent sensing investigation of adsorption and antigen binding behavior of monoclonal antibody showed that the ZnO FBAR biosensor operated at about 1.5 GHz could reach 2 kHz cm^{-2} ng^{-1} during the detection of the human prostate-specific antigen [326]. Furthermore, the sensitivity could be enhanced through increasing the sensing area using the nanomaterials (e.g., nanotips or nanowires) and selecting the shear mode bulk wave. A film of the functionalized ZnO nanotips on FBAR using the magnesium zinc oxide ($Mg_x Zn_{1-x} O$) as piezoelectric material were reported to increase the mass sensitivity higher than 103 Hz·cm^2 ng^{-1} for selectively immobilizing DNA [327]. Beside the longitudinal mode, the shear mode was used for the FBAR biosensor with a inclined angle of 16° in the ZnO piezoelectric film to measure the model system avidin/anti-avidin, resulting in a sensitivity up to 585 Hz·cm^2 ng^{-1} [328].

Considering the flexible design of the microfluidic structure, the ZnO FBAR biosensor array was placed and sealed in a flow cell consisting of an acrylic glass chamber with an inflow and outflow that were connected to a fluidic system [328]. A microfluidic channel was integrated with the ZnO

FBAR biosensor to keep the quality factor up to 150 in the liquid environment [326]. The competitive adsorption and exchange behavior of proteins were monitored among the proteins of albumin (Alb), immunoglobulin (IgG), and fibrinogen (Fib); and a minimum detectable mass was reached for 1.35 ng cm^{-2} [329].

The ZnO FBAR biosensors are also used to detect the enzymatic reactions as well as the model system of the immune reactions (e.g., avidin/anti-avidin, IgE/IgG). The odorant-binding protein were reported to detect the N,N-diethyl-metatoluamide using a ZnO FBAR biosensor with a resonant frequency of about 1.5 GHz [326]. The sensing layer was made by immobilizing the acetylcholinesterase enzyme on one of the faces of the shear mode ZnO FBAR resonator with a resonant frequency of 1.47 GHz and Q-factors of 411 in air and 298 in dimethylformamide liquid, which could detect the organophosphorous pesticides with a very low concentration of 4.1×10^{-11} M [330].

The gases can also be detected using the ZnO FBAR sensors at room temperature combining the sensing materials. For instance, a frequency downshift of 131 kHz was observed with a response time of 12 s under the exposure of 1.4 ppm [331]. The nerve gas could be monitored through coating the poly(vinylidene fluoride) on the top of the ZnO FBAR with a W/SiO$_2$ Bragg reflector, which showed the gas sensitivity of 718 kHz ppm^{-1} and a good linear correlation in the range of 10–50 ppm between the frequency shifts and the concentrations of the nerve gas [332]. When a Pd thin film was deposited on the piezoelectric ZnO film as the electrode and sensing layer, the ZnO FBARs with an A/W Bragg reflector and resonant frequency of 2.39 GHz could capture hydrogen to reach a detection limit of 0.05% at room temperature [333].

As described previously about the fundamental properties, the photoconductivity of the ZnO films is prone to changing with the defects. The ZnO FBARs can be applied as the optical sensors to monitor the UV light and infrared (IR) light. The UV sensing with ZnO FBARs were demonstrated that the resonant frequency was increased under the exposure to the UV illumination [334]. The sensitivity could reach 9.8 kHz for the 365-nm UV light with the intensity of 600 μW cm^{-2} [334]. Further studies of the effects of temperature, relative humidity, and reducing gases on the UV response of ZnO-based FBAR showed that the response of the ZnO FBAR UV sensor degraded with the increases in the temperature and relative humidity; reducing the gases (e.g., acetone) also degraded the UV sensitivity [335]. The reason was that the adsorbed oxygen was influenced by the temperature, water molecular, and gas molecular [335]. Thickness field excitation FBAR and lateral field excitation FBAR were reported to detect the IR light, corresponding to the detection limits of 0.7 and 2 μW mm^{-2}, respectively. The sensing principle was explained based on the temperature-dependent Young's modulus of the ZnO film, which subsequently resulted in the resonant frequency shifts [334].

5.7   SUMMARY AND FUTURE TRENDS

This chapter presents a comprehensive review of the ZnO thin films and nanostructures for the acoustic wave microfluidic and sensing applications. The fundamental properties of the ZnO thin film and nanostructures are dependent on the growth techniques and growth conditions. The good piezoelectric ZnO thin films possess large electromechanical coupling coefficient that could be fabricated for the ZnO SAW devices with a better acoustic performance. The SAWs can be excited to mix, stream, pump, eject, and atomize the liquid. Therefore, the ZnO SAW devices are very attractive to be integrated into a LOC system where the SAWs can transport bioliquids to the desired area, mix the extracted DNA or proteins, and detect the changes of the signals. The ZnO SAW devices in combination with different sensing layers could be also used to successfully detect the gas, UV light, and biochemicals with remarkable sensitivity.

Although the studies of ZnO thin films and nanostructures have obtained great achievements and large success, the huge space of the research and development for future is still left. The multifunctional integration is promising for the ZnO SAW devices, such as combining with surface plasma resonance, electrowetting dielectrics, and phononic crystal structures. The novel design of the acoustic wave devices are interesting to enhance the microfluidic and sensing applications. They can excite new ideas for the various possible applications based on in-depth understanding the acoustic properties. The coupling mechanism of the acoustic wave with the fluids is sensitive to the deformation of the interface, which is difficult to predict; accurate interpretations of the interaction between the SAW and fluids is urgently needed for the design of novel microfluidic devices.

The ZnO FBAR sensor exhibited an excellent sensitivity for monitor the changes of the mass, bioreactions, gases, UV, and IR lights. The high resonant frequency and Q-factor are vital to the performance of the sensors, which will be improved with the new designs of the structures. The multiple functions are the future trends of the ZnO FBAR sensors highly integrated with different sensing materials and MEMS technique. However, some environmental perturbations still need to be overcome such as the temperature, humidity, optical response and the vibration from the holders. The mechanism of the special sensing processes is also not interpreted well when the physical parameters are monitored. The low cost, portable, high throughput, and miniature ZnO FBAR sensors are pursued with the development of the fabrication techniques in future.

ACKNOWLEDGMENTS

This work is supported by the University's development fund (grant number 201343) and Doctoral research fund (grant number 2014QDJ017) from Xi'an University of Science and Technology, the National Natural Science Foundation of China (NSFC grant numbers 11504291 and 11504292), UK Engineering Physics and Science Research Council (EPSRC grant number EP/P018998/1), Newton Mobility Grant (grant number IE161019) through Royal Society and NFSC, Royal Society of Edinburgh, Carnegie Trust Funding, the Royal Society-Research Grant (grant number RG090609), and Scottish Sensor System Centre (SSSC).

KEYWORDS

- **zinc oxide**
- **thin film**
- **nanostructure**
- **piezoelectric**
- **microfluidics**
- **surface acoustic wave**
- **lab on chip**

REFERENCES

1. Hull, R.; Jagadish, C.; Osgood, Jr.; R. M.; Parisi, J.; Wang, Z.; Warlimont, H. *Zinc Oxide from Fundamental Properties towards Novel Applications*; Springer-Verlag: Berlin, Heidelberg, 2010; p 2.

2. Honerlage, B.; Levy, R.; Grun, J. B.; Klingshirn, C.; Bohnert, K. *Phys. Rep.* **1985**, *124*, 161.

3. Brown, H. E. *Zinc Oxide, Properties and Applications*; Pergamon Press: New York, 1976.

4. Ding, Y.; Kong, X. Y.; Wang, Z. L. *Phys. Rev. B: Condens. Matter* **2004**, *70*, 235408.

5. Fan, Z.; J. Lu, G. *J. Nanosci. Nanotechnol.* **2005**, *5*, 1561.

6. Schmidt-Mende, L.; MacManus-Driscoll, J. L. *Mater. Today* **2007**, *10*, 40.

7. Janotti, A.; Van de Walle, C. G. *Rep. Prog. Phys.* **2009**, *72*, 126501.

8. [8] Wang, Z. L. *J. Phys.: Condens. Matter* **2004**, *16*, R829.

9. McCluskey, M. D.; Jokela, S. J. *J. Appl. Phys.* **2009**, *106*, 071101.

10. Zhang, Q. F.; Dandeneau, C. S.; Zhou, X. Y.; Cao, G. Z. *Adv. Mater.* **2009**, *21*, 4087.

11. Djurišić, A. B.; Ng, A. M. C.; Chen, X. Y. *Prog. Quant. Electron.* **2010**, 34, 191.

12. Y. Q. Fu, Luo, J. K.; Du, X. Y.; Flewitt, A. J.; Y. Li, Markx, G. H.; Walton, A. J.; Milne, W. I. *Sens. Actuators, B* **2010**, *143*, 606.

13. Abgrall, P.; Gue, A.-M. *J. Micromech. Microeng.* **2007**, *17*, R15.

14. Friend, J.; Yeo, L. Y. *Rev. Mod. Phys.* **2011**, *83*, 647.

15. Haeberle, S.; Zengerle, R. *Lab Chip* **2007**, *7*, 1094.

16. Luo, J. K.; Fu, Y. Q.; Li, Y.; Du, X. Y.; Flewitt, A. J.; Walton, A. J.; Milne, W. I. *J. Micromech. Microeng.* **2009**, *19*, 054001.

17. Yeo, L. Y.; Friend, J. R. *Biomicrofluidics* **2009**, *3*, 012002.

18. Wixforth, A. *JALA* **2006**, *11*, 399.

19. Campanella, H. *Acoustic Wave and Electromechanical Resonators: Concept to Key Applications*; Artech House: Boston/London, 2010; pp 38–46.

20. Wise, K. D. *Sens. Actuators, A* **2007**, *136*, 39.

21. Wei, A.; Pan, L.; Huang, W. *Mater. Sci. Eng. B* **2011**, *176*, 1409.

22. Yan, Z.; Zhou, X. Y.; Pang, G. K. H.; Zhang, T.; Liu, W. L.; Cheng, J. G.; Song, Z. T.; Feng, S. L.; Lai, L. H.; Chen, J. Z.; Wang, Y. *Appl. Phys. Lett.* **2007**, *90*, 143503.

23. Voiculescu, I.; Nordin, A. N. *Biosen. Bioelectron.* **2012**, *33*, 1.

24. Rossler, U. *Phys. Rev.* **1969**, *184*, 733.

25. Vogel, D.; Krüger, P.; Pollmann, J. *Phys. Rev. B: Condens. Matter* **1995**, *52*, R14316.

26. Schroer, P.; Kruger, P.; Pollmann, J. *Phys. Rev. B: Condens. Matter* **1993**, *47*, 6971.

27. Varshni, Y. P. *Physica* **1967**, *34*, 149.

28. Ohtomo, A.; Kawasaki, M.; Koida, T.; Masubuchi, K.; Koinuma, H.; Sakurai, Y.; Yoshida, Y.; Yasuda, T.; Segawa, Y. *Appl. Phys. Lett.* **1998**, *72*, 2466.

29. Makino, T.; Segawa, Y.; Kawasaki, M.; Ohtomo, A.; Shiroki, R.; Tamura, K.; Yasuda, T.; Koinuma, H. *Appl. Phys. Lett.* **2001**, *78*, 1237.

30. Bagnall, D. M.; Chen, Y. F.; Zhu, Z.; Yao, T.; Koyama, S.; Shen, M. Y.; Goto, T. *Appl. Phys. Lett.* **1997**, *70*, 2230.

31. Manbachi, A.; Cobbold, R. S. C. *Ultrasound* **2011**, *19*, 187.

32. Corso, A. D.; Posternak, M.; Resta, R.; Balderschi, A. *Phys. Rev. B: Condens. Matter* **1994**, *50*, 10715.

33. Pang, H. F.; Garcia-Gancedo, L.; Fu, Y. Q.; Porro, S.; Gu, Y. W.; Luo, J. K.; Zu, X. T.; Placido, F.; Wilson, J. I. B.; Flewitt, A. J.; Milne, W. I. *Phys. Status Solidi (a)* **2013**, *210*, 1575.

34. Luo, J. T.; Pan, F.; Fan, P.; Zeng, F.; Zhang, D. P.; Zheng, Z. H.; Liang, G. X. *Appl. Phys. Lett.* **2012**, *101*, 172909.

35. Chen, G.; Peng, J. J.; Song, C.; Zeng, F.; Pan, F. *J. Appl. Phys.* **2013**, *113*, 104503.

36. Emanetoglu, N. W.; Muthukumar, S.; Wu, P.; Wittstruck, R.; Chen, Y.; Lu, Y. *IEEE Trans. Ultrason., Ferroelect. Freq. Contr.* **2003**, *50*, 537.

37. Chen, Y.; Emanetoglu, N. W.; Saraf, G.; Wu, P.; Lu, Y. *IEEE Trans. Ultrason., Ferroelect. Freq. Contr.* **2005**, *52*, 1161.

38. Xiang, H. J.; Yang, J.; Hou, J. G.; Zhu, Q. *Appl. Phys. Lett.* **2006**, *89*, 223111.

39. Agrawal, R.; Espinosa, H. D. *Nano Lett.* **2011**, *11*, 786.

40. Park, W. I.; Jun, Y. H.; Jung, S. W.; Yi, G.-C. *Appl. Phys. Lett.* **2003**, *82*, 964.
41. Grabowska, J.; Meaney, A.; Nanda, K. K.; Mosnier, J.-P.; Henry, M. O.; J.-R. Duclère, McGlynn, E. *Phys. Rev. B: Condens. Matter* **2006**, *71*, 115439.
42. Wu, K.; He, H.; Lu, Y.; Huang, J.; Ye, Z. *Nanoscale* **2012**, *4*, 1701.
43. Main, K.; Shimada, R.; Fujita, Y.; Neogi, A. *Phys. Status Solidi (RRL)* **2013**, *7*, 1089.
44. Lin, B.; Fu, Z.; Jia, Y. *Appl. Phys. Lett.* **2001**, *79*, 943.
45. Zhang, D. H.; Xue, Z. Y.; Wang, Q. P. *J. Phys. D: Appl. Phys.* **2002**, *35*, 2837.
46. Zeng, H.; Duan, G.; Li, Y.; Yang, S.; Xu, X.; Cai, W. *Adv. Funct. Mater.* **2010**, *20*, 561.
47. Bagnall, D. M.; Chen, Y. F.; Zhu, Z.; Yao, T.; Koyama, S.; Shen, M. Y.; Goto, T. *Appl. Phys. Lett.* **1997**, *70*, 2230.
48. Tang, Z. K.; Wong, G. K. L.; Yu, P.; Kawasaki, M.; Ohtomo, A.; Koinuma, H.; Segawa, Y. *Appl. Phys. Lett.* **1998**, *72*, 3270.
49. Dong, H.; Liu, Y.; Lu, J.; Chen, Z.; Wang, J.; Zhang, L. *J. Mater. Chem. C* **2013**, *1*, 202.
50. [50] Djurišić, A. B.; Leung, Y. H.; Choy, W. C. H.; Cheah, K. W.; Chan, W. K. *Appl. Phys. Lett.* **2004**, *84*, 2635.
51. Mollow, E. In: *Proceedings of the Photoconductivity Conference*; Breckenridge, R. G.; Wiley: New York, 1954; p 509.
52. Miller, P. H. In: *Proceedings of the Photoconductivity Conference*; Breckenridge, R. G.; Wiley: New York, 1954; p 287.
53. Zhang, D. H. *J. Phys. D: Appl. Phys.* **1995**, *28*, 1273.
54. Studenikin, S. A.; Golego, N.; Cocivera, M. *J. Appl. Phys.* **2000**, *87*, 2413.
55. Sharma, P.; Mansingh, A.; Sreenivas, K. *Appl. Phys. Lett.* **2002**, *80*, 553.
56. Reyes, P. I.; Ku, C.-J.; Duan, Z.; Xu, Y.; Garfunkel, E.; Lu, Y. *Appl. Phys. Lett.* **2012**, *101*, 031118.
57. Studenikin, S. A.; Cocivera, M. *J. Appl. Phys.* **2002**, *91*, 5060.
58. Tomm, J. W.; Ullrich, B.; Qiu, X. G.; Segawa, Y.; Ohtomo, A.; Kawasaki, M.; Koinuma, H. *J. Appl. Phys.* **2000**, *87*, 1844.
59. Heiland, G. *Surf. Sci.* **1969**, *13*, 72.
60. Look, D. C. *Surf. Sci.* **2007**, *601*, 5315.
61. Schmidt, O.; Kiesel, P.; Van de Walle, C. G.; Johnson, N. M.; Nause, J.; Döhler, G. H. *Jpn. J. Appl. Phys.* **2005**, *44*, 7271.
62. Chen, C.-Y.; Chen, M.-W.; Ke, J.-J.; Lin, C.-A.; Retamal, J. R. D.; He, J.-H. *Pure Appl. Chem.* **2010**, *82*, 2055.
63. Mtangi, W.; Nel, J. M.; Auret, F. D.; Chawanda, A.; Diale, M.; Nyamhere, C. *Physica B* **2012**, *407*, 1624.
64. Krusemeyer, H. J. *Phys. Rev.* **1959**, *114*, 655.
65. Heiland, G.; Ibach, H. *Sol. Stat. Commun.* **1966**, *4*, 353.
66. Klingshirn, C. F.; Waag, A.; Hoffmann, A.; Geurts, J. *Zinc Oxide From Fundamental Properties Towards Novel Applications*. Springer-Verlag: Berlin and Heidelberg, 2010.
67. Link, M.; Schreiter, M.; Weber, J.; Gabl, R.; Pitzer, D.; Primig, R.; Wersing, W.; Assouar, M. B.; Elmazria, O. *J. Vac. Sci. Technol. A* **2006**, *24*, 218.
68. Milyutin, E.; Gentil, S.; Mural, P. *J. Appl. Phys.* **2008**, *104*, 084508.
69. Qin, L.; Wang, Q.-M. *J. Appl. Phys.* **2010**, *108*, 104510.
70. Pang, H. F.; Fu, Y. Q.; Hou, R.; Kirk, K.; Hudson, D.; Zu, X. T.; Placido, F. *Ultrasonics* **2013**, *53*, 1264.
71. Lee, Y. E.; Kim, S. G.; Kim, Y. J.; Kim, H. J. *J. Vac. Sci. Technol. A* **1997**, *15*, 1194.
72. Bensmaine, S.; Le Brizoual, L.; Elmazria, O.; Fundenberger, J. J.; Benyoucef, B. *Phys. Status Solidi (a)* **2007**, *204*, 3091.
73. Pang, H. F.; Zhang, G. A.; Tang, Y. L.; Fu, Y. Q.; Wang, L. P.; Zu, X. T.; Placido, F. *Appl. Surf. Sci.* **2012**, *259*, 747.
74. Yanagitani, T.; Morisato, N.; Takayanagi, S.; Matsukawa, M.; Watanabe, Y. *IEEE Trans. Ultrason., Ferroelect., Freq. Contr.* **2011**, *58*, 1062.

75. Zhang, H.; Kosinski, J. A. *IEEE Trans Ultrason Ferroelectr Freq Control* **2012**, *59*, 2831.
76. Yoshino, Y.; Inoue, K.; Takeuchi, M.; Makino, T.; Katayama, Y.; Hata, T. *Vacuum* **2000**, *59*, 403.
77. Novotný, M.; Čížek, J.; Kužel, R.; Bulíř, J.; Lančok, J.; Connolly, J.; McCarthy, E.; Krishnamurthy, S.; Mosnier, J.-P.; Anwand, W.; Brauer, G. *J. Phys. D: Appl. Phys.* **2012**, *45*, 225101.
78. Misra, P.; Kukreja, L. M. *Thin Solid Films* **2005**, *485*, 42.
79. Khranovskyy, V.; Minikayev, R.; Trushkin, S.; Lashkarev, G.; Lazorenko, V.; Grossner, U.; Paszkowicz, W.; Suchocki, A.; Svensson, B. G.; Yakimova, R. *J. Cryst. Growth* **2007**, *308*, 93.
80. Phan, D.-T.; Suh, H.-C.; Chung, G.-S. *Microelectr. Eng.* **2011**, *88*, 105.
81. Kim, J. H.; Kim, E.-M.; Andeen, D.; Thomson, D.; DenBaars, S. P.; Lange, F. F. *Adv. Funct. Mater.* **2007**, *17*, 463.
82. Nakahata, H.; Fujii, S.; Higaki, K.; Hachigo, A.; Kitabayashi, H.; Shikata, S.; Fujimori, N. *Semicond. Sci. Technol.* **2003**, *18*, S96.
83. Pang, H. F.; Garcia-Gancedo, L.; Y. Fu, Q.; Porro, S.; Y. Gu, W.; Luo, J. K.; Zu, X. T.; Placido, F.; Wilson, J. I. B.; Flewitt, A. J.; Milne, W. I. *Phys. Status Solidi (a)* **2013**, *210*, 1575.
84. Garcia-Gancedo, L.; Pedros, J.; Zhu, Z.; Flewitt, A. J.; Milne, W. I.; Luo, J. K.; Ford, C. J. B. *J. Appl. Phys.* **2012**, *112*, 014907.
85. Gulino, A.; Lupo, F.; Fragalà, M. E. *J. Phys. Chem. C* **2008**, *112*, 13869.
86. Tang, K.; Wang, L.; Huang, J.; Xu, R.; Lai, J.; Wang, J.; Min, J.; Shi, W.; Xia, Y. *Plasma Sci. Technol.* **2009**, *11*, 587.
87. Amaike, H.; Hazu, K.; Sawai, Y.; Chichibu, S. F. *Appl. Phys. Express* **2009**, *2*, 105503.
88. Koster, G.; Rijnders, G. J. H. M.; Blank, D. H. A.; Rogalla, H. *Appl. Phys. Lett.* **1999**, *74*, 3729.
89. Hwang, D.-K.; Bang, K.-H.; Jeong, M.-C.; Myoung, J.-M. *J. Cryst. Growth* **2003**, *254*, 449.
90. Hur, T.-B.; Hwang, Y.-H.; Kim, H.-K.; Lee, I. J. *J. Appl. Phys.* **2006**, *99*, 064308.
91. Shi, J.; Wang, X. *J. Phys. Chem. C* **2010**, *114*, 2082.
92. Tsiaoussis, I.; Khranovskyy, V.; Dimitrakopulos, G. P.; Stoemenos, J.; Yakimova, R.; Pecz, B. *J. Appl. Phys.* **2011**, *109*, 043507.
93. Kim, S.-W.; Fujita, S.; Fujita, S. *Appl. Phys. Lett.* **2002**, *81*, 5036.
94. Lu, J. G.; Ye, Z. Z.; Zhang, Y. Z.; Liang, Q. L.; Fujita, S.; Wang, Z. L. *Appl. Phys. Lett.* **2006**, *89*, 023122.
95. Kluth, O.; Schope, G.; Hüpkes, J.; Agashe, C.; Müller, J.; Rech, B. *Thin Solid Films* **2003**, *442*, 80.
96. Calnan, S.; Upadhyaya, H. M.; Thwaites, M. J.; Tiwari, A. N. *Thin Solid Films* **2007**, *515*(15), 6045–6050.
97. Li, F. M.; Bernhard Bayer, C.; Hofmann, S.; Speakman, S. P.; Ducati, C.; Milne, W. I.; Flewitt, A. J. *Phys. Status Solidi (b)* **2013**, *250*, 957.
98. Bae, S. H.; Lee, S. Y.; Jin, B. J.; Im, S. Appl. *Surf. Sci.* **2000**, *154–155*, 458.
99. Jagadish, C.; Pearton, S. *Zinc Oxide Bulk, Thin Films and Nanostructures*; Elsevier: Oxford, 2006; p 88.
100. Hwang, D.-K.; Oh, M.-S.; Lim, J.-H.; Park, S.-J. *J. Phys. D: Appl. Phys.* **2007**, *40*, R387.
101. Pant, P.; Budai, J. D.; Aggarwal, R.; Narayan, R. J.; Narayan, J. *J. Phys. D: Appl. Phys.* **2009**, *42*, 105409.
102. Yoshida, T.; Tachibana, T.; Maemoto, T.; Sasa, S.; Inoue, M. Appl. Phys. A **2010**, *101*, 685.
103. Ohtomo, A.; Tamura, K.; Saikusa, K.; Takahashi, T.; Makino, T.; Segawa, Y.; Koinuma, H.; Kawasaki, M. *Appl. Phys. Lett.* **1999**, *75*, 2635.

104. Tsukazaki, A.; Ohtomo, A.; Yoshida, S.; Kawasaki, M.; Chia, C. H.; Makino, T.; Segawa, Y.; Koida, T.; Chichibu, S. F.; Koinuma, H. *Appl. Phys. Lett.* **2003**, *83*, 2784.
105. Pant, P.; Budai, J. D.; Narayan, J. *Acta Mater.* **2010**, *58*, 1097.
106. Duclere, J.-R.; McLoughlin, C.; Fryar, J.; R. O'Haire, Guilloux-Viry, M.; Meaney, A.; Perrin, A.; McGlynn, E.; Henry, M. O.; Mosnier, J.-P. *Thin Solid Films* **2006**, *500*, 78.
107. Chen, Y. F.; Hong, S.; Ko, H. *Appl. Phys. Lett.* **2000**, *76*, 559.
108. Sha, Z. D.; Wang, J.; Chen, Z. C.; Chen, A. J.; Zhou, Z. Y.; Wu, X. M.; Zhuge, L. J. *Physica E* **2006**, *33*, 263.
109. Wei, X.; Zhao, R.; Shao, M.; Xu, X.; Huang, J. *Nanoscale Res. Lett.* **2013**, *8*, 112.
110. Rogers, D. J.; Look, D. C.; Teherani, F. H.; Minder, K.; Razeghi, M.; Largeteau, A.; Demazeau, G. *Phys. Status Solidi (c)* **2008**, *5*, 3084.
111. Yata, S.; Nakashima, Y.; Kobayashi, T. *Thin Solid Films* **2003**, *445*, 259.
112. Kim, T. H.; Nam, S. H.; Park, H. S.; Song, J. K.; Park, S. M. Appl. *Surf. Sci.* **2007**, *253*, 8054.
113. Cho, A. Y.; Arthur, J. R. *Prog. Solid State Chem.* **1975**, *10*, 157.
114. Henini, M. *Molecular Beam Epitaxy: From Research to Mass Production*; Elsevier Inc.: Oxford, 2012; p 369.
115. Opel, M.; Geprägs, S.; Althammer, M.; Brenninger, T.; Gross, R. *J. Phys. D: Appl. Phys.* **2014**, *47*, 034002.
116. Zhou, H.; Wang, H.-Q.; Liao, X.-X.; Zhang, Y.; Zheng, J.-C.; Wang, J.-O.; Muhemmed, E.; Qian, H.-J.; Ibrahim, K.; Chen, X.; Zhan, H.; Kang, J. *Nanoscale Res. Lett.* **2012**, *7*, 184.
117. Tsukazaki, A.; Saito, H.; Tamura, K.; Ohtani, M.; Koinuma, H.; Sumiya, M.; Fuke, S.; Fukumura, T.; Kawasaki, M. *Appl. Phys. Lett.* **2002**, *81*, 235.
118. Emanetoglu, N. W.; Gorla, C.; Liu, Y.; Liang, S.; Lu, Y. *Mat. Sci. Semicond. Proc.* **1999**, *2*, 247.
119. Wang, X. Q.; Sun, H. P.; Pan, X. Q. *Appl. Phys. Lett.* **2010**, *97*, 151908.
120. Cho, M. W.; Setiawan, A.; Ko, H. J.; Hong, S. K.; Yao, T. *Semicond. Sci. Technol.* **2005**, *20*, S13.
121. Wang, H.-C.; Liao, C.-H.; Chueh, Y.-L.; Lai, C.-C.; Chou, P.-C.; Ting, S.-Y. *Opt. Mater. Express* **2013**, *3*, 295.
122. Choi, Y. S.; Kang, J. W.; Hwang, D. K.; Park, S. J. *IEEE Trans. Electron. Dev.* **2010**, *57*, 26.
123. Przezdziecka, E.; Wierzbicka, A.; Reszka, A.; Goscinski, K.; Droba, A.; Jakiela, R.; Dobosz, D.; Krajewski, T. A.; Kopalko, K.; Sajkowski, J. M.; Stachowicz, M.; Pietrzyk, M. A.; Kozanecki, A. *J. Phys. D: Appl. Phys.* **2013**, *46*, 035101.
124. Barnes, T. M.; Olsen, K.; Wolden, C. A. *Appl. Phys. Lett.* **2005**, *86*, 112112.
125. Yao, B.; Shen, D. Z.; Zhang, Z. Z.; Wang, X. H.; Wei, Z. P.; Li, B. H.; Lv, Y. M.; Fan, X. W.; Guan, L. X.; Xing, G. Z.; Cong, C. X.; Xie, Y. P. *J. Appl. Phys.* **2006**, *99*, 123510.
126. Wang, L. G.; Zunger, A. *Phys. Rev. Lett.* **2003**, *90*, 256401.
127. Maksimov, O. Rev. *Adv. Mater. Sci.* **2010**, *24*, 26.
128. Jagadish, C.; Pearton, S. *Zinc Oxide Bulk, Thin Films and Nanostructures*; Elsevier: Oxford, 2006; p 448.
129. Janotti, A.; Van de Walle, C. G. *Rep. Prog. Phys.* **2009**, *72*, 126501.
130. Zheng, C. C.; Xu, S. J.; Ning, J. Q.; Bao, W.; Wang, J. F.; Gao, J.; Liu, J. M.; Zhu, J. H.; Liu, X. L. *Semicond. Sci. Technol.* **2012**, *27*, 035008.
131. Kashiwaba, Y.; Haga, K.; Watanabe, H.; Zhang, B. P.; Segawa, Y.; Wakatsuki, K. *Phys. Status Solidi (b)*, **2002**, *229*, 921.
132. Kirchner, C.; Gruber, T.; Reuss, F.; Thonke, K.; Waag, A.; Giessen, C.; Heuken, M. *J. Cryst. Growth* **2003**, *248*, 20.
133. Haga, K.; Abe, S.; Takizawa, Y.; Yubuta, K.; Shishido, T. *J. Phys.: Conf. Ser.* **2013**, *417*, 012059.

134. Jagadish, C.; Pearton, S. *Zinc Oxide Bulk, Thin Films and Nanostructures.* Elsevier: Oxford, 2006; p 448.
135. Hwang, D.-K.; Oh, M.-S.; Lim, J.-H.; Park, S.-J. *J. Phys. D: Appl. Phys.* **2007,** *40,* R387.
136. Liang, H.; Gordon, R. G. *J. Mater. Sci.* **2007,** *42,* 6388.
137. Seki, S.; Onodera, H.; Sekizawa, T.; Sakuma, M.; Haga, K.; Seki, Y.; Sawada, Y.; Shishido, T. *Phys. Status Solidi (c),* **2010,** *7,* 1565.
138. Hussaina, M.; Hussainb, S. T. *Chem. Eur. J.* **2010,** *1* (2), 96.
139. Nicolay, S.; Benkhaira, M.; Ding, L.; Escarre, J.; Bugnon, G.; Meillaud, F.; Ballif, C. *Solid Energy Mater. Sol. C* **2012,** *105,* 46.
140. Znaidi, L. *Mater. Sci. Eng. B* **2010,** *174,* 18.
141. Viswanatha, R.; Sapra, S.; Satpati, Satyam, P. V.; Dev, B. N.; Sarma, D. D. *J. Mater. Chem.* **2004,** *14,* 661–668.
142. Kubo, R. *J. Phys. Soc. Jpn.* **1962,** *17,* 975.
143. Fonoberov, V. A.; Balandin, A. A. *Phys. Rev. B: Condens. Matter* **2004,** *70,* 195410.
144. Fonoberov, V. A.; Balandin, A. A. *J. Nanoelectron. Optoelectron.* **2006,** *1,* 19.
145. Cheng, H.-M.; Lin, K.-F.; Hsu, H.-C.; Hsieh, W.-F. *Appl. Phys. Lett.* **2006,** *88,* 261909.
146. He, R.; Tsuzuki, T. *J. Am. Ceram. Soc.* **2010,** *93,* 2281.
147. Asok, A.; Gandhi, M. N.; Kulkarni, A. R. *Nanoscale* **2012,** *4,* 4943.
148. Han, L.-L.; Cui, L.; Wang, W.-H.; Wang, J.-L.; Du, X.-W.; *Semicond. Sci. Technol.* **2012,** *27,* 065020.
149. Hsu, S.; Lin, Y. Y.; Huang, S.; Lem, K. W.; Nguyen, D. H.; Lee, D. S. *Nanotechnology* **2013,** *24,* 475102.
150. Sreeja, R.; John, J.; Aneesh, P. M.; Jayaraj, M. K. *Opt. Commun.* **2010,** *283,* 2908.
151. Son, D. I.; Kwon, B. W.; Park, D. H.; Seo, W.-S.; Yi, Y.; Angadi, B.; Lee, C.-L.; Choi, W. K.; *Nature Nanotechnol.* **2012,** *7,* 465.
152. Maikhuri, D.; Purohit, S. P.; Mathur, K. C. *AIP Adv.* **2012,** *2,* 012160.
153. Yuan, Q.; Hein, S.; Misra, R. D. K. *Acta Biomater.* **2010,** *6,* 2732.
154. Forleo, A.; Francioso, L.; Capone, S.; Siciliano, P.; Lommens, P.; Hens, Z. *Sens. Actuators B* **2010,** *146,* 111.
155. Shao, D.; Sun, X.; Xie, M.; Sun, H.; Lu, F.; George, S. M.; Lian, J.; Sawyer, S. *Mater. Lett.* **2013,** *112,* 165.
156. Tan, S. T.; Sun, X. W.; Zhang, X. H.; Chen, B. J.; Chu, S. J.; Yong, A.; Dong, Z. L.; Hu, X. *J. Cryst. Growth* **2006,** *290,* 518.
157. Zhang, X.; Kobayashi, K.; Tomita, Y.; Maeda, Y.; Kohno, Y. *Phys. Status Solidi (c)* **2013,** *10,* 1576.
158. Lu, J. G.; Ye, Z. Z.; Huang, J. Y.; Zhu, L. P.; Zhao, B. H.; Wang, Z. L.; Fujita, S. *Appl. Phys. Lett.* **2006,** *88,* 063110.
159. Rani, S.; Suri, P.; Shishodia, P. K.; Mehra, R. M. Sol. *Energy Mater. Sol. C* **2008,** *92,* 1639.
160. Zhang, Q.; Dandeneau, C. S.; Zhou, X.; Cao, G. *Adv. Mater.* **2009,** *21,* 4087.
161. Rai, P.; Yu, Y.-T. *Sens. Actuators, B* **2012,** *173,* 58.
162. Xiong, H.-M.; *Adv. Mater.* **2013,** *25,* 5329.
163. Aleshin, A. N.; Shcherbakov, I. P.; Petrov, V. N.; Titkov, A. N. *Org. Electron.* **2011,** *12,* 1285.
164. Hanley, C.; Thurber, A.; Hanna, C.; Punnoose, A.; Zhang, J.; Wingett, D. G. *Nanoscale Res Lett.* **2009,** *4,* 1409.
165. Zhao, M. H.; Wang, Z. L.; Mao, S. X. *Nano Lett.* **2004,** *4,* 587.
166. Chen, C. Q.; Shi, Y.; Zhang, Y. S.; Zhu, J.; Yan, Y. J. *Phys. Rev. Lett.* **2006,** *96,* 075505.
167. Fan, J.; Lee, W.; Hauschild, R.; Alexe, M.; Rhun, G. L.; Scholz, R.; Dadgar, A.; Nielsch, K.; Kalt, H.; Krost, A.; Zacharias, M.; Gosele, U. *Small* **2006,** *2,* 561.
168. Yang, P.; Yan, H.; Mao, S.; Russo, R.; Johnson, J.; Saykally, R.; Morris, N.; Pham, J.; He, R.; Choi, H.-J.; *Adv. Funct. Mater.* **2002,** *12,* 323.

169. Tak, Y.; Yong, K. J.; *J. Phys. Chem. B* **2005**, *109*, 19263.
170. Tian, J.-H.; Hu, J.; Li, S.-S.; Zhang, F.; Liu, J.; Shi, J.; Li, X.; Tian, Z.-Q.; Chen, Y. *Nanotechnology* **2011**, *22*, 245601.
171. Wagner, R. S.; Ellis, W. C. *Appl. Phys. Lett.* **1964**, *4*, 89.
172. Yi, G.-C, Wang, C.; Park, W. I. *Semicond. Sci. Technol.* **2005**, *20*, S22.
173. Zhu, Z.; Chen, T.-L.; Gu, Y.; Warren, J.; Osgood, R. M. *Chem. Mater.* **2005**, *17*, 4227.
174. Yoo, J.; Hong, Y.-J.; An, S.; Yi, G.-C.; Chon, B.; Joo, T.; Kim, J.-W.; Lee, J.-S. *Appl. Phys. Lett.* **2006**, *89*, 043124.
175. Huang, M. H.; Mao, S.; Feick, H.; Yan, H.; Wu, Y.; Kind, H.; Weber, E.; Russo, R.; Yang, P. *Science* **2001**, *292*, 1897.
176. Banerjee, D.; Rybczynski, J.; Huang, J. Y.; Wang, D. Z.; Kempa, K.; Ren, Z. F. *Appl. Phys. A* **2005**, *80*, 749.
177. Wang, X.; Song, J.; Li, P.; Ryou, J. H.; Dupuis, R. D.; Summers, C. J.; Wang, Z. L. *J. Am. Chem. Soc.* **2005**, *127*, 7920.
178. Fan, H. J.; Zacharias, M. *J. Mater. Sci. Technol.* **2008**, *24*, 589.
179. Wang, X.; Summers, C. J.; Wang, Z. L. *Nano Lett.* **2004**, *4*, 423.
180. Soman, P.; Darnell, M.; Feldman, M. D.; Chen, S. *J. Nanosci. Nanotechnol.* **2011**, *11*, 1.
181. Wen, L.; Wong, K. M.; Fang, Y.; Wu, M. Lei, Y. *J. Mater. Chem.* **2011**, *21*, 7090.
182. Huang, M. H.; Wu, Y.; Feick, H.; Tran, N.; Weber, E.; Yang, P. *Adv. Mater.* **2001**, *13*, 113.
183. Qin, Y.; Yang, R.; Wang, Z. L. *J. Phys. Chem. C* **2008**, *112* (48), 18734.
184. Chik, H.; Liang, J.; Cloutier, S. G.; Kouklin, N.; Xu, J. M. *Appl. Phys. Lett.* **2004**, *84*, 3376.
185. Yi, G.-C. *Semiconductor Nanostructures for Optoelectronic Devices*; Springer: Berlin, 2012; p 12.
186. Miao, L.; Ieda, Y.; Tanemura, S.; Cao, Y. G.; Tanemura, M.; Hayashi, Y.; Toh, S.; Kaneko, K. Sci. Tech. *Adv. Mater.* **2007**, *8*, 443.
187. Park, W. I.; Yi, G. C.; Kim, M. Y.; Pennycook, S. *J. Adv. Mater.* **2002**, *14*, 1841.
188. Greene, L.; Law, M.; Goldberger, J.; Kim, F.; Johnson, J. C.; Zhang, Y. F.; Saykally, R. J.; Yang, P. D. *Angew. Chem., Int. Ed.* **2003**, *42*, 3031.
189. Unalan, H. E.; Hiralal, P.; Rupesinghe, N.; Dalal, S.; Milne, W. I.; Amaratunga, G. A. J. *Nanotechnology* **2008**, *19*, 255608.
190. Baruah, S.; Dutta, J. *Sci. Technol. Adv. Mater.* **2009**, *10*, 013001.
191. Joo, J.; Chow, B. Y.; Prakash, M.; Boyden, E. S.; Jacobson, J. M. *Nat. Mater.* **2011**, *10*, 596.
192. Liu, Z.; Zhu, R.; Zhang, G. *J. Phys. D: Appl. Phys.* **2010**, *43*, 155402.
193. Yeo, J.; Hong, S.; Wanit, M.; Kang, H. W.; Lee, D.; Grigoropoulos, C. P.; Sung, H. J.; Ko, S. H. *Adv. Funct. Mater.* **2013**, *23*, 3316.
194. Mahpeykar, S. M.; Koohsorkhi, J.; Ghafoori-Fard, H. *Nanotechnology* **2012**, *23*, 165602.
195. Wu, G. S.; Xie, T.; Yuan, X. Y.; Li, Y.; Yang, L.; Xiao, Y. H.; Zhang, L. D. *Sol. Stat. Commun.* **2005**, *134*, 485.
196. Htay, M. T.; Tani, Y.; Hashimoto, Y.; Ito, J. *J. Mater. Sci.* **2009**, *20*, 341.
197. Fan, Z.; Dutta, D.; Chien, C.-J.; Chen, H.-Y.; Brown, E. C.; Chang, P.-C.; Lu, J. G. *Appl. Phys. Lett.* **2006**, *89*, 213110.
198. Ramírez, D.; Gómez, H.; Lincot, D. *Electrochim. Acta* **2010**, *55*, 2191.
199. Yang, K. X.; Wang, Z. L. *Appl. Phys. Lett.* **2004**, *84*, 975.
200. Ding, Y.; Wang, Z. L. *J. Phys. Chem. B* **2004**, *108*, 12280.
201. Wang, Z. L. *J. Phys.: Condens. Matter* **2004**, *16*, R829.
202. Wang, Z. L. *J. Mater. Chem.* **2005**, *15*, 1021.
203. Wei, Y.; Ding, Y.; Li, C.; Xu, S.; Ryo, J.-H.; Dupuis, R.; Sood, A. K.; Polla, D. L.; Wang, Z. L. *J. Phys. Chem. C* **2008**, *112* (48), 18935.
204. Zhao, M. H.; Wang, Z. L.; Mao, S. X. *Nano Lett.* **2004**, *4*, 587.

205. Wang, W. Z.; Zeng, B. Q.; Yang, J.; Poudel, B.; Huang, J. Y.; Naughton, M. J.; Ren, Z. F. *Adv. Mater.* **2006,** *18* (24), 3275.
206. Cao, B. Q.; Liu, Z. M.; Xu, H. Y.; Gong, H. B.; Nakamura, D.; Sakai, K.; Higashihata, M.; Okada, T. *CrystEngComm* **2011,** *13,* 4282.
207. Sadek, A. Z.; Choopun, S.; Wlodarski, W.; Ippolito, S. J.; Kalantar-Zadeh, K. *IEEE Sen. J.* **2007,** *7,* 919.
208. Li, F.; Ding, Y.; Gao, P.; Xin, X.; Wang, Z. L. *Angew. Chem.* **2004,** *116,* 5350.
209. Peng, Y.; Xu, A.-W.; Deng, B.; Antonietti, M.; Cölfen, H. *J. Phys. Chem. B* **2006,** *110,* 2988.
210. Hussain, S.; Liu, T.; Kashifc, M.; Miao, B.; He, J.; Zeng, W.; Zhang, Y.; Hashimc, U.; Pan, F. *Mater. Lett.* **2014,** *118,* 165.
211. Chen, H.; Wu, X.; Gong, L.; Ye, C.; Qu, F.; Shen, G. *Nanoscale Res. Lett.* **2009,** *5,* 570.
212. Xu, C. X.; Sun, X. W.; Dong, Z. L.; Yu, M. B. *Appl. Phys. Lett.* **2004,** *85,* 3878.
213. Ding, G. Q.; Shen, W. Z.; Zheng, M. J.; Fan, D. H. *Appl. Phys. Lett.* **2006,** *88,* 103106.
214. Reiss, P.; Protiere, M.; Li, L. *Small* **2009,** *5* (2), 154.
215. Pugliese, D.; Bella, F.; Cauda, V.; Lamberti, A.; Sacco, A.; Tresso, E.; Bianco, S. *ACS Appl. Mater. Interfaces* **2013,** *5,* 11288.
216. Cheng, C.; Fan, H. J. *Nano Today* **2012,** *7,* 327.
217. Fu, M.; Zhou, J.; Xiao, Q.; Li, B.; Zong, R.; Chen, W.; Zhang, J. *Adv. Mater.* **2006,** *18,* 1001.
218. Chandra, D.; Mridha, S.; Basak, D.; Bhaumik, A. *Chem. Commun.* **2009,** *7,* 2384.
219. Li, Y.; Cai, W.; Duan, G.; Cao, B.; Sun, F.; Lu, F. *J. Colloid Interface Sci.* **2005,** *287,* 634.
220. Polarz, S.; Orlov, A. V.; Schüth, F.; Lu, A. H. *Chemistry* **2007,** *13,* 592.
221. Huang, K.-M.; Ho, C.-L.; Chang, H.-J.; Wu, M.-C. *Nanoscale Res. Lett.* **2013,** *8,* 306.
222. Liu, H. L.; Wu, J. H.; Min, J. H.; Zhang, X. Y.; Kim, Y. K. *Mater. Res. Bull.* **2013,** *48,* 551.
223. Wang, K.; Chen, J. J.; Zeng, Z. M.; Tarr, J.; Zhou, W. L.; Jiang, C. S.; Pern, J.; Mascarenhas, A. *Appl. Phys. Lett.* **2010,** *96,* 123105.
224. Zeng, H.; Cai, W.; Hu, J.; Duan, G.; Liu, P.; Li, Y. *Appl. Phys. Lett.* **2006,** *88,* 171910.
225. Zhou, T.; Lu, M.; Zhang, Z.; Gong, H.; Chin, W. S.; Liu, B. *Adv. Mater.* **2010,** *22,* 403.
226. Zhu, D.; Li, W.; Ma, L.; Lei, Y. RSC Adv. **2014,** *4,* 9372.
227. Zhu, Y. F.; Fan, D. H.; Shen, W. Z. *J. Phys. Chem. C* **2008,** *112,* 10402.
228. Saha, S.; Sarkar, S.; Pal, S.; Sarkar, P. *J. Phys. Chem. C* **2013,** *117* (31), 15890.
229. Mayoa, D. C.; Marvinney, C. E.; Bililign, E. S.; McBride, J. R.; Mud, R. R.; Haglund, R. F. *Thin Solid Films* **2014,** *553,* 132.
230. Miao, J.; Yang, H. B.; Khoo, S. Y.; Liu, B. *Nanoscale* **2013,** *5,* 11118.
231. Chen, Y.; Wei, L.; Zhang, G.; Jiao, J. *Nanoscale Res Lett.* **2012,** *7* (1), 516.
232. Law, M.; Greene, L. E.; Radenovic, A.; Kuykendall, T.; Liphardt, J.; Yang, P. *J. Phys. Chem. B* **2006,** *110* (45), 22652.
233. Thierry, R.; Perillat-Merceroz, G.; Jouneau, P. H.; Ferret, P.; Feuillet, G. *Nanotechnology* **2012,** *23,* 085705.
234. Yang, P.; Xiao, X.; Li, Y.; Ding, Y.; Qiang, P.; Tan, X.; Mai, W.; Lin, Z.; Wu, W.; Li, T.; Jin, H.; Liu, P.; Zhou, J.; Wong, C. P.; Wang, Z. L. *ACS Nano* **2013,** *7* (3), 2617.
235. Elias, J.; Levy-Clement, C.; Bechelany, M.; Michler, J.; Wang, G.-Y.; Wang, Z.; Philippe, L. *Adv. Mater.* **2010,** *22,* 1607.
236. Li, C.; Li, G.; Shen, C.; Hui, C.; Tian, J.; Du, S.; Zhang, Z.; Gao, H.-J. *Nanoscale* **2010,** *2,* 2557.
237. Shi, R.; Yang, P.; Wang, J.; Zhang, A.; Zhu, Y.; Cao, Y.; Ma, Q. *CrystEngComm* **2012,** *14,* 5996.
238. McPeak, K. M.; Le, T. P.; Britton, N. G.; Nickolov, Z. S.; Elabd, Y. A.; Baxter, J. B. *Langmuir* **2011,** *27,* 3672.
239. Zhang, H.; Yang, D.; Ma, X.; Ji, Y.; J. Xu, Que, D. *Nanotechnology* **2004,** *15,* 622.
240. Pan, A.; Yu, R.; Xie, S.; Zhang, Z.; Jin, C.; Zou, B. *J. Cryst. Growth* **2005,** *282,* 165.

241. Baxter, J. B.; Aydil, E. S. *Appl. Phys. Lett.* **2005,** *86*, 053114.
242. Jiang, C. Y.; Sun, X. W.; Lo, G. Q.; Kwong, D. L.; Wang, J. X. *Appl. Phys. Lett.* **2007,** *90*, 263501.
243. He, F.-Q.; Zhao, Y.-P. *Appl. Phys. Lett.* **2006,** *88*, 193113.
244. Lao, J. Y.; Wen, J. G.; Ren, Z. F. *Nano Lett.* **2002,** *2* (11), 1287.
245. Fan, H. J.; Scholz, R.; Kolb, F. M.; Zacharias, M. *Appl. Phys. Lett.* **2004,** *85*, 4142.
246. Xu, X.; Wu, M.; Asoro, M.; Ferreira, P. J.; Fan, D. L. *Cryst. Growth Des.* **2012,** *12* (10), 4829.
247. Deng, J.; Yu, B.; Lou, Z.; Wang, L.; Wang, R.; Zhang, T. *Sens. Actuators, B* **2013,** *184*, 21.
248. Ko, S. H.; Lee, D.; Kang, H. W.; Nam, K. H.; Yeo, J. Y.; Hong, S. J.; Grigoropoulos, C. P.; Sung, H. J. *Nano Lett.* **2011,** *11*, 666.
249. Yeo, L. Y.; Friend, J. R. *Annu. Rev. Fluid Mech.* **2014,** *46*, 379.
250. Büyükköse, S.; Vratzov, B.; van der Veen, J.; Santos, P. V.; van der Wiel, W. R. *Appl. Phys. Lett.* **2013,** *102*, 013112.
251. Smith, W. R.; Gerard, H. M.; Collins, J. H.; Reeder, T. M.; Shaw, H. J. *IEEE Trans. Microwave Theory Technol.* **1969,** *17*, 856.
252. Du, X. Y.; Fu, Y. Q.; Tan, S. C.; Luo, J. K.; Flewitt, A. J.; Milne, W. I.; Lee, D. S.; Park, N. M.; Park, J.; Choi, Y. J.; Kim, S. H.; Maeng, S. *Appl. Phys. Lett.* **2008,** *93*, 094105.
253. Mitsuyu, T.; Ono, S.; Wasa, K. *J. Appl. Phys.* **1980,** *51*, 2464.
254. Mortet, V.; Williams, O. A.; Haenen, K. *Phys. Status Solidi (a)* **2008,** *205*, 1009.
255. Luo, J. K.; Fu, Y. Q.; Milne, S. B.; Le, H. R.; Williams, J. A.; Spearing, S. M.; Flewitt, A. J.; Milne, W. I. *J. Micromech. Microeng.* **2007,** *17*, S147.
256. Fujii, S. *Phys. Status Solidi (a)* **2011,** *208*, 1072.
257. Fujii, S.; Shikata, S.; Uemura, T.; Nakahata, H.; Harima, H. *IEEE Trans. Ultrason. Ferroelectr. Freq. Control.* **2005,** *52*, 1817.
258. Campbell, C. K. *Proc. IEEE,* **1989,** *77* (10), 1453–1484.
259. Jin, H.; Zhou, J.; He, X.; Wang, W.; Guo, H.; Dong, S.; Wang, D.; Xu, Y.; Geng, J.; Luo, J. K.; Milne, W. I. *Sci. Rep.* **2013,** *3*, 2140.
260. Ippolito, S. J.; Kandasamy, S.; Kalantar-Zadeh, K.; Wlodarski, W.; Galatsis, K.; Kiriakidis, G.; Katsarakis, N.; Suchea, M. *Sens. Actuators, B* **2005,** *111–112*, 207.
261. Pang, H.-F.; Fu, Y.-Q.; Li, Z.-J.; Li, Y.-F.; Placido, F.; Walton, A. J.; Zu, X.-T. *Sens. Actuators A* **2013,** *193*, 87.
262. Singh, D.; Narasimulu, A. A.; Garcia-Gancedo, L.; Fu, Y. Q.; Hasan, T.; Lin, S. S.; Geng, J.; Shao, G.; Luo, J. K. *J. Mater. Chem. C* **2013,** *1*, 2525.
263. Grudkowski, T. W.; Black, J. F.; Reeder, T. M.; Cullen, D. E.; Wagner, R. A. *Appl. Phys. Lett.* **1980,** *37*, 993.
264. Hashimoto, K.; *RF Bulk Acoustic Wave Filters for Communications*; Artech House: Norwood, MA/London, 2009; p 52.
265. Qin, L.; Chen, Q.; Cheng, H.; Wang, Q. M. *IEEE Trans Ultrason. Ferroelectr. Freq. Control* **2010,** *57* (8), 1840.
266. Link, M.; Schreiter, M.; Weber, J.; Primig, R.; Pitzer, D.; Gabl, R. *IEEE Trans Ultrason. Ferroelectr. Freq. Control* **2006,** *53* (2), 492.
267. Chen, D.; Wang, J.; Xu, Y.; Li, D.; Zhang, L.; Liu, W. *J. Micromech. Microeng.* **2013,** *23*, 095032.
268. García-Gancedo, L.; Pedrós, J.; Iborra, E.; Clement, M.; Zhao, X. B.; Olivares, J.; Capilla, J.; Luo, J. K.; Lu, J. R.; Milne, W. I.; Flewitt, A. J. *Sens. Actuators, B* **2013,** *183*, 136.
269. Shung, K. K. *Diagnostic Ultrasound: Imaging and Blood Flow Measurements*; CRC Press: Boca Raton, FL, 2006, Vol 3; p 46.
270. Hashimoto, K.; Suzuki, H.; Yamamoto, M.; Yamaguchi, M. *Proc. IEEE Ultrason. Symp.* **1989,** 339–342.
271. Yamamoto, M.; Hashimoto, K.; Rajendran, V.; Yamaguchi, M. Jpn. *J. Appl. Phys.* **1990,** *29*, 53.

272. Cannata, J. M.; Williams, J. A.; Zhou, Q. F.; Sun, L.; Shung, K. K.; Yu, H.; Kim, E. S. *J. Appl. Phys.* **2008,** *103* (8), 084109.
273. Feng, G.-H.; Sharp, C. C.; Zhou, Q. F.; Pang, W.; Kim, E. S.; Shung, K. K. *J. Micromech. Microeng.* **2005,** *15*, 586.
274. Martin, P. M.; Good, M. S.; Johnston, J. W.; Posakony, G. J.; Bond, L. J.; Crawford, S. L. *Thin Solid Films* **2000,** *379* (1–2), 253.
275. Zhou, Q.; Lau, S.; Wu, D.; Shung, K. K. *Prog. Mater. Sci.* **2011,** *56*, 139.
276. Ito, Y.; Kushida, K.; Sugawara, K.; Takeuchi, H. *IEEE Trans Ultrason Ferroelectr Freq Control* **1995,** *42*, 316.
277. Jen, C.-K.; Sreenivas, K.; Sayer, M. *J. Acoust. Soc. Am.* **1988,** *84*, 26.
278. Pang, H. F.; Fu, Y. Q.; Hou, R.; Kirk, K. J.; Hutson, D.; Zu, X. T.; Placido, F. *Ultrasonics* **2013,** *53*, 1264.
279. Yanagitani, T.; Morisato, N.; Takayanagi, S.; Matsukawa, M.; Watanabe, Y. *IEEE Trans. Ultrason. Ferroelectr. Freq. Control* **2011,** *58*, 1062.
280. Ding, X.; Li, P.; Lin, S.-C. S.; Stratton, Z. S.; Nama, N.; Guo, F.; Slotcavage, D.; Mao, X.; Shi, J.; Costanzo, F.; Huang, T. J. *Lab Chip* **2013,** *13*, 3626.
281. Du, X. Y.; Fu, Y. Q.; Luo, J. K.; Flewitt, A. J.; Milne, W. I. *J. Appl. Phys.* **2009,** *105*, 024508.
282. Alghane, M.; Chen, B. X.; Fu, Y. Q.; Li, Y.; Desmulliez, M. P. Y.; Mohammed, M. I.; Walton, A. *J. Phys. Rev. E* **2012,** *86*, 056304.
283. Alghane, M.; Chen, B. X.; Fu, Y. Q.; Li, Y.; Luo, J. K.; Walton, A. J. *J. Micromech. Microeng.* **2011,** *21*, 015005.
284. Pang, H. F.; Fu, Y. Q.; Garcia-Gancedo, L.; Porro, S.; Luo, J. K.; Placido, F.; Wilson, B.; Flewitt, A. J.; Milne, W. I.; Zu, X. T. *Microfluids Nanofluids* **2013,** *15*, 377.
285. Fu, Y. Q.; Li, Y.; Zhao, C.; Placido, F.; Walton, A. *J. Appl. Phys. Lett.* **2012,** *101*, 194101.
286. Reboud, J.; Bourquin, Y.; Wilson, R.; Pall, G. S.; Jiwaji, M.; Pitt, A. R.; Graham, A.; Waters, A. P.; Cooper, J. M. *PNAS* **2012,** *109* (38), 15162.
287. Hoummadyy, M.; Campitelliyz, A.; Wlodarskiyz, W. Smart Mater. Struct. **1997,** *6*, 647.
288. Sadek, A.; Wlodarski, W.; Li, Y.; Yu, W.; Yu, X.; Kalantar-Zadeh, K.; Li, X. *Thin Solid Films* **2007,** *515*, 8705.
289. Tasaltin, C.; Ebeoglu, M. A.; Ozturk, Z. Z. *Sensors* **2012,** *12*, 12006.
290. Ippolito, S. J.; Kandasamy, S.; Kalantar-Zadeh, K.; Wlodarski, W. *Sens. Actuators, B* **2005,** *108*, 553.
291. Ippolito, S. J.; Ponzoni, A.; Kalantar-Zadeh, K.; Wlodarski, W.; Comini, E.; Faglia, G.; Sberveglieri, G. *Sens. Actuators, B* **2006,** *117*, 442.
292. Huang, F.-C.; Chen, Y.-Y.; Wu, T.-T. *Nanotechnology* **2009,** *20*, 065501.
293. Jakubik, W. P. *Thin Solid Films* **2007,** *515*, 8345.
294. Nagai, H.; Kawai, S.; Ito, O.; Oizmi, T.; Tsuji, T.; Takeda, N.; Yamanaka, K. Possibility for Sub-ppm Hydrogen Detection with the Ball SAW Sensor. In: *The 20th International Congress on Acoustics*, Sydney, Australia, 23–27 August, 2010.
295. García-González, D. L. *Aparicio R. Grasas y Aceites, 53. Fasc.* **2002,** *1*, 96.
296. Dasgupta, D.; Sreenivas, K. *J. Appl. Phys.* **2011,** *110*, 044502.
297. Emanetoglu, N. W.; Zhu, J.; Chen, Y.; Zhong, J.; Chen, Y.; Lu, Y. *Appl. Phys. Lett.* **2004,** *85*, 3702.
298. Wei, C.-L.; Chen, Y.-C.; Cheng, C.-C.; Kao, K.-S.; Cheng, D.-L.; Cheng, P.-S. *Thin Solid Films* **2010,** *518*, 3059.
299. Wu, T.-T.; Wang, W.-S.; Chou, T.-H.; Chen, Y.-Y. *J. Acoust. Soc. Am.* **2008,** *123*, 3377.
300. Sharma, P.; Sreenivas, K. *Appl. Phys. Lett.* **2003,** *83*, 3617.
301. Kumar, S.; Sharma, P.; Sreenivas, K. *Semicond., Sci. Technol.* **2005,** *20*, L27.
302. Phan, T.; Chung, G.-S. *Curr. Appl. Phys.* **2012,** *12*, 521–524.
303. Kumar, S.; Kim, G.-H.; Sreenivas, K.; Tandon, R. P. *J. Electroceram.* **2009,** *22*, 198.
304. Peng, W.; He, Y.; Wen, C.; Ma, K. *Sens. Actuators A* **2012,** *184*, 34.

305. Phan, D.-T.; Chung, G.-S. *Curr. Appl. Phys.* **2012**, *12*, 210.
306. Wang, W.-S.; Wu, T.-T.; Chou, T.-H.; Chen, Y.-Y. *Nanotechnology* **2009**, *20*, 135503.
307. Chivukula, V.; Ciplys, D.; Shur, M.; Dutta, P. *Appl. Phys. Lett.* **2010**, *96*, 233512.
308. Water, W.; Jhao, R.-Y.; Ji, L.-W.; Fang, T.-H.; Chen, S.-E. *Sens. Actuators A* **2010**, *161*, 6.
309. Chu, S. Y.; Water, W.; Liaw, J. T. *Ultrasonics* **2003**, *4*, 133.
310. Chang, R.-C.; Chu, S.-Y.; Hong, C.-S.; Chuang, Y.-T. *Thin Solid Films* **2006**, *498*, 146.
311. Länge, K.; Rapp, B. E.; Rapp, M. *Anal. Bioanal. Chem.* **2008**, *391*, 1509.
312. Voiculescu, I.; Nordin, A. N. *Biosen. Bioelectron.* **2012**, *33*, 1.
313. Rocha-Gaso, M.-I.; March-Iborra, C.; Montoya-Baides, Á.; Arnau-Vives, A. *Sensors* **2009**, *9*, 5740.
314. Kalantar-Zadeh, K.; Wlodarski, W.; Chen, Y. Y.; Fry, B. N.; Galatsis, K. *Sens. Actuators, B* **2003**, *91*, 143.
315. Krishnamoorthy, S.; Iliadis, A. A.; Bei, T.; Chrousos, G. P. *Biosens. Bioelectron.* **2008**, *24*, 313.
316. Luo, J.; Luo, P.; Xie, M.; Du, K.; Zhao, B.; Pan, F.; Fan, P.; Zeng, F.; Zhang, D.; Zheng, Z.; Liang, G. *Biosens. Bioelectron.* **2013**, *49*, 512.
317. Tigli, O.; Bivona, L.; Berg, P.; Zaghloul, M. E. *IEEE Trans. Biomed. Circuits Syst.* **2010**, *4*, 62.
318. Nirschl, M.; Schreiter, M.; Voros, J. *Sens. Actuators A* **2011**, *165*, 415.
319. Lin, R.-C.; Chen, Y.-C.; Chang, W.-T.; Cheng, C.-C.; Kao, K.-S. *Sens. Actuators, A* **2008**, *147*, 425.
320. Mai, L.; Kim, D.-H.; Yim, M.; Yoon, G. *Microw. Opt. Technol. Lett.* **2004**, *42*, 505.
321. Johnston, M. L.; Kymissis, I.; Shepard, K. L. *IEEE Sensors J.* **2010**, *10*, 1042.
322. Nirschl, M.; Schreiter, M.; Vörös, J. *Sens. Actuators, A* **2011**, *165*, 415.
323. Link, M.; Weber, J.; Schreiter, M.; Wersing, W.; Elmazria, O. Alnot, P. *Sens. Actuators B* **2007**, *121*, 372.
324. Garcia-Gancedo, L.; Pedros, J.; Zhao, X. B.; Ashle, G. M.; Flewitt, A. J.; Milne, W. I.; Ford, C. J. B.; Lu, J. R.; Luo, J. K. *Biosens. Bioelectron.* **2012**, *38*, 369.
325. Pottigari, S. S.; Kwon, J. W. *IEEE International Conference on Solid-State Sensors, Actuators and Microsystems*, Denver, CO, June 21–25, 2009; pp 156–159.
326. Zhao, X.; Pan, F.; Ashley, G. M.; Garcia-Gancedod, J.; Luo, J.; Flewittd, A. J.; Milne, W. I.; Lu, J. R. *Sens. Actuators, B* **2014**, *190*, 946.
327. Chen, Y.; Reyes, P. I.; Duan, Z.; Saraf, G.; Wittstruck, R.; Lu, Y.; Taratula, O.; Galoppini, E. *J. Electron. Mater.* **2009**, *38* (8), 1605–1611.
328. Weber, J.; Albers, W. M.; Tuppurainen, J.; Link, M.; Gabl, R.; Wersing, W.; Schreiter, M. *Sens. Actuators, A* **2006**, *128*, 84.
329. Xu, W.; Zhang, X.; Choi, S.; Chae, J. *J. Microelectromech. S* **2011**, *20*, 213.
330. Chen, D.; Wang, J.; Xu, Y.; Li, D. *Sens. Actuators B* **2012**, *171–172*, 1081.
331. Wang, Z.; Qiu, X.; Shi, J.; Yu, H. *J. Electrochem. Soc.* **2012**, *159*, J13.
332. Chen, D.; Wang, J.; Li, D.; Liu, Y.; Song, H.; Li, Q. *J. Micromech. Microeng.* **2011**, *21*, 085017.
333. Chen, D.; Wang, J.; Liu, Q.; Xu, Y.; Li, D.; Liu, Y. *J. Micromech. Microeng.* **2011**, *21*, 115018.
334. Wang, Z.; Qiu, X.; Chen, S. J.; Pang, W.; Zhang, H.; Shi, J.; Yu, H. *Thin Solid Films* **2011**, *519*, 6144.
335. Qiu, X.; Tang, R.; Zhu, J.; Oiler, J.; Yu, C.; Wang, Z.; Yu, H. *Sens. Actuators B* **2011**, *151*, 360.

CHAPTER 6

DILUTED MAGNETIC SEMICONDUCTING OXIDES

NGUYEN HOA HONG*

Department of Physics and Astronomy, Seoul National University, Seoul, South Korea

*E-mail: nguyenhong@snu.ac.kr

CONTENTS

ABSTRACT

More than a decade ago, it was suggested theoretically that room temperature ferromagnetism (FM) could be obtained in various semiconductors such as ZnO, GaAs, GaN, etc. if we dope Mn transition-metal (TM) into these systems. The magnetic ordering in those compounds was thought to be induced by the Ruderman–Kittel–Katsuya–Yosida interaction of localized moments of dopants via 2p holes or 4s electrons. Being excited by this idea, many experimental groups have tried to dope TMs into many oxides such as ZnO, TiO_2, SnO_2, In_2O_3, ZrO_2, HfO_2, etc. aiming to obtain room temperature FM in semiconductors, in order to be able to exploit both charge and spin in same compounds. Indeed, room temperature FM was found; however, the phenomenon is not the same as what theorists proposed. The finding of Coey's group in 2004 about room temperature FM of pristine HfO_2 has given some warning to the magnetism community to be more careful when judging the role of TM dopings. More recently, experimental observations of FM for various oxides such as TiO_2, HfO_2, In_2O_3, ZnO, CeO_2, Al_2O_3, and MgO in nanometer materials have confirmed that FM is certainly possible for pristine semiconducting oxides. It is very likely that FM could stem from oxygen vacancies and/or defects that were formed more comfortably at the surface and interfaces.

Research on very thin films and nanoparticles of diluted magnetic semiconducting oxides (DMSO) has shown that downscaling magnetic oxide semiconductors to nanometer size must be a key point, in order to make them ferromagnetic. Exploiting the bright side of nanoworld in the field would help to open a new door for spintronic applications.

The domain of DMSO research still demands a huge of efforts of leading research groups toward higher levels with hope to make it able to approach closer to a realization of spintronic devices based on DMSO materials.

6.1 INTRODUCTION

For devices using semiconductors, taking advantage of the charge of electrons is very important. Additionally, if semiconducting materials are ferromagnetic, they can also be exploited with respects to electron spin. If we use both charges and spin of electrons in the same material, then the spin of electrons, which carries the information, can serve an added degree of freedom in novel electronic devices. Thus, functional ferromagnetic semiconductors can be applied for promising devices for the future.

Fourteen years ago, Dietl et al. theoretically pointed out that room temperature ferromagnetism (FM) could be obtained in p-type diluted magnetic semiconductors (DMSs) made by doping few percent of Mn into ZnO [1]. Their calculations showed that FM could be created in a DMS, which is based on a wide band-gap semiconductor. In the same year, Katayama-Yoshida group in Osaka had stated that in order to introduce room temperature FM, the whole series of transition-metals (TMs) except Mn, could be used to partially dope into ZnO [2].

One year later, the first experimental evidence was reported: room temperature FM was observed in Co-doped TiO_2 thin films. However, the observed magnetic moment is very modest (as of 0.3 μ_B per Co atom) [3]. After this proof, plenty of studies have been carried out, whether to look for a new compound which might be simply ferromagnetic at room temperature, or to find materials with larger magnetic moments [4,5]. The most important point for applications is that such room temperature FM must be intrinsic, that is, it should really originate from the doped matrices but not from dopant clusters. Since the mechanism that governs the magnetic interactions in this system is still unclear, and in fact quite controversial, many groups had been trying to verify these issues. It had been found that the induced magnetic properties in DMS systems do not simply result only from the Ruderman–Kittel–Kasuya–Yosida (RKKY) interaction, but it most probably come from defects and/or oxygen vacancies [6–8]. Obviously, the behaviors of semiconducting oxides such as ZnO, TiO_2, SnO_2, In_2O_3, etc. are quite different from other traditional semiconductors such as GaAs or GaN. Recently, people have been rather prudent when discussing magnetic properties of semiconducting oxides, then have separated this as a particular family of materials that is so-called DMSO in order to distinguish it with other classical semiconductors that do not have problems relating to defects or oxygen vacancies. Since the discovery of FM in pristine oxides such as HfO_2, TiO_2, In_2O_3, SnO_2, CeO_2, MgO [6,9–13], people have given special attention to magnetism due to defects and/or oxygen vacancies. It is sure that we should be careful when rejudging the actual role of 3d element doping in introducing FM in a semiconducting host: Does the doping indeed introduce FM? Or does it just enhance the magnetism that already exists in the oxide hosts under a nanometered-size form? Moreover, it was shown that the observed FM could be seen only in low-dimension systems, pointing that one cannot ignore the role of confinement effects.

In this chapter, we will review the results of this research field, from its theoretical background and experimental results for magnetic oxide semiconductors, to newest models that have been proposed, in order to understand better the mechanism of these materials. We hope that our review could help the readers to have an overall picture about this material then may have some ideas for the future development of spintronic applications using ferromagnetic semiconducting oxide.

We do not plan to make our chapter to be a complete report about ferromagnetic oxides but only want to express our own viewpoint about standing issues, based on working experiences of the authors in the domain.

6.2 THEORETICAL BACKGROUND

In this paragraph, we recall about the previous theories of magnetism in DMS systems that actually have led experimental research over more than a decade. These days we are putting a lot of questions if all are not right based on the fact that experiments do not support those theoretical ideas, even if from first looks, it looked like "really believable."

The group of Dietl was the first group that has opened this direction [1], suggesting that by doping Mn into ZnO, one might induce FM at high temperatures, along with a requirement that the doped compound must be p-type. Working along this line, Sato and coworkers had developed this theory to a higher level, viewing more fully for the whole range of doping of TMs [2,14]. We then just try to report their ideas below.

The Fermi level (E_F) of ZnO lies in the majority impurity band. The significant energy gain arises from the broadening of the impurity band with increasing concentration c. When the concentration is increased from a lower value to a higher value, the density of state (DOS)-weight is transferred from around E_F to lower energies, leading to an energy gain, which stabilizes the ferromagnetic state. This energy gain is proportional to the bandwidth W of the impurity band, which is proportional to the square root of the concentration. This is known as Zener's double exchange [15,16].

According to the electronic structure of TM impurity in semiconductors [17], fivefold degenerated d states of TM impurity are split into doubly degenerated d_γ states and threefold degenerated d_ε states in the tetrahedral coordination. Two d_γ states have the symmetry of $3z^2-r^2$ and x^2-y^2. Three d_ε states have the symmetry of xy, yz, and zx. The wave functions of these d_ε states are extended to anions; thus, d_ε states can hybridize with O-2p states which make host valence band; therefore, the bonding states (t^b) and the anti-bonding counterparts (t^a) are created. Additionally, the wave functions of d_γ states are extended to interstitial region, so that the hybridization of d_γ states with host valence band is weak and d_γ states are remained as nonbonding states.

In ZnO-based DMS, the magnetic properties are decided by the competition between ferromagnetic double exchange and antiferromagnetic super-exchange. In order to ensure the ferromagnetic state, there must be itinerant electrons. It was found that the ferromagnetic state can be stable when delocalized t^a states are partially occupied. By this rule, the induced magnetism in DMS could be explained. Remembering that TM may have either 2^+ or 3^+ charge state in II–VI [28] or III–V [22], respectively, then their 3d-electron configurations are expected.

In Figure 6.1, computed total energy difference per one formula unit between the ferromagnetic state and the spin-glass state is shown for V-, Cr-, Mn-, Fe-, Co-, and Ni-doped ZnO [2]. The positive energy difference indicates that the ferromagnetic state is more stable than the spin-glass state. For the Mn-doped ZnO, the disordered local moment state is most stable, while for V- and Cr-doped ZnO, the ferromagnetic states are more stable than the disordered local moment states.

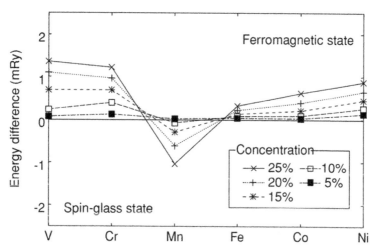

FIGURE 6.1 Stability of the ferromagnetic state in TM-doped ZnO systems (reprinted with permission from Ref. [14]). (Reprinted with permission from Sato, K.; et al. Computational Materials Design of ZnO-Based Semiconductor Spintronics. In: *Magnetism in Semiconducting Oxides*; Hong, N. H., Ed.; Transworld Research Network, 2007.)

Indeed with only few percent of doping (in case of DMSO), dopants are indeed quite isolated with no magnetic nearest neighbors; therefore, the conventional picture of magnetism in insulators and magnetic semiconductors cannot well explain anything for what really happens in DMSO that we will discuss in the next section.

6.3   EXPERIMENTAL RESULTS

6.3.1   DILUTED MAGNETIC SEMICONDUCTING OXIDES (OR TM-DOPED SEMICONDUCTING OXIDES)

Since 2001, many studies have been carried out in the field of DMSO, whether to look for a new compound which might be simply room temperature ferromagnetic or to discover some materials which should have a large magnetic moment [4,5]. It is very crucial for applications that such room temperature FM must indeed come from the doped matrices, not from dopant clusters. Since the mechanism of the induced magnetism in this system is still unclear, many groups have been trying to elucidate these issues. It is likely that the magnetism in DMS systems does not simply result from the RKKY interaction, but it seems to stem from defects and/or oxygen vacancies that are formed during the films' growth [6,8]. In order to clarify these standing points, systematic investigations on TM-doped TiO_2, SnO_2, In_2O_3, and ZnO thin films were done.

The most popular example for the observed FM in DMSO is TM-doped TiO_2 thin films [18]. All of TM-doped TiO_2 films deposited within the chosen range of growth conditions are room temperature ferromagnetic. The magnetization versus element of dopant is replotted in Figure 6.2. Room temperature FM was confirmed by the $[M(H)]$ curves taken at 300 K that demonstrate very well-defined hysteresis loops, and T_C is above 400 K [18]. The observed magnetic moments are quite large, and the largest value as of 4.2 μ_B per atom was found in V-doped TiO_2 films (see inset of Fig. 6.2). This finding is worth to be noticed since vanadium itself is known to be nonmagnetic. Therefore, the observed FM could be supposed to be intrinsic. This remark is enforced by the magnetic force microscopy (MFM) measurements that were performed at room temperature on a film of V-doped TiO_2 (Fig. 6.3). The V-doped TiO_2 sample shows a uniform MFM image that was distinctly different from the atomic force microscopy (morphology) image indicating a global ferromagnetic effect that rules out the cause of clusters. (More details can be seen in Ref. [18].)

The chemical trend for TM-doped TiO_2 films is rather similar to what the theory depicted for TM-doped ZnO single crystals, except one point is that we have obtained FM in Mn-doped TiO_2, while according to theories, antiferromagnetic must be the ground state of Mn-doped ZnO [2]. This could be explained by the difference of the hosts (TiO_2 in this case), and of the growth conditions, which were hard to maintain exactly the same for different runs.

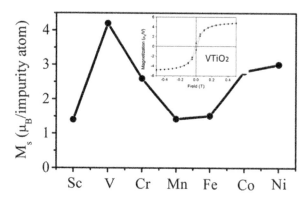

FIGURE 6.2 Saturated magnetization (at 300 K) versus element for TM-doped TiO₂ thin films grown on LaAlO₃ substrate at 650°C. The inset shows the M-H curve taken at 300 K for a V-doped TiO₂ film (Field was applied parallel to the film plane).

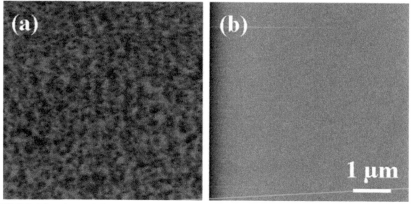

FIGURE 6.3 (a) and (b) AFM and MFM images taken at room temperature for a V-doped TiO2 film grown on LaAlO3 substrate at 650°C and the tip was magnetized parallel to the film plane.

Till now, in comparison to the work on TiO₂ and ZnO systems, there have been few reports on TM-doped SnO₂ thin films. While Kimura et al. reported about the absence of FM in Mn:SnO₂ [19], other groups obtained FM above room temperature with rather large magnetic moments in Co:SnO₂ thin films (7.5 μ_B Co⁻¹) and Fe:SnO₂ thin films (1.8 μ_B Fe⁻¹) [20,21]. It has been supposed that such a large magnetic moment was probably resulted from kind of unquenched orbital contributions [20].

Cr/Ni/V-doped SnO₂ films were grown on (0 0 1) LaAlO₃ (LAO), (0 0 1) SrTiO₃ (STO) and R-cut Al₂O₃ substrates with optimal conditions. Except the V:SnO$_2$ films that were deposited on Al$_2$O$_3$ substrates show paramagnetic behavior (note that only these films are amorphous, while all others are well formed as SnO$_2$ structure), Cr:SnO$_2$ and Ni:SnO$_2$ films grown on three types of substrates, as well as V:SnO$_2$ films on both LAO and STO substrates are room temperature ferromagnetic. All films have T_C higher than 400 K. A very large magnetic moment of 6 μ_B per atom was obtained in Cr-doped SnO$_2$ films [22–24]. This huge value cannot come from Cr metal clusters, because Cr metal is paramagnetic at high temperatures and antiferromagnetic below 308 K [25]. Since the LAO substrate is diamagnetic, it is also impossible to attribute the large ferromagnetic signals of the Cr:SnO$_2$ films on LAO to the substrates. Additionally, the magnetic moment of 2–6 μ_B per impurity atom in TM:SnO$_2$ films on LAO are too large to be attributed to any precipitations. Consequently, it must be explained that in these cases, the orbital quenching effects are most probably absent.

Another thing to be aware about TM-doped SnO$_2$ films is substrate effect. Films that were grown under the same fabrication conditions on STO and Al$_2$O$_3$ substrates have a magnetic moment of one order smaller than that of films grown on LAO substrates. The [M–(T)] curves of these films on STO and Al$_2$O$_3$ show a rise up at very low temperatures suggesting an existence of clusters, because if there are some antiferromagnetic precipitations, their contributions could cause a reduction of magnetic moment [22]. The reason should be the big difference in substrate's morphology that strongly influences the interface. The LAO substrate has a very smooth and flat surface while the STO substrate is very rough with lots of steps on it. This can be the reason to make the film's morphology, and the structural nature of the compound as well, to be modified at the interface (see topology images for substrate in Ref. [22]). As consequences, some precipitations could be favorably formed when films were deposited on STO substrates (also similarly on Al$_2$O$_3$ substrates), in comparison with films grown on LAO substrates.

Since In$_2$O$_3$ is a transparent, wide-band-gap semiconductor with a cubic structure, which is rather complicated and different from the other host oxides that have been investigated so far, it is not easy to obtain FM in this type of oxides. In this section, only selected results on the V/Cr/Fe/Co/Ni-doped In$_2$O$_3$ thin films grown on MgO and Al$_2$O$_3$ substrates will be discussed.

V/Cr/Fe/Co/Ni-doped In$_2$O$_3$ thin films grown by appropriate conditions are well crystallized, and all showed room temperature FM (T_C higher than 400 K) [26]. As regarding to V and Cr doping cases, it is impossible to suspect if the FM in the films could stem from dopant metal clusters, because in fact, V and Cr metals themselves are known to be nonmagnetic. As for Fe and Co doping, the saturation magnetization (M_s) is very modest (0.4 and 0.5 μ_B atom^{-1}, respectively), ruling out the assumption if the FM in those films could originate from Fe and Co metal particles/clusters, since for Fe and Co metals, M_s should be 2.2 and 1.7 μ_B, respectively [25]. As for Ni doping, the value of M_s of the Ni:In$_2$O$_3$ film as of 0.7 μ_B atom^{-1} also does not match the value of M_s for Ni metal as of 0.6 μ_B atom^{-1} [25]. Note that to calculate the values of magnetization from magnetic moment, we used very precise numbers of atoms that were strictly determined from the Rutherford Back-scattering Spectroscopy (RBS) data, by supposing that all the dopant atoms contributed to the magnetism of the films. For the films grown on Al$_2$O$_3$ substrates, we also obtained room temperature FM with similar values of M_s, and the maximal M_s that could be obtained is also 0.7 μ_B for the Ni:In$_2$O$_3$ films [26]. There is almost no difference for M_s of TM:In$_2$O$_3$ films grown on MgO and Al$_2$O$_3$ substrates. In general, the chemical trend is very similar to the theoretical trend of the TM:ZnO system [2], and that of TM:TiO$_2$ films that was mentioned earlier [18].

Even though our work on TM:ZnO thin films have showed that room temperature FM could be obtained in V/Mn/Cr/Fe-doped ZnO, the main goal of our study on the ZnO-based system is not to find the FM but to understand the origin of its FM. In comparison to the other host oxides such as TiO$_2$ or SnO$_2$ that we have discussed earlier, doping TMs in ZnO does not result in a very big magnetic moment (in fact, it is very modest, i.e., one order smaller than that of TM:TiO$_2$ or TM:SnO$_2$ films). However, these compounds are found to be quite sensitive to defects and/or oxygen vacancies. Thus, we expected that this case might help to justify the nature of magnetism in DMS compounds.

For V:ZnO films, it was found that by changing the substrate temperature of only 50°C, the magnetization could be changed by 1 order of magnitude [27], then it implied that growth conditions in fact play a very important role in tuning FM in ZnO-based systems. Thereafter, we have exploited its influences to try to deal with the most controversial case in the field: Mn-doped ZnO. A theoretical work claimed that antiferromagnetic is the ground state of Mn-doped ZnO [28], and Mn doping alone cannot produce FM in ZnO system; therefore, one must codope with Cu [29]. Our work on Mn-doped ZnO showed that oxygen vacancies could play a more important role than that of additional carriers [30]. It was found that doping Mn alone does not result in room temperature FM if inappropriate conditions were applied (i.e., 400°C and an oxygen pressure of 10^{-6} Torr). However, under appropriate ones (i.e., 650°C and an oxygen pressure of 0.1 Torr), it could be absolutely possible. The substrate temperature and oxygen pressure during the growth process might create defects and/or necessary oxygen vacancies that are similar to an n-type doping. This hypothesis is in accord with the explanations in Ref. [6] for the magnetism in HfO_2 films, which has become agreed later by a theoretical work supposing that vacancies can be necessary ingredients to create additional bands inside the semiconducting gap that is responsible for such FM [31]. Our work on Cr-doped ZnO films have shown that an oxygen annealing could certainly improve the film's crystallinity [8], but simultaneously destroys the ferromagnetic ordering (see Fig. 6.9 in Ref. [8]). It is understood that in this system, a perfect crystallinity does not go

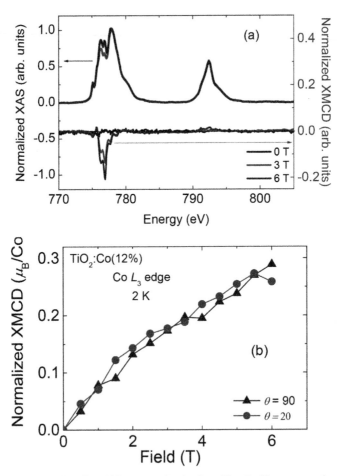

FIGURE 6.4 (a) XAS and XMCD spectra for the $Co_{0.12}Ti_{0.88}O_2$ film measured at the Co $L_{2,3}$ edges (taken at 2 K, 6 T when the magnetic field applied at an angle of 20° to the film plane) and (b) XMCD versus magnetic field curves taken at 2 K for the $Co_{0.12}Ti_{88}O_2$ film when the magnetic field applied at an angle of 20° and 90° to the film plane.

along with FM, and filling up oxygen vacancies enormously degrades FM. In general, defects and oxygen vacancies should play a very important role in tuning FM in DMSO [8].

The wonder about an intrinsic nature of FM in TM-doped semiconducting thin films discussed above has suggested us to re-examine the role that a 3d element dopant indeed plays. A series of X-ray absorption spectra (XAS) and X-ray magnetic circular dichroism measurements (XMCD) on Cr-, Mn-, and Co-doped TiO_2 films was carried out at the Cr, Mn, and Co $L_{2,3}$ edges. A typical example of XAS and XMCD spectra is shown in Figure 6.4 for the Co $L_{2,3}$ edges of a TiO_2 film doped with 12% of Co. One can clearly see that the dependence of the XMCD signals on magnetic field is well paramagnetic. The magnetic moment at remanence (i.e., $H =$ 0) is estimated to be below 0.02 μ_B atom^{-1} by applying the magnetooptical sum rules [32]. A very similar feature was also observed for the Mn- and Cr-doped TiO_2 film. These results reveal that the main contribution to the ferromagnetic signal basically comes from the pristine TiO_2 host matrix. The same results were also obtained for Co-doped ZnO. Co contribution to magnetic properties of Co-doped ZnO is indeed paramagnetic [33], or in other words, FM does not come from RKKY interaction of localized moments of impurities. These facts explain why the FM (T_C, etc.) in DMSO does not depend much on the type and concentration of dopant. It is obvious here that new theories are required in order to understand this novel class of materials.

Conclusions: Doping TMs in TiO_2, SnO_2, In_2O_3, and ZnO under thin film forms with appropriate growth conditions have shown to result in FM above room temperature. In lots of cases, a large magnetic moment could be obtained. This work not only presents many promising candidates for spintronics application but also helps to better understand about these systems from a viewpoint of fundamental research. Besides, the RKKY interaction that might play some role in introducing magnetism in these compounds, there must be another source such as structural defects and/or oxygen vacancies (which are originated basically from thin film forms and nanoparticles) that play a more important role in tuning FM. A big question arises here if the TM doping is indeed important for introducing FM into nonmagnetic oxides, or it just acts as a catalyst to add some contribution to the magnetic ordering already exists in the host?

6.3.2 ROLE OF DEFECTS AND OXYGEN VACANCIES IN THE INDUCED FM OF PRISTINE OXIDES

In 2004, Dublin group first reported about the observed FM in HfO_2 thin films grown on sapphire or silicon substrates [6]. This report has immediately attracted special attentions of the magnetism community about a new phenomenon, so-called d^0 magnetism. Indeed, the thin film form (i.e., 2-dimension-structure) might be another key point here, since defects and/or oxygen vacancies that are formed during the film growth can also become a source for magnetism. The fabrication conditions can actually create oxygen vacancies, which play a similar role to an n-type doping. Working along this direction, many experimental works have shown that defects certainly can tune the magnetic properties of diluted magnetic oxide thin films. For example, it was obvious that defects could intentionally introduce FM into ZnO system [7]. In some other cases, it was found that a good crystallinity could indeed destroy the ferromagnetic ordering. Additionally, it was also found that filling up oxygen vacancies could enormously degrade the magnetic moment of those DMSO [8]. One theory group has performed simulation on HfO_2 and assumed that isolated cation vacancies in HfO_2 could form high-spin defect states, and as the results, they could be ferromagnetically coupled with a rather short-range magnetic interaction leading to a ferromagnetic ground state [31]. We ourselves obtained FM in HfO_2 thin films grown on yttrium-stabilized zirconia (YSZ) substrates [9]; however, the corresponding XMCD data do not show any magnetic signal on the Hf site due to big ratio of noises. Later, Coey group stated that the FM of their HfO_2 films is not stable but has a strong aging effect [34]. Another group proposed their model but it does not explain well experimental results [35]. All of these questionable issues have suggested us to experimentally investigate the magnetic properties of several types of undoped oxides. We could indeed confirm that pristine oxides could be certainly room temperature ferromagnetic in nanostructured forms. Additionally, a TM doping in fact does not play any important role in inducing FM in those hosts.

Pure TiO_2, HfO_2, ZnO, In_2O_3, SnO_2 films were deposited by a pulsed-laser deposition system (KrF, 248 nm) from ceramic targets on (1 0 0) LAO, (1 0 0) YSZ, C-cut Al_2O_3, (0 0 1) MgO substrates, respectively. The typical thickness is 220 nm for TiO_2, HfO_2, and SnO_2 films, 600 nm for In_2O_3 films on MgO, and 375 nm for ZnO films on Al_2O_3. Magnetic moment data were basically taken when the magnetic field was applied parallel to the film plane. From Figure 6.5, it is shown that all films are room temperature ferromagnetic. While the magnetic moments for HfO_2 and TiO_2 films are quite large (M_s is almost 30 emu cm^{-3} for HfO$_2$ and 20 emu cm^{-3} for TiO$_2$), it is one order smaller for In$_2$O$_3$ films on MgO and ZnO films on Al$_2$O$_3$ (i.e., only about few emu cm^{-3}). Later, a group in Northwestern Univ. also reported similar results for their laser ablated TiO$_2$ films [36], and the TiO$_2$ films made by spin coating in-air by our group also showed room temperature FM [37]. Thus, we can certainly confirm that the observed phenomenon is not mistaken. It is not possible to attribute such a large value of magnetic moment in the case of HfO$_2$ and TiO$_2$ films to any kind of impurity. What can be the source for magnetism here? For the TiO$_2$ case, neither Ti^{4+} nor O^{2-} is magnetic. Also for the HfO$_2$ case, neither Hf^{4+} nor O^{2-} is magnetic. An initial assumption is that it is due to impurities. From the viewpoint of purity of the targets, we must say that such possibility is very small, since impurities of less than 10^{-2} wt% could not create such huge magnetic moments. From the viewpoint of the structural properties of the deposited films, it is found that there is no trace of impurities that could be seen from XRD and films are single phase [23].

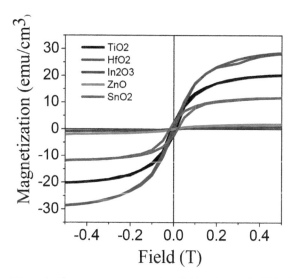

FIGURE 6.5 Magnetization versus magnetic field taken at 300 K for pristine 220-nm-thick-TiO$_2$ grown on LaAlO$_3$, 220-nm-thick HfO$_2$ grown on YSZ, 600-nm-thick In$_2$O$_3$ grown on MgO, 375-nm-thick ZnO grown on Al$_2$O$_3$, and 220-nm-thick-SnO$_2$ grown on LaAlO$_3$ films.

One should notice that films of In$_2$O$_3$ on MgO are room temperature ferromagnetic but with a rather modest magnetic moment. However, an important feature that needs to be noted here is that In$_2$O$_3$ films fabricated under the same conditions on Al_2O_3 substrates are diamagnetic; even if the films were well crystallized [9]. There has been no report so far about In_2O_3 that could be magnetic, since In^{3+} could not be the source of magnetism. Even though In_2O_3 tends to create oxygen vacancies [38], the fact that FM is observed on only one type of substrate but not on the other implies some sort of defects that might cause such magnetism. Similarly, in the ZnO case, neither Zn^{2+} nor O^{2-} is magnetic; thus, according to the conventional concepts, there is no source for magnetism in pure ZnO.

As for pristine SnO_2 films, there is no reason to attribute the introduction of FM to any dopant, and moreover, there is no 3d electron involved, so that one cannot think of any interaction that may originate from that. Some groups reported that their SnO_2 films are diamagnetic, while ours is certainly ferromagnetic. Hays et al. reported that their nanoparticles of SnO_2 are nonferromagnetic [39], while Rao's group has confirmed that their SnO_2 nanoparticles were weakly ferromagnetic with some paramagnetic components [12]. To explain the observed FM in our films, we must say that most probably, oxygen vacancies formed during the growth are a key factor here, beside the presence of confinements. However, growth conditions and how to make them precisely controllable must be a standing issue. We would like to recall Ref. [12] stating that thermal treatments could drastically influence the magnetic properties of SnO_2 nanoparticles. This explained well why we could get a very pronounced FM in our SnO_2 films, while in the case of Ref. [12], the paramagnetic phase is still more dominant than the coexisted ferromagnetic one.

We studied the possible formation and segregation of oxygen vacancies near the surface of SnO_2 thin films from oxygen K-edge X-ray emission and absorption spectra and discovered that the distribution of O 2p unoccupied states for ferromagnetic SnO_2 thin films is completely different from that of postannealed SnO_2 films under oxygen atmosphere showing diamagnetic behavior. This spectroscopic result implied that oxygen vacancies should be the main source of the surface-induced magnetism in SnO_2 thin films. This possibility was then clarified by calculating the lowest energy levels of the structural defects (impurities or neutral vacancies) with two localized carriers near the surface of SnO_2 film using a quantum–mechanical approach combined with the image charge method. A magnetic triplet state is determined to be the ground state of those defects in the vicinity of the SnO_2 surface, whereas the nonmagnetic singlet is the ground state of the bulk SnO_2. Surface-induced ferromagnetic order can certainly appear at room temperature via two-dimensional magnetic percolation, if the vacancy concentration is greater than 3×10^{16} m^{-2} (see Figure 6.6 and refer to Ref. [40]).

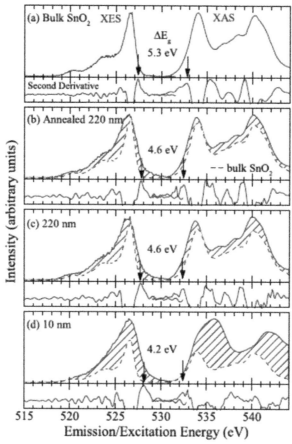

FIGURE 6.6 O K-edge X-ray emission ($OK\alpha$ XES) and absorption (O 1s XAS) spectra for bulk SnO_2 (a), the 200-nm-thick SnO_2 thin film with O_2 postannealing and (b), the 200-nm-thick as-grown film (c), and the 10-nm-thick as-grown film (d) (reprinted with permission from Ref. [40]). (Reprinted from Chang, G.-S.; Forrest, J.; Kurmaev, E. Z.; Morozovska, A. N.; Glinchuk, M. D.; McLeod, J. A.; Moewes, A.; Surkova, T. P.; Hong, N. H. Oxygen-vacancy-induced ferromagnetism in undoped SnO2 thin films. *Phys. Rev. B* 2012, *85*, 165319.© 2012 American Physical Society.)

Recently, very similar results were reported for FM in undoped CeO_2 and MgO [12,13]. All come to the same conclusion that samples of pristine semiconducting oxides can be ferromagnetic if

(1) they have nanometer-size configurations (such as thin films and nanoparticles);
(2) there are defects and/or oxygen vacancies.

In Figure 6.7a,b, it is shown that bulks of TiO_2, HfO_2, ZnO, In_2O_3, SnO_2 (pieces cut from targets from which films were made) are diamagnetic. It implies that indeed FM was induced only in the thin film form, or in other words, it is unique for low-dimension structures (also seen in Ref. [12] for nanoparticles and Ref. [37] for films made by spin coating). One should assume that confinement effects should play some role here. We see from Figure 6.7c that the ultrathin film such as 5-nm-thick has a much larger magnetic moment than the 200-nm-thick one. It suggests that besides the reason of confinement effects in low-dimension systems, the observed FM must be induced by defects and/or oxygen vacancies that are mostly formed at and near to the surface and/or interface (between film and substrate).

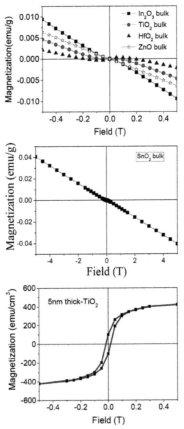

FIGURE 6.7 Magnetization versus magnetic field taken at 300 K for (a) and (b) bulks of TiO_2, HfO_2, In_2O_3, ZnO, and SnO_2 and (c) for a 5-nm-thick TiO_2 film (field was applied parallel to the film plane).

The assumption about FM due to defects and oxygen vacancies are enforced by data shown in Figure 6.8. Figure 6.8a shows M–H curves for the as-grown TiO_2 film as well as for the TiO_2 post-annealed films in oxygen atmosphere. One can see that after 2 h of annealing in oxygen atmosphere, the magnetic moment of the as-grown film reduces about one order of magnitude. When it was annealed for longer time, up to 8–10 h, the ferromagnetic ordering was completely destroyed. It seems that the observed FM is related directly to oxygen vacancies, because when the oxygen vacancies are filled, the FM is degraded [9].

Another example for FM induced by defects can be seen from the case of carbon-doped SnO_2 thin films (see Fig. 6.8b). Even pristine SnO_2 is ferromagnetic already, when C was doped, the magnetic moment was increased significantly. Doping C is known to create defects in the host; therefore, the relationship between defect and the enhancement of magnetization in this case is also confirmed [41].

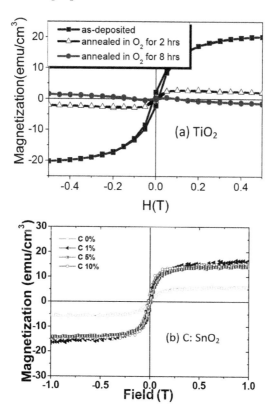

FIGURE 6.8 Magnetization versus magnetic field taken at 300 K for (a) as-grown and oxygen post-annealed TiO_2 films and (b) for $C_xSn_{1-x}O_2$ films where $x = 0\%$, 1%, 5%, and 10%.

To ensure about above remarks, the chemical and orbital selectivity of X-ray absorption spectroscopy and X-ray magnetic circular dichroism were exploited, to prove that the observed FM in laser-ablated TiO_2 films is an intrinsic property of this material, directly originating from the O-2p and, to a lesser extent, from the Ti-3d electrons. This FM stems from a surface region, of a thickness of several nanometers, which is likely to be rich in oxygen vacancies [42].

In order to find some model that fit better to explain our experimental results that are quite far from what theory of Dietl et al. could cover, Huong proposed a model of FM in oxide thin films of HfO_2 and TiO_2 to investigate the possibility of magnetism due to oxygen vacancies and the confinement effect [42]. Due to HfO_2 and TiO_2 structure, each oxygen atom is surrounded by three or four Hf (or Ti) atoms and the local symmetry is D_{3h} and D_{4h} accordingly. An oxygen vacancy would result in the loosing of bonding of two d-electrons in the outer shell of the Hf (Ti) atom. It was assumed that in the local D_{3h} or D_{4h} molecular orbital high symmetry, the two electrons do not really leave their own shell and remain as the d-electrons, becoming a d^2 impurity center. The exchange interaction of these two d-electrons with each other and with the molecular orbital field leads to splitting of energy level of the impurity band around the vacancy, so that produces a large magnetic moment. Beside the orbital local symmetry, the electrons here are considered to be confined in the two-dimensional confinement. The thin film configuration would have two effects on the structure of the impurity band: it enhances the formation of the oxygen vacancy and enhances the coupling between the d-electrons and the local field as well. The boundary conditions at the surfaces of the thin film in the strong confinement approximation make the matrix elements for the exchange interaction much larger in comparison with the bulk case. In the strong local field, the interaction of each d-electron with the local tetragonal orbital field is considered first to get the eigenstates and the splitting of the fivefold degenerate energy level of the single d-impurity center into e- and t_2-orbitals. Then, the interaction between the two d-electrons is calculated according to irreducible representation of the product of E and T_2 representations, along with the spin part and the Pauli principle. By diagonalizing the Coulomb exchange matrices, the splitting of all energy levels and the states of the impurity band with a high-spin ground state could be obtained [43].

The tight binding calculation was used for HfO_2, TiO_2, and In_2O_3 thin films to consider the effect of the molecular orbital field and the impurity band created by the exchange interaction of the electrons created by the oxygen vacancy and trapped around the vacancy, while taking into account the two-dimensional confinement effect of the thin films. The splitting of energy levels of this impurity band has been obtained and the two-dimensional confinement results in large exchange interaction matrix element, which is very different in comparison with the bulk's case. It was found that a high-spin state with a magnetic moment per vacancy of 3.18 μ_B for TiO_2, 3.05 μ_B for HfO_2, and 0.16 μ_B for In_2O_3. If the vacancy concentration about 3%, we would have the magnetic moment comparable to the experimental values. According to this model, FM is quite possible for thin film configurations of undoped semiconducting oxides, especially for the configurations that favor a formation of oxygen vacancies (thin films and nanoparticles) [43].

In many semiconducting oxides, if the observed FM is due to defects, then it would be very difficult to control and manipulate for application purposes. Thus, the search for a more suitable candidate is pursuit. So far, there is only one case that likely show an intrinsic induced FM, it is cubic Mn-doped ZrO_2 films. As seen in Figure 6.9, undoped Mn-doped ZrO_2 is paramagnetic, and doping Mn really introduces FM into ZrO_2 host. One cannot blame that FM in Mn-doped ZrO_2 is due to defects, since it shows clearly that it comes only after Mn doping [44]. To be able to generalize this case, we have investigated the magnetic properties of Fe/Co/Ni-doped ZrO_2

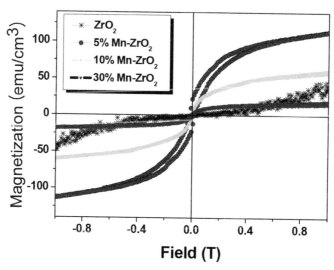

FIGURE 6.9 Magnetization versus magnetic field (applied parallel to the film plane) taken at 300 K for 80-nm-thick $Mn_xZr_{1-x}O_2$ films (where $x = 0\%$, 5%, 10%, and 30%). The inset shows the $M–H$ curve measured in magnetic field perpendicular to the film plane for the 5% Mn-doped ZrO_2 film taken at 300 K.

laser-ablated thin films in comparison with the known results of Mn-doped ZrO_2. It is found that doping with a TM can really induce room temperature FM in "fake diamond." Theoretical analysis based on density functional theory has confirmed the experimental data, by revealing that the magnetic moments of Mn- and Ni-doped ZrO_2 thin films are much larger than that of Fe- or Co-doped ZrO_2 thin films. Most importantly, our calculations confirm that Mn- and Ni-doped ZrO_2 show a ferromagnetic ground state in comparison to Co- and Fe-doped ZrO_2, which favor an antiferromagnetic ground state (Fig. 6.10). This might give an important guide for material research [45].

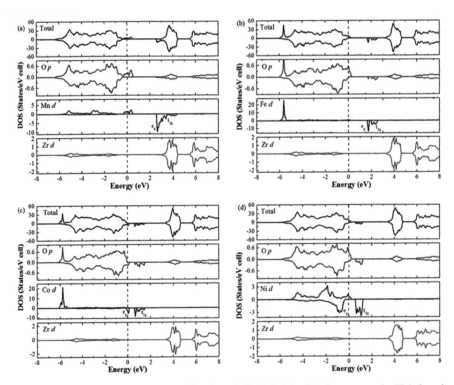

FIGURE 6.10 The total and partial spin-polarized DOSs for ferromagnetic TM-doped ZrO_2: (a) Mn, (b) Fe, (c) Co, and (d) Ni. Spin up and spin down correspond to positive and negative values, respectively. The vertical dashed line denotes the Fermi level (reprinted with permission from Ref. [45]). (Reprinted with permission from Hong, N. H.; Kanoun, M. B.; Goumri-Said, S.; Song, J.-H.; Chikoidze, E.; Dumont, Y.; Ruyter, A.; Kurisu, M. The origin of magnetism in transition metal-doped ZrO_2 thin films: experiment and theory. *J. Phys.: Phys.* Condens. 2013, *25*, 436003. http://iopscience.iop.org/article/10.1088/0953-8984/25/43/436003/meta © 2013 IOP Publishing Ltd.)

Conclusion: Many experimental results have shown that FM could be obtained in undoped semiconducting oxide thin films and nanoparticles. The observed FM is most probably due to oxygen vacancies and/or defects. The assumption for FM due to oxygen vacancies/defects in TiO_2 thin films is strongly confirmed by XMCD: There is a presence of XMCD signals at both O K and Ti $L_{2,3}$ edges. It implies that the FM in TiO_2 films stems from both O-2p and Ti-3d electrons. Our theoretical model also suggests that confinement effects must play a key role in tuning magnetic properties of thin films, or more generally, of low dimension structured systems.

This finding may open a new direction for searching for a novel class of materials that are promising for spintronic applications. By down-scaling semiconducting oxides to nanosize, under appropriate conditions that may create oxygen vacancies/defects, room temperature ferromagnets can be obtained. And as the results, spin and charge could be manipulated at the same time in the same device. Making these properties controllable must be the next step in order to be able to exploit them for future applications.

6.4 MODELS

The model of Dietl [1] for DMS is indeed Heisenberg model considering dopant cations to have well-localized magnetic moments, and those couple with each other ferromagnetically (i.e., long-range ordered interaction) via 2p holes or 4s electrons. This unlikely works well, as researchers have seen through experimental facts. It is not proven experimentally that dopants in DMSO contribute any ferromagnetic component, and they give paramagnetic contribution instead. This model assumed that the long-range order interaction must exist, but indeed with a very dilute amount of dopants (few percent). It is really difficult to imagine how the exchange interaction could take place. Additionally, Dielt's theoretical work did not care about dimensionality (confinement effect) as well as oxygen and cation defects that certainly exist in the compounds. Naturally, it fails to explain the mechanism of DMSO. It is figured out that, basically the observed Tc does not depend on dopant concentration, and it means that indeed the RKKY interaction cannot be the main source of FM in DMSO, but oxygen vacancies and other defects must play a more important role.

The Coey's model as of "bound magnetic polaron" (2005), also considered that the interaction between localized moments of dopant via intermediates plays a principal role, however, in here the "intermediates" are "electrons associated with defects" [46] leading to an impurity band. But still, this one cannot explain how T_C in DMSO could be too high, or how undoped oxides (i.e., without 3d electron) could be ferromagnetic.

In order to understand the induced FM in undoped semiconducting oxides, one may recall the model of Elfimov [47] for surface magnetism due to defects at the surface/interface, including surfaces between grain boundaries. However, this model cannot explain how a high T_C could occur as we have seen in DMSO, because according to Heisenberg models, T_C should be limited very much due to spin wave excitations.

Very recently, Coey introduced another model, so-called charge-transfer ferromagnetism (CTF) which is based on defect-based impurity band [48] but moreover assumed that there must be a charge reservoir in the system to facilitate the hopping of electrons to from the impurity band leading to spin splitting. In DMSO, this reservoir goes along with dopant ions. This may somehow explain for FM due to defects in doped DMSO such as Mn-doped ZnO or TM-doped TiO_2 or TM-doped SnO_2. However, it still cannot explain how FM occur in undoped HfO_2, TiO_2, ZnO, SnO_2, etc. because in those pristine oxides, where is the charge reservoir? Moreover, in many TM-doped oxides, even with dopants, there is no charge reservoir if the compound is not conductive but insulating. Actually, DMSO can be metallic, semiconducting, and insulating, and the CTF model cannot cover all those.

The observed FM in DMSO appears to be very hard to be explained well by any single model so far. If one can explain one behavior, then fails to explain the others. If one can explain well for one system (such as TM-doped DMSO), then fails to explain other systems (such as undoped oxides), or vice versa. Therefore, the field is still very interesting to be discovered, because once we can understand well the mechanism, we can manipulate the materials to meet our needs for applications.

6.5   CONCLUSIONS AND PERSPECTIVES

Studies on diluted magnetic semiconducting oxide (DMSO) systems have been done by many efforts of many laboratories around the world, bringing in many interesting results but also raising many questions. New phenomena have attracted many theorists to look for suitable models with hope to be able to explain well their mechanism. There are many issues that may cause confusions in the domain such as

(1) Different laboratories have used different growth conditions and techniques that lead to samples with different qualities, including concentrations of defects and oxygen vacancies, conductivities (n-type or p-type, insulating or conductive), homogeneity (clusters or not), etc.
(2) Dimensionality (e.g., bulks, films, and particles behave differently), etc.

It is certain that RKKY interaction is not the main cause for the induced FM in DMSO, and defects should play a big role here. However, even though it is very interesting to study, practically, in order to bring DMSO into the device world, we need to make it clear if those defects could be stabilized, controlled, and manipulated, because otherwise, there would be no sense for applications.

Research on very thin films and nanoparticles of DMSO has interestingly shown that downscaling materials of this family to nanometer-scale can be a key issue, to turn DMSO from diamagnetic (as in bulks) to become ferromagnetic (as in low-dimension structures). It opens a door to exploit the great side of nanoworld in this field of spintronics.

The field of DMSO still requires big efforts of leading research groups, both experimentally and theoretically, to pursuit toward a higher level of research in order to be able to bring it closer to a realization of spintronic devices based on DMSO materials.

ACKNOWLEDGMENTS

The author would like to thank J. Sakai, A. Hassini, A. Ruyter, N. Poirot. V. Brize, N. Q. Huong, A. Barla, C-K. Park, and J-H. Song for their cowork and then contributions in experiments that leading to our results in the DMSO field. The work on ZrO_2 was supported by project 3348-20100041 of The National Research Foundation of Korea. The work on C-doped SnO_2 was supported by project 3348-20100016 of SNU R&D Foundation. The author finally thanks project 2017060271 of the NRF of Korea for its partial support for her work in 2017.

KEYWORDS

- **ferromagnetism**
- **magnetic oxide semiconductors**
- **spintronic applications**
- **magnetization**
- **dopant**

REFERENCES

1. Dietl, T.; Ohno, H.; Matsukura, F.; Cibert, J.; Ferrand, D. *Science* **2000**, *287*, 12019.
2. Sato, K.; Katayama-Yoshida, H. *Jpn. J. Appl. Phys.* **2000**, *39* (Part 2), L555.
3. Matsumoto, Y.; Murakami, M.; Shono, T.; Hasegawa, T.; Fukumura, T.; Kawasaki, M.; Ahmet, P.; Chikyow, T.; Koshihara, S.; Koinuma, H. *Science* **2001**, *291*, 854.
4. Prellier, W.; Fouchet, A.; Mercey, B. *J. Phys.: Condens. Matter* **2003**, *15*, R1583.
5. Ranish, R.; Gopal, P.; Spaldin, N. A. *J. Phys.: Condens. Matter* **2005**, *17*, R657.
6. Venkatesan, M.; Fitzgerald, C. B.; Coey, J. M. D. *Nature* **2004**, *430*, 630.
7. Schwartz, D. A.; Gamelin, D. R. *Adv. Mater.* **2004**, *16*, 2115.
8. Hong, N. H.; Sakai, J.; Huong, N. T.; Poirot, N.; Ruyter, A. *Phys. Rev. B* **2005**, *72*, 45336.
9. Hong, N. H.; Sakai, J.; Poirot, N.; Brizé, V. *Phys. Rev. B* **2006**, *73*, 132404.
10. Hong, N. H.; Sakai, J.; Huong, N. T.; Ruyter, A.; Brizé, V. *J. Phys.: Condens. Matter* **2006**, *18*, 6897.
11. Hong, N. H.; Poirot, N.; Sakai, J. *Phys. Rev. B* **2008**, *77*, 33205; Chang, G. S.; Forrest, J. Kurmaev, E. Z.; Morozovska, A. N.; Glinchuk, M. D.; McLeod, J. A.; Moewes, A.; Surkova, T. P.; Hong, N. H. *Phys. Rev. B* **2012**, *85*, 165319.
12. Sundaresan, A.; Bhagravi, B.; Rangarajan, N.; Siddesh, U.; Rao, C. N. R. *Phys. Rev. B* **2006**, *74*, 161306 (R).
13. Martínez-Boubeta, C.; Beltrán, J. I.; Balcells, L.; Konstantinović, Z.; Valencia, S.; Schmitz, D.; Arbiol, J.; Estrade, S.; Cornil, J.; Martínez, B. *Phys. Rev. B* **2010**, *82*, 024405.
14. Sato, K.; et al. Computational Materials Design of ZnO-Based Semiconductor Spintronics. In: *Magnetism in Semiconducting Oxides*; Hong, N. H., Ed.; Transworld Research Network, 2007.
15. Sato, K.; Dederichs, P. H.; Katayama-Yoshida, H.; Kudrnovsky, J. *J. Phys.: Condens. Matter* **2004**, *16*, S5491.
16. Sato, K.; Dederichs, P. H.; Katayama-Yoshida, H. *Europhys. Lett.* **2003**, *61*, 403.
17. Zunger, A. *Solid State Phys.* **1986**, *39*, 276.
18. Hong, N. H.; Sakai, J.; Prellier, W.; Hassini, A.; Ruyter, A.; Gervais, F. *Phys. Rev. B* **2004**, *70*, 195204.
19. Kimura, H.; Fukumura, T.; Kawasaki, M.; Inaba, K.; Hasegawa, T.; Koinuma, H. *Appl. Phys. Lett.* **2002**, *80*, 94.

20. Ogale, S. B.; Choudhary, R. J.; Buban, J. P.; Lofland, S. E.; Shinde, S. R.; Kale, S. N.; Kulkarni, V. N.; Higgins, J.; Lanci, C.; Simpson, J. R.; Browning, N. D.; Das Sarma, S.; Drew, H. D.; Greene, R. L.; Venkatesan, T. *Phys. Rev. Lett.* **2003**, *91*, 77205.

21. Coey, J. M. D.; Douvalis, A. P.; Fitzgerald, C. B.; Venkatesan, M. *Appl. Phys. Lett.* **2004**, *84*, 1332.

22. Hong, N. H.; Sakai, J.; Prellier, W.; Hassini, A. *J. Phys.: Condens. Matter* **2005**, *17*, 1697.

23. Hong, N. H.; Ruyter, A.; Prellier, W.; Sakai, J.; Huong, N. T. *J. Phys.: Condens. Matter* **2005**, *17*, 6533.

24. Hong, N. H.; Sakai, J. *Physica B* **2005**, *358*, 265.

25. Kittel, C. *Introduction to Solid State Physics*, 7th ed., John Wiley: New York, 1996.

26. Hong, N. H.; Sakai, J.; Huong, N. T.; Ruyter, A.; Brizé, V. *J. Phys.: Condens. Matter* **2006**, *18*, 6897.

27. Hong, N. H.; Sakai, J.; Hassini, A. *J. Phys.: Condens. Matter* **2005**, *17*, 199.

28. Wang, Q.; Sun, Q.; Rao, B. K.; Sena, P. *Phys. Rev. B* **2004**, *69*, 233310.

29. Spaldin, N. A. *Phys. Rev. B* **2004**, *69*, 125201.

30. Hong, N. H.; Brizé, V.; Sakai, J. *Appl. Phys. Lett.* **2005**, *86*, 82505.

31. Pemmaraju, D. P.; Sanvito, S. *Phys. Rev. Lett.* **2005**, *94*, 217205.

32. Hong, N. H.; Barla, A.; Sakai, J.; Huong, N. Q. *Phys. Status Solidi (C)* **2007**, *4*, 4461.

33. Barla, A.; Schmerber, G.; Beaurepaire, E.; Dinia, A.; Bieber, H.; Colis, S.; Scheurer, F.; Kappler, J.-P.; Imperia, P.; Nolting, F.; Wilhelm, F.; Rogalev, A.; Muller, D.; Grob, J.-J. *Phys. Rev. B* **2007**, *76*, 125201.

34. Coey, J. M. D.; Venkatesan, M.; Stamenov, P.; Fitzgerald, C. B.; Dorneles, L. S. *Phys. Rev. B* **2005**, *72*, 24450.

35. Bouzerar, G.; Ziman, T. *Phys. Rev. Lett.* **2006**, *96*, 207602.

36. Yoon, S. D.; Chen, Y.; Yang, A.; Goodrich, T. L.; Zuo, X.; Arena, D. A.; Ziemer, K.; Vittoria, C.; Harris, V. G. *J. Phys.: Condens. Matter* **2006**, *18*, L355.

37. Hassini, A.; Sakai, J.; Lopez, J. S.; Hong, N. H. *Phys. Lett. A* **2008**, *372*, 3299.

38. Hartnagel, H. L.; Dawar, A. L.; Jain, A. K.; Jagadish, C. *Semiconducting Transparent Thin Films.* IOP Publishing: Bristol, Philadelphia, 1995.

39. Hays, J.; Punnoose, A.; Baldner, R.; Engelhard, M. H.; Peloquin, J.; Reddy, K. M. *Phys. Rev. B* **2005**, *72*, 075203.

40. Chang, G.-S.; Forrest, J.; Kurmaev, E. Z.; Morozovska, A. N.; Glinchuk, M. D.; McLeod, J. A.; Moewes, A.; Surkova, T. P.; Hong, N. H. *Phys. Rev. B* **2012**, *85*, 165319.

41. Hong, N. H.; Song, J.-H.; Raghavender, A. T.; Asaeda, T.; Kurisu, M. *Appl. Phys. Lett.* **2011**, *99*, 052505.

42. Barla, A.; Hong, N. H.; et al. *Unpublished XMCD results on TiO₂ system*, 2007–2012.

43. Huong, N. Q. Magnetism Due to Oxygen Vacancies in Undoped Oxide Thin Films. In: *Magnetism in Semiconducting Oxides*; Hong, N. H.; Ed.; Transworld Research Network, 2012.

44. Hong, N. H.; Park, C.-K.; Raghavender, A. T.; Ruyter, A.; Chikoidze, E.; Dumont, Y. *J. Mag. Mag. Mater.* **2012**, *324*, 3013.

45. Hong, N. H.; Kanoun, M. B.; Goumri-Said, S.; Song, J.-H.; Chikoidze, E.; Dumont, Y.; Ruyter, A.; Kurisu, M. *J. Phys.: Phys. Condens.* **2013**, *25*, 436003.

46. Coey, J. M. D.; Venkatesan, M.; Fitzgerald, C. B. *Nat. Mater.* **2005**, *4*, 173.

47. Elfimov, I. S.; Yunoki, S.; Sawatzky, G. A. *Phys. Rev. Lett.* **2002**, *89*, 216403.

48. Alaria, J.; Cheval, N.; Rode, K.; Venkatesan, M.; Coey, J. M. D. *J. Phys. D: Appl. Phys.* **2008**, *41*, 135004.

CHAPTER 7

FIRST PRINCIPLES CALCULATIONS IN EXPLORING THE MAGNETISM OF OXIDE-BASED DMS

YIREN WANG and JIABAO YI*

School of Materials Science and Engineering, UNSW, Kensington 2052, NSW, Australia

Corresponding author. E-mail: Jiabao.yi@unsw.edu.au

CONTENTS

ABSTRACT

Spintronics device is a future device using spin rather than charge to realize device functions, having advantage over traditional devices, such as low power, high speed, and flow-a-spin current without dissipation. Diluted magnetic semiconductor (DMS) is one of the promising materials for the applications of spintronics device since it has both spin and semiconductor behavior, easily integrated to current semiconductor technology. Mn-doped GaAs is one of the model systems for III–V-based diluted magnetic semiconductor. However, its low Curie temperature makes it unable for practical applications. Oxide semiconductor such as ZnO has been predicted to be one of the promising semiconductor host for DMSs with high Curie temperature. The following research on oxide semiconductor (i.e., ZnO, TiO_2, SnO_2, In_2O_3)-based DMSs have shown that high Curie temperature has been observed in many systems. However, there are disputes on the origin of the ferromagnetism. In Chapter 6, oxide-based DMSs have been introduced. In this chapter, we will introduce the study of magnetism in oxide-based DMSs using first principles calculations. First, we introduce the basics for first principles calculations. Then, we introduce the study of different oxide-based systems by first principles calculations.

7.1   INTRODUCTION

Conventional semiconductors devices, such as field-effect transistor, computer processor, and random access memories, are based on the manipulation of charge transport. The charge flow on/off controlled by the gate voltage represents the digital states 1/0 of the devices. Different functions are realized via the integrated circuit composed of transistors, capacitors, resistors, and diodes. Semiconducting electronics have been developed rapidly since 1970s, which are popularly considered as following the Moore's law. The Moore's law has successfully predicted that the power of these semiconductor devices doubles in approximately 18 months for several decades, while now it is almost reaching the limits.

For the continuous shrink of the chip size and limitations of the current microelectronic technique, spintronics device is proposed to be a good candidate for substituting the established electronic devices. Spin instead of charge is used as the logical unit for devices. For one electron spin, it has two states, spin up and spin down. In ferromagnetic (FM) materials, the spins can be aligned in one direction, and a small magnetic field can manipulate the spin state from up to down. These two states of up and down can be used as the logic "on" and "off" of a semiconductor device. Combined both charge and spin information of electrons, spintronics devices are expected to have broader properties and applications, since the spin devices will have many advantages over conventional semiconductor devices, such as low power, high speed, and flow-a-spin current without dissipation.

Diluted magnetic semiconductor (DMS), which possesses both semiconductor and magnetic behavior, is one of the most ideal and promising materials for fabricating spintronics devices. The semiconductor behavior is for easy integration into the current semiconductor production and the magnetic behavior is for the realization of spin behavior. DMS was initially fabricated via doping transition metals (TMs) into nonmagnetic semiconductors [1]. The materials for producing DMS have been extensively investigated both experimentally and theoretically. Mn-doped GaAs is one of the model examples, which has shown every aspect promising for spintronics devices. However, its low Curie temperature limits its practical applications. In 2000, Dietl et al. [2] predicted that ZnO and GaN can be two host candidates to achieve room temperature ferromagnetism (RTFM) when doped with TMs. Following the prediction, many researchers have reported RTFM in TM-doped GaN or ZnO systems. Further research has extended the DMS based on wide gap oxide semiconductor systems, such as TiO_2, SnO_2, In_2O_3, etc. RTFM has also been extensively reported in these systems.

However, clusters and secondary phases have also been proposed to be the origin of the ferromagnetism in many cases including III–V and II–VI semiconductor-based DMSs, especially for those thin-film DMSs fabricated under a low oxygen partial pressure [3]. This direct interaction between local moments of magnetic clusters can be one of the origins of the ferromagnetism. These clusters and secondary phases induced ferromagnetism is not intrinsic, which is not suitable for realizing spintronics device functions. Ideally, the ferromagnetism in DMS should be carrier-mediated, as explained by Dietl et al. using mean-field theory based on Zener model, which makes the spin manipulation possible by an external electric field to realize spin functions.

To avoid the ferromagnetism induced by magnetic clusters or secondary phases, nonmagnetic elements have been purposely introduced into the oxide host to generate RTFM. The ferromagnetism is supposed to be originated from defects, such as vacancies, which makes the origins of RTFM in DMSs very complicated. Until now, none of the existing theory can explain the magnetic mechanisms of different kinds of DMSs satisfactorily.

Due to the limitation of the current experiment techniques, and the great development in high-performance computers as well as the increasing progress in algorithmic methods, theoretical studies are reaching unprecedented high levels of precision. Based on density functional theory (DFT), first principles calculation is able to explain the given properties of a certain material and predict the possible properties as well. In this chapter, we make efforts to summarize the recent progress of theoretical investigations in oxide-based DMS. The chapter is organized as follows. In the following two sections, we will first give a brief introduction of the computational theory and the basic information of the FM interactions in DMS materials. Second, detailed descriptions of first principles studies in several oxide-based DMS will be given, including ZnO, TiO_2, In_2O_3, SnO_2, and HfO_2 with certain dopants. Both TM and nonmagnetic elements are taken into consideration. The last section of the chapter will describe the issues and prospective in current computational studies as well as some possible applications in the future.

7.2 COMPUTATIONAL THEORY

7.2.1 FIRST PRINCIPLES CALCULATION

First principles calculation, or ab-initio calculation, is an algorithmic method which is based on the interaction between atomic nuclei, electrons, and their motion patterns. It starts to calculate physical properties from established laws of physics without additional assumptions such as empirical or fitted parameters. Derived from the molecular orbital theory [4], there are only five exclusive constants included in ab-initio method, which are electron mass m_0, electron charge e, Planck constant h, velocity of light c, and Boltzmann constant K_b. However, in order to get satisfactory results, some less responsible factors will be reasonably ignored. Typical first principles calculations can deal with the atomic structure of materials by applying a series of approximations, such as systematic energy (entropy, enthalpy, and Gibbs free energy) to solve the Schrödinger's equation instead of fitting experimental data. This computational approach provides a direct prediction toward physical, chemical, and mechanical properties. Moreover, it does not require any input parameter, or a minimal set, and the results are mainly self-converged by energy or force on an atom. Consequently, first principles calculation can reach a high precision in physical state. For instance, lattice constants and bulk modulus obtained from computations have been shown in good agreement with experiments. The relative error is only a few percentages. With the great enhancement in computing capacity, first principles calculation is regarded as a fundamental and critical technique in computational materials.

7.2.2 DENSITY FUNCTIONAL THEORY

In 1964, Hohenberg and Kohn proposed the electron DFT [5], and Kohn and Sham developed its primary fulfillment in 1965 [6]. Using the electron density as the fundamental variable instead of the electron wave function, DFT has become a convincing quantum mechanical modeling method in exploring the electronic structure of many-body systems.

7.2.3 HOHENBERG–KOHN THEOREM AND KOHN–SHAM EQUATION

Originated from the work Thomas and Fermi made in 1927 [7], the starting point of DFT is that we could describe the ground-state physical properties of atoms, molecules and solids in terms of the density function. In order to perform the research on inhomogeneous electron gas, Hohenberg and Kohn introduced two remarkable theorems.

For a system in an external potential with electrons of given number N, the Hamilton operator $\hat{H}$ of the system can be defined as

$$\hat{H} = \hat{T} + \hat{V} + \hat{W} \tag{7.1}$$

The kinetic operator $\hat{T}$ equals to

$$\hat{T} = -\frac{1}{2}\int \nabla \Psi^+(r) \cdot \nabla \Psi(r) dr \tag{7.2}$$

The potential operator $\hat{V}$ equals to

$$\hat{V} = \int v(r) \Psi^+(r) \Psi(r) dr \tag{7.3}$$

And the Coulomb interaction operator $\hat{W}$ is defined as

$$\hat{W} = \frac{1}{2}\int \frac{1}{|r-r'|} \Psi^+(r) \Psi(r) \Psi^+(r') \Psi(r') drdr'. \tag{7.4}$$

Here, $\Psi^+ (r)$ and $\Psi(r)$ are operators which represent for one electron created (+) or annihilated in position r. The electron density function $\rho(r)$ is written as below, where ψ stands for the wave function at ground state:

$$\rho(r) = \psi \,|\, \Psi^* (r) \Psi(r) \,|\, \psi \tag{7.5}$$

Therefore, Hohenberg and Kohn can draw inference from the functions above: $v(r)$ is the unique functional of $\rho(r)$. If the ground-state density is given, so that the $v(r)$ is given, then the Hamilton operator can be determined uniquely. That is to say, the ground-state density of identical electrons can determine the external potential with a constant, and the ground-state energy is the only functional. For instance, if two systems of electrons trapped in potentials $v_1(\vec{r})$ and $v_2(\vec{r})$ separately, while these two systems share the same ground-state density, we could obtain that $v_1(\vec{r}) - v_2(\vec{r}) = A$, where A is a constant.

In systems containing certain number of particles, the density $\rho(r)$ determines everything. Therefore, we could set a universal functional at ground state which is irrelevant to external potential:

$$F[\rho] = T[\rho] + W[\rho] \tag{7.6}$$

Here, $T[\rho]$ and $W[\rho]$ are functionals that give for any ground-state density, the kinetic energy of the electrons and the electron–electron repulsion energy.

The electronic energy functional is given as

$$E[\rho] = T[\rho] + W[\rho] + V[\rho] = F[\rho] + V[\rho] \tag{7.7}$$

$$V[\rho] = \int v(r)\, \rho(r)\, dr \tag{7.8}$$

With the restriction of the numbers of electrons in system and combined with Eqs. (7.5) and (7.6), Levy [8] determined the value of $F[\rho]$:

$$F[\rho] = \min_{\psi \Rightarrow \rho} \psi \,|\, \hat{T} + \hat{W} \,|\, \psi \tag{7.9}$$

$F[\rho]$ is the minimal value of all antisymmetric wave functions which assign the expectation value ($\hat{T} + \hat{W}$) to density ρ.

Derived from Eqs. (7.7) and (7.8), for given $v(r)$, the energy functional $E[\rho]$ is set as

$$E[\rho] \equiv \psi \,|\, \hat{T} + \hat{W} \,|\, \psi + \int v(r)\rho(r)dr \tag{7.10}$$

According to the variation principle, taking the ground-state density as a variable, $E[\rho]$ will reach the minimum at the ground-state density. Subsequently, we could determine the minimal energy functional of the system, which is equivalent to energy in ground state.

Hohenberg–Kohn theorems point out that the density function is the basic variable in exploring the physical properties of many-body system in ground state. However, problems still remained until Kohn and Sham introduced a new way to evaluate the system by replacing the kinetic energy with a combination of $T_s[\rho]$ which is foregone in a noninteracting ($W[\rho] = 0$) system, and a minor correction as well [6]. The universal functional is, therefore, shown as

$$F[\rho] = T_s[\rho] + E_{xc}[\rho] + J[\rho] \tag{7.11}$$

$$T_s[\rho] = \sum_k^N \left\langle \varphi_k \left| -\frac{\nabla^2}{2} \right| \varphi_k \right\rangle. \tag{7.12}$$

Here, $J[\rho]$ is the functional of classic Coulomb interaction. $E_{xc}[\rho]$ is the function including both correlation and exchange effects. $E_{xc}[\rho]$ can be regarded as the consistence of the two components: the difference between the real kinetic energy $T[\rho]$ and the energy of a reference system $T_s[\rho]$; the second part is the correction of electrons self-interaction $W[\rho]$ versus Coulomb interaction $J[\rho]$. The expression of the exchange–correlation energy is

$$E_{xc}[\rho] = T[\rho] - T_s[\rho] + W[\rho] - J[\rho] \tag{7.13}$$

Eq. (7.7) is therefore resolved into following parts:

$$\begin{aligned} E[\rho] &= T_s[\rho] + V[\rho] + E_{xc}[\rho] + J[\rho] \\ &= T_s[\rho] + E_{xc}[\rho] + J[\rho] + \int v(r)\rho(r)dr \end{aligned} \tag{7.14}$$

Here,

$$\rho = \sum_k^N |\varphi_k|^2 \tag{7.15}$$

φ_k is known as the "Kohn–Sham orbitals," and the sum of φ_k of all the single particle states equals to the amount of the ground-state charge density. Concerning from ground-state charge density, it seems the wave function is likely to be a single Slater determinant built from φ_k. However, it is not a fact. We can obtain the Kohn–Sham equation by varying the calculation to the total energy from single particle orbital

$$\hat{F}_{ks}\varphi_k = \varepsilon_k^{KS}\varphi_k \qquad (7.16)$$

Here

$$\hat{F}_{ks} = \hat{T}_s + \hat{V}_{eff} \qquad (7.17)$$

$$V_{eff}(r) = V_{ne}(r) + V_{Coulomb}\left[\rho(r)\right] + V_{xc}\left[\rho(r)\right] \qquad (7.18)$$

And the electrostatics, $V_{ns}(r)$ is a potential caused by the electrons distribution, $V_{Coulomb}[\rho(r)]$ is the Coulomb potential between the electrons, and $V_{xc}[\rho(r)]$ represents the exchange–correlation potential. The meaning of the eigenvalue ε_k^{KS} is not easy to be established though it is believed to be physically significant [9].

The central idea of Kohn–Sham equation is to construct the single-particle potential in such a way that the density of the auxiliary noninteracting system equals the density of the interacting system of interest. It should be noted that DFT does not give any hint on how to construct $E_{xc}[\rho]$. It only holds the promise that $E_{xc}[\rho]$ is a universal functional of the density.

7.2.4 APPROXIMATIONS OF EXCHANGE–CORRELATION ENERGY FUNCTIONAL

As afore discussion, DFT itself does not give the solution for the construction of $E_{xc}[\rho]$. In order to solve the Kohn–Sham equation, we must give a specific form to $E_{xc}[\rho]$.

7.2.4.1 LOCAL DENSITY APPROXIMATION

If electrons density is changing smoothly, Kohn and Sham [6] suggest an idea of using the density $\varepsilon_{xc}[\rho]$ of every electron in homogeneous electron gas replacing the density of inhomogeneous gas.

$$E_{xc}[\rho] = \int \rho(r)\varepsilon_{xc}[\rho(r)]dr \qquad (7.19)$$

The $\varepsilon_{xc}[\rho]$ can be divided into the exchange and correlation parts,

$$\varepsilon_{xc}[\rho] = \varepsilon_x[\rho] + \varepsilon_c[\rho] \qquad (7.20)$$

In Thomas–Fermi–Dirac Theory [10], the exchange energy is

$$\varepsilon_x[\rho] = -C_x \rho(r)^{\frac{1}{3}} \quad C_x = \frac{3}{4}\left(\frac{3}{\pi}\right)^{\frac{1}{3}} \tag{7.21}$$

In 1980, Ceperley and Alder [11] obtained the exact value of correlation energy using a quantum Monte Carlo method:

$$\varepsilon_c[\rho] = E[\rho] - E_x[\rho] - T_s[\rho] \tag{7.22}$$

Eq. (7.19) is the local density approximation (LDA), the corresponding exchange–correlation potential is

$$V_{XC}(\rho) = \frac{\delta E_{XC}[\rho]}{\delta\rho} = \varepsilon_{xc}[\rho(r)] + \rho(r)\frac{\delta\varepsilon_{xc}[\rho]}{\delta\rho} \tag{7.23}$$

Considering the spin polarization, then the potential of local spin density approximation (LSDA) will be

$$V_{XC}^{\delta}(\rho) = \frac{\delta E_{XC}[\rho_+\rho_-]}{\delta\rho(r)} \tag{7.24}$$

From Eqs. (7.16), (7.23), and (7.24), the localized Kohn–Sham equation is built as

$$[T_s[\rho] + V_{ne}(r) + V_{\text{Coulomb}} + V_{XC}^{L(S)DA}(\rho)]\varphi_k = \varepsilon_k^{KS}\varphi_k \tag{7.25}$$

LDA works quite well in nonuniform systems with spatially slowly varying density, such as solid. First principles computations under the framework of L(S)DA have achieved great success in electronic structure research based on DFT. Cooperated with the band structure calculations, we can give accurate descriptions of properties of semiconductors and metals at ground state. The calculated lattice constant, crystal formation energy, and mechanical property, for instance, have been authenticated in accord with the empirical data. However, L(S)DA cannot deal with systems with d or f electrons. The predicted band gap value of semiconductor materials is always underestimated. Therefore, certain appropriate corrections are needed to be made to the L(S)DA theory.

7.2.4.2   GENERALIZED GRADIENT APPROXIMATION

The idea of generalized gradient approximation (GGA) is to make a gradient expansion of $E_{xc}[\rho]$ since the actual exchange–correlation effect is nonlocalized, while L(S)DA carries a localized approximation. With the expansion, better descriptions can be given to the nonuniform systems.

One of the widely used improvements of LSDA is presented by Perdew and Wang in 1986 [12]: The exchange and correlation energy is not simply dependent on electron density but the gradient density as well. The exchange energy is therefore written as

$$\varepsilon_X^{PW86} = \varepsilon_X^{LSDA}\left(1 + ax^2 + bx^4 + cx^5\right), x = \frac{|\nabla\rho|}{\rho^{\frac{4}{3}}} \tag{7.26}$$

Here, x is the dimension of gradient variable; a, b, and c are constants. This method is known as GGA-I or PW81.

In 1991, Perdew and Wang proposed the GGA-II, or PW91 method [13]. In this approximation method, they introduced some half empirical parameters to calculate more precisely.

$$E_X^{PW91}\left[\rho\uparrow,\rho\downarrow\right] = \frac{1}{2}E_X^{PW91}\left[2\rho\uparrow\right] + \frac{1}{2}E_X^{PW91}\left[2\rho\downarrow\right] \tag{7.27}$$

$$E_X^{PW91}\left[\rho\right] = \int\rho\varepsilon_x\left(r_s,0\right)F\left(s\right)d^3r \tag{7.28}$$

$$\varepsilon_x\left(r_s,0\right) = -\frac{3k_F}{4\pi} \tag{7.29}$$

PW91 has shown the advantages of LDA. However, problems still exist [14]: (1) PW91 depends on extreme complicated inferences; (2) the analytic function f is rather complicated and fuzzy; (3) f depends too much on parameterization; and (4) those parameters do not fit themselves well. When doing the dimensionless reduced density gradient, no matter in small or large scale, the above-mentioned problems will lead to abnormal fluctuations of exchange–correlation potential $\dfrac{\delta E_{xc}}{\delta n[r]}$, which as a consequence, makes it hard to build the GGA-based pseudopotential.

After a period of time, the PW91–GGA pseudopotential was improved by Perdew, Burke, and Ernzerhof, namely the GGA–PBE approximation [15].

PBE has shown higher precision and is more practical than GGA. When the density changes smoothly, the relevant exchange and correlation energy can be described as

$$E_X^{PBE}\left[n(r)\right]=\int d^3 rn\left[\varepsilon_c^0\left(r_s,\xi,t\right)+H\left(r_s,\xi,t\right)\right] \tag{7.30}$$

$$E_C^{PBE}=\int d^3 rn\varepsilon_C^0\left(n\right)F_x\left(s\right) \tag{7.31}$$

Here, r_s is the localized Seitz radius $(n=\dfrac{3}{4}\pi r_s^3=\dfrac{k_F^3}{3\pi^2})$; ξ is the spin polarization coefficient; $t=\dfrac{|\nabla n|}{2\phi k s''}$ is the dimensionless density gradient;

$$\phi(\xi)=\frac{\sum(1+\xi)^{\frac{2}{3}}+\left[1-\xi\dfrac{2}{\beta}\right]}{2}$$

stands for the optional ratio;

$k_s=\sqrt{\dfrac{4k_F}{\pi a_0}}$ $(a_0=\dfrac{\hbar}{me^2})$ is the shield coefficient raised by Thomas–Fermi;

$\varepsilon_C^0\left(n\right)=-\dfrac{3e^2 k_F}{4\pi}$ is the exchange energy in uniform density; $F_x\left(s\right)$ is the spin-polarized strengthen coefficient which satisfies $F_x\left(s\right)=1+k-\dfrac{k}{1+\dfrac{\mu s^2}{k}}$, where $k=0.085$.

7.3 PLANE WAVE FUNCTION AND PSEUDOPOTENTIAL METHOD

7.3.1 PLANE WAVE METHOD FOR BASIS FUNCTION

One of the crucial points of carrying on the first principles calculation is to choose appropriate basis function to expand the Kohn–Sham wave function φ_k. The basis function must match following conditions:

(1) It should be or almost be a complete set. Hence, it could be used to expand any wave function.
(2) It should have correct approximation relation with the described system, so that less basis is required for the wave function.
(3) It is easy to calculate the orbit integration defined by this basis, especially the multi-centered integration. Hence the self-convergence can be achieved fast.

The common basis functions are plane wave function (PW); molecular atomic orbital, such as Slater-type orbitals, Gaussian-type orbitals; natural atomic orbitals; pseudo-atomic orbitals; etc. The PW basis is the basic orthogonal and complete function group [16]. It has the following characteristics:

(1) It has a preferable analyses form: orthogonal in most cases and the Hamilton matrix can be expressed in a simple formula.
(2) More wave functions can be added to improve the properties of basic functions.
(3) At the same time, it is nonlocalized, suggesting that it may not be dependent on the position of atoms.

If we expand φ_k under the PW basis as follows:

$$\varphi_k = \frac{1}{\sqrt{\Omega}} \sum_G c_{nkG} exp^{i(k+G)r} \tag{7.32}$$

where Ω is the cell volume, G is the reciprocal grid vector, and all the k points are located in Brillouin zone.

If we combine Eq. (7.33) with the single electron Kohn–Sham equation, we can obtain a group of linear equations. After diagonalization, the eigenvalue of energy will be determined as well as the eigenfunction. In principle, the mentioned linear equations should have infinite order, while only limited numbers of equations can be used for practical calculations. Normally, we set a cutoff energy together with the following pseudopotential method.

7.3.2 PSEUDOPOTENTIAL OF EXTERNAL POTENTIAL $V_{ext}(r)$

In solid substances, the motion status of valence electrons changes at extreme extend, while the electrons of inner shell change smoothly. As Figure 7.1 depicts the wave function of valence electrons $\Psi(Z/r)$ change rather smoothly in areas above the cutoff radius r_c; while $\Psi(Z/r)$ fluctuates greatly in inner ion cores. The strong variation reflects the orthogonality with the inner core electrons wave function. The request to be orthogonal is acting as a repulsion potential, which, to some extent, cancels out the inner attraction $V(Z/r)$. Scientists made the primary proof based on the orthogonalized PW function of band calculation then brought out the concept of pseudopotential. Pseudopotential is a hypothetical potential in the ion cores, which is used to replace the real one when solving the wave equation.

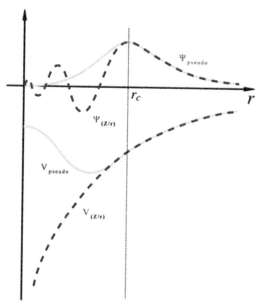

FIGURE 7.1 Comparison of a wave function in the Coulomb potential of the nucleus (blue) to the one in the pseudopotential (yellow). The real and the pseudo wave function and potentials match above a certain cutoff radius r_c.

There are three pseudopotentials frequently used: the norm-conserving pseudopotentials (NC-pp) proposed in late 1970s [17], the ultrasoft pseudopotentials (US-pp) in 1990 [18], and the pseudopotential come from the projector-augmented wave (PAW-pp) method in 1994 [19]. The NC pseudopotential requires an equal numbers of the total charges of all electrons wave function and pseudopotential wave function within the cutoff radius r_c, while the wave function above r_c will be kept, as shown in Figure 7.1. Therefore, these functions can completely equal besides the cutoff radius, and the precision of the pseudopotential can be greatly enhanced. However, the NC-pp is quite hard which means it needs several PW functions to expand and therefore the amount of calculation is large. The US-pp relaxes the demand of equal mode and allows the wave function to be soft enough to lower the cutoff energy largely. Therefore, less PW basis functions are needed for US-pp. The PAW-pp combines the benefits of both NC-pp and US-pp, and it is much easier to produce basis set than US-pp. Moreover, PAW-pp is more accurate in calculating the magnetic materials, alkali metals, rare-earth metals, elements with 3d electrons, lanthanide, and actinide series elements.

DFT provides a way to systematically map the many-body problem, with external potential $\hat{V}$, onto a single-body problem, without $\hat{V}$. Considering the pseudopotentials, plane-wave basis sets, and lattice geometry, DFT is now recognized as a conventional tool for first principles simulations in magnetic materials.

7.4 FERROMAGNETIC INTERACTIONS IN DMS MATERIALS

Magnetism is a kind of physical property that materials respond to the magnetic field. Originated from quantum mechanics of angular momentum of electrons, there are two major sources of magnetism: electron motion and the nuclear magnetic moments of atomic nuclei. Regarding electrons motion, it can be described as orbital magnetic moment and spin magnetic moment. For the magnetic moment in TMs, spin moment is the major contribution since the orbital moment is usually quenched.

Generally, the magnetic moments (both orbital and spin) of numerous electrons in a material would cancel out, due to the result of the Pauli exclusion principle that paired electrons produce opposite spin-magnetic moments, or due to the zero net orbital motion caused by the tendency of electrons filling subshells. However, there are diverse factors that can affect the magnetic moments, like thermal vibration energy, external magnetic field, and sometimes the individual magnetic moment can spontaneously couple each other.

Ferromagnetism is a property that certain materials, such as iron, show spontaneous magnetic moments without external magnetic field. The magnetic moments line up parallel due to the internal crystal field. The magnetic moments normally decrease with increasing temperature due to the thermal vibration energy, which randomizes the lined magnetic moments and decreases abruptly to zero at certain temperature called Curie temperature T_c.

Numerous researches have been performed to study the magnetic properties of DMS materials. The FM interactions in DMS can be divided into two fundamental groups: the direct exchange and the indirect exchange [20]. It is generally known that the observed ferromagnetism is contributed by the coexistence of two exchange interactions where several mechanisms compete and interplay with each other. However, the mechanisms of ferromagnetism in DMS still remain debated since none of the existing theories can explain all the experimental observations.

7.4.1 DIRECT EXCHANGE

Direct exchange interaction is regarded as the main origin of the ferromagnetism in simple magnetic materials such as Fe, Co, and Ni. It is originated from the overlapping of the wave functions of two neighboring atomic orbitals that induce the interactions, which can be expressed by the Heisenberg model as

$$H_{i,j} = -J_{i,j} S_i S_j \qquad (7.33)$$

where J is the exchange integral. S_i and S_j are the spins.

The potential and kinetic exchanges determine the magnitude J to be either positive or negative. The magnetic alignments vary with the distances of the neighboring atoms. For atoms close to each other, like a H_2 molecule, the exchange interactions are contributed by the Coulomb repulsion of electrons. Pauli's exclusion principle forces the electrons spins to align in opposite orientation, which contributes to negative exchange and antiferromagnetism. Increasing the interatomic distances, the moments of the electrons in atoms like Co will be aligned in parallel, resulting in ferromagnetism and a positive exchange.

Figure 7.2 shows the Bethe–Slater curve of several TMs as functions of the ratio of interatomic distance (a) to the radius of 3d shell of electron (r). The exchange integral can explain the FM (positive J) or anti-FM (negative J) nature of certain elements. However, the direct coupling is short ranged which depends greatly on the distances of ions, while the long-range magnetic ordering needs a better explanation.

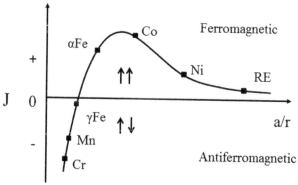

FIGURE 7.2   The Bethe–Slater curves of transition metals (TM) and rare-earth (RE) elements.

7.4.2 INDIRECT EXCHANGE

Indirect exchange via carrier-mediated interaction is raised to complete a better understanding in DMS. Generally, the localized spins of the doped TM can be mediated by the carriers (electrons and holes) in the semiconducting host thus generating ferromagnetism. The magnetic properties of such materials mainly depend on the concentration of doping elements and carriers. At high-carrier density, the itinerant carriers like electrons or holes will be coupled with the magnetic dopants, and this mediation interaction leads to ferromagnetism.

7.4.2.1 RKKY MECHANISM

This theory was initially brought out by Ruderman and Kittel, and then completed by Kasuya and Yosida. It describes a long-range exchange interaction of the magnetic moments in nuclei or the electron spins of localized inner d or f electrons through the metallic electron gas. The coefficient $J(r_{ij})$, which characterizes the spin polarization of conduction electrons, can be expressed as

$$H_{i,j} = \sum_{i,j} J_{(r_{ij})} J_i J_j \tag{7.34}$$

There are diverse factors that influence the coupling strength $J(r_{ij})$, especially the Bloch wave vector at the Fermi energy like Fermi wave vector k_F, and its form is

$$J_{(r_{ij})} = \frac{2mk_F^4}{\pi h^2} J_{ex}^2 F(2k_F r_{ij}) \tag{7.35}$$

where J_{ex} stands for the interaction between the itinerant electrons and impurity atom, and the function $F(x)$ is given as

$$J_{ex} = \frac{x\cos(x) - \sin(x)}{x^4} \tag{7.35}$$

7.4.2.2 DOUBLE EXCHANGE

This interaction usually occurs in systems with mixed valence ions which display different oxidation states. The spins of the ions with different states can be aligned in FM ordering via hoping between the electrons shells due to local Hund's coupling.

7.4.2.3 MEAN-FIELD ZENER MODEL

This model derives from the mean-field theory and RKKY interaction and was used to explain hole-mediated ferromagnetism by Dietl et al. [21]. It is one of the most widely accepted models for ferromagnetism in (Ga,Mn) As and related DMS materials. Three different interactions are included to describe ferromagnetism in TMs like Fe, Ni, and Co: the direct exchange between d shells; the exchange coupling between conduction electrons and d shells and kinetic energy of the conduction electrons.

7.4.2.4 BOUND MAGNETIC POLARON MODEL

Bound magnetic polaron (BMP) model is usually used to describe the ferromagnetism mechanism in compound semiconductors such as ZnO, which contains certain point defects such as zinc interstitials and oxygen vacancies. A BMP is the alignment of the impurity charge carrier with the localized spins. At low carrier density, a percolation network will form. When the pairs of BMPs begin to overlap with each other and the magnetic impurities tend to form polarons, FM interaction therefore generates around the clusters [22]. The origin of the ferromagnetism in DMS is always a combination of one or more of the above mechanisms. Even in a certain mechanism, various types of the magnetisms can be present: such as paramagnetism, ferromagnetism, antiferromagnetism, and sometimes a combination of them. Therefore, the study of the ferromagnetism in DMS can be very complicated but fascinating.

7.5 FIRST PRINCIPLES CALCULATIONS IN OXIDE-BASED DMS

Ferromagnetism has been widely reported in TM-doped DMS. As mentioned before, there are two major theories in explaining the magnetism. The first theory derived from the general mean-field theory, which believes that the substitution of TM elements in bulk materials can somehow develop into a random alloy with the host in a form like (Zn, TM)O. The local magnetic moments of TM elements can be mediated by the free carriers like electrons or holes in the host and therefore produce the ferromagnetism. The mean-field approximation is also applied on the spin–spin coupling, which is suggested to be a long-range interaction. However, researchers suggest that it is the magnetic element alone that generates the ferromagnetism via forming magnetic clusters, and the ferromagnetism is very localized, determined by the magnetic dopant itself.

Since the preparation environment can affect the properties of samples intensively, it is hard to verify the mechanisms from experiments. First principles calculations have been regarded as a strong and efficient tool in predicting potential DMS materials and investigating their electronic structures and magnetic properties since no experimental results are required. First principle calculations have predicted stable RTFM in several wide-gap semiconductor-based DMSs. In addition, carrier-mediated ferromagnetism has also been achieved by of the variation of charge carriers [23]. Among the various kinds of materials that have been investigated, oxide-based DMS, in particular ZnO, TiO_2, In_2O_3, and SnO_2-based materials, have attracted extensive attention due to their possible high Curie temperatures above room temperature.

7.5.1 ZnO

Zinc oxide (ZnO) has attracted considerable attention in both academic and industry fields due to its numerous applications in optoelectronics, sensing, and transparent electronics in the past decades. ZnO has a hexagonal wurtzite structure with lattice constants of $a = b = 3.25$ Å, $c = 5.21$ Å, as shown in Figure 7.3 [24]. It is a native n-type semiconductor that exhibits a wide band-gap of about 3.4 eV.

It well known that ZnO is enriched with different lattice defects, such as oxygen vacancies (V_O), oxygen interstitials (O_{int}), Zn vacancies (Zn_O), Zn interstitials (Zn_{int}), etc. Those defects have been found to have great influence on the properties of the ZnO materials. The original n-type behavior of undoped ZnO is mainly due to the native defects such as oxygen vacancies (V_O) and Zn interstitials (Zn_{int}). However, which one is the dominant donor is difficult to determine from experiments. First principles calculations [25,26] show that all intrinsic defects exhibit rather low-concentration shallow donor properties, and V_O is unlikely to be the dominant donor compared to Zn_{int}, since the calculations suggest that it has relatively high ionization energy of about 30 ± 5 meV [27].

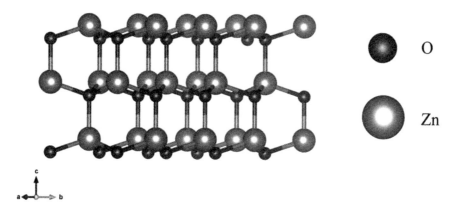

FIGURE 7.3   Crystal structure of ZnO supercell.

Experimentally, a variety of magnetic element-doped ZnO DMSs have been reported and RTFM have been observed in these magnetic element-doped ZnO DMSs. However, the functionalities of the doping elements are not yet fully understood. Theoretically, many models have been established to investigate the electronic structure and magnetic properties of ZnO-based DMSs. These works have considered the factors that may influence the doping like intrinsic defects and doping concentration.

7.5.1.1   MAGNETISM IN TM-DOPED ZnO

Based on the mean-field theory, Zener's p–d exchange model theory is proposed to describe the carrier-induced ferromagnetism by Dietl et al. and Ohno et al. [2,21]. Based on the Zener model, ZnO is predicted to be a candidate for the host semiconductor to achieve RTFM and its magnetic order is favored by p-type conductivity. Ever since then, there have been plenty of studies focusing on the ZnO-based DMS.

Numerous fabrications of TM-doped ZnO have been reported, such as Mn-doped ZnO [28–31], Cu-doped ZnO [30–32], Ti-doped ZnO [30,31,33], Co-doped ZnO [34–36], etc. These materials can be both FM and non-FM under different preparation conditions. The Curie temperatures of them can vary from 2 K to over room temperature. Works on the III–V semiconductor-based DMSs have shown that FM ordering can be formed via the mediation of holes. However, situations in TM-doped oxides can be rather complicated due to the involvement of magnetic clusters, secondary phases as well as intrinsic defects, especially for those fabricated under a poor oxygen environment.

Initially, the TM-doped ZnO was reported to be non-FM [29,30]. Subsequently, it has been reported from experiments that Mn-doped ZnO thin films [28] can be possible DMS material. Later, Co-doped ZnO thin films fabricated by pulsed laser deposition (PLD) technique [37] exhibit RTFM. Since then, hundreds of researches on ZnO-based DMS have been reported.

Theoretically, Sato and Katayama–Yoshida [38,39] applied the band-structure calculation to predict that Mn, V, Cr, Fe, Co, and Ni-doped ZnO can be FM. They performed the Korringa–Kohn–Rostoker Green's function method, combined with the coherent potential approximation to explore the disordered systems. Calculation of total energy difference of TM-doped ZnO without considering the defects showed that Mn-doped ZnO preferred a spin-glass state, while the FM ordering of Mn magnetic moments could be induced by hole doping, such as introducing nitrogen doping. They also found Fe-, Co-, and Ni-doped ZnO preferred the FM state without carriers, where the magnetic moments of TM atoms coupled with each other and generated the ferromagnetism.

In DMS systems, the 3d orbitals of the doping elements are partially occupied. When the magnetic moments of neighboring TM ions become parallel, the 3d electron can hop to the neighboring FM state. This interaction is governed by the double-exchange mechanism. Sato calculated the total density of states (DOSs) and partial density of d states in (Zn, TM)O systems and believed that the double-exchange mechanism could direct the stabilization of the ferromagnetism. Moreover, they predicted (Zn, Mn, Fe)O, (Zn, Mn, Co)O, or (Zn, Mn, Ni)O may show carrier-induced ferromagnetism under electron doping by tuning the ratio of Mn to Fe, Co, or Ni and may achieve high Curie temperature as well.

Later, Spaldin [40] presented results of a computational study of ZnO in the presence of Co and Mn-substitutional impurities. She imposed high concentration of Zn vacancies to simulate p-type doping in ZnO. As a result, ferromagnetism can be induced by Co or Mn substitution.

In 2005, Mofor et al. observed the existence of RTFM in Mn-doped ZnO [41], which exhibits similar magnetic properties with II–VI magnetic semiconductors [28,29,42]. The d–d transition of Mn ions absorption can induce spin-glass magnetic behaviors and a large magnetoresistance at low temperature, which suggests a long range magnetic ordering.

Besides the magnetic element such as Ni, Fe, or Co-doped DMS, other TM-element-doped oxides have also been found to be FM. The initial researches focus on the magnetic properties of Cu-doped ZnO. Incipiently,

Sato and Katayama-Yoshida [38] predicted that 25% Cu-doped ZnO is nonmagnetic according to their ab-initio calculations. However, this is because of the relatively small supercell performed in this study, which leads to high concentration of Cu in a unit. While later works of Park and Min in 2003 [36] and Chien and Chiou et al. in 2004 [43] found that ZnO doped with lower concentrations (6.25% and 3.125% separately) of Cu are FM from DFT using different approximations. According to these calculations, the separation of Cu atoms with each other is required to be wider than the adjacent basal planes for achieving ferromagnetism. Subsequently, Feng [44] explored the electronic structures of (Zn, Cu)O with the B3LYP hybrid density functional. Different Cu–Cu distances and the related magnetic ordering states were investigated. In Feng's calculation, when the Cu–Cu distance is 5.205 Å and the two Cu ions are aligned in the direction of the c-axis, Cu-doped ZnO favors the FM state. The material tends to be anti-ferromagnetic (AFM) when the Cu–Cu distance is 3.249 Å, which is the shortest distance between two cations in the ab-plane of ZnO. However, in 12.5%Cu-doped ZnO, the total energy of the FM state is lower than AFM state with about 23 meV. Therefore, Cu-doped ZnO is likely to be FM. Further researches by Ye et al. [45] which applied accurate full-potential linearized augmented plane–wave method and DMol3 calculations based on DFT, agreed well with the previous works. They also predicted the possibility of ferromagnetism in both n-type and p-type Cu-doped ZnO. Experimentally, magnetic measurements carried by Hou et al. [46] indicated the existence RTFM in n-type Cu-doped ZnO thin films and their results also indicated that itinerant electrons were responsible for the ferromagnetism. Experimentally and theoretically, dopant clusters have also been claimed to be the origin of the ferromagnetism. Park et al. [34] indicated that the observed ferromagnetism in Co-doped ZnO films originated from the nano-meter-sized Co clusters. Spaldin [40] performed a 32-atom-ZnO model doped with both Co and Cu using LSDA based on DFT. She found that Cu would induce FM interactions between the Co ions. Experimentally, Ando and Saito et al. [47] proved 0.3%Cu-doped ZnO thin films can be a candidate of DMS. Han, Song et al. [48] achieved ferromagnetism in (Zn, Fe)O system by doping with a certain concentration of Cu and the Curie temperature is up to 550 K. Based on Han's work, Shim et al. [49] investigated the origin of the ferromagnetism using the zero-field Fe nuclear magnetic resonance and neutron diffraction. And they found some of the Fe ions can form a secondary phase therefore leading to the ferromagnetism. The origin of the ferromagnetism in TM-doped DMS is still an open topic.

7.5.1.2   MAGNETISM IN NONMAGNETIC ELEMENT-DOPED ZnO

From above discussion, magnetic secondary phases or nanoclusters are possible origins of the ferromagnetism in oxide-based DMSs. However, the kind of ferromagnetism is not suitable for practical applications since to realize spin functions, the ferromagnetism should be able to be manipulated by an electric field, whereas intrinsic FM DMS is following Zener model-based mean field theory, the ferromagnetism of which is carrier-mediated and be able to be manipulated by an electric field.

To avoid the formation of secondary phases and magnetic clusters, researchers begin to explore the possibility of DMS fabricated by doping nonmagnetic element. Intrinsic RTFM was observed in carbon-doped ZnO systems by Pan et al. [50] from both experimental and theoretical view. Based on their calculations, a magnetic moment of 2.02 μ_B is expected per carbon atom. The magnetism mainly resided on the carbon p orbitals (0.85 μ_B), together with small contribution made by the neighboring Zn atoms (0.11 μ_B), and second nearest neighboring oxygen atoms (0.05 μ_B), as depicted in Figure 7.4. From the figure, we can also found a strong interaction between the carbon p orbitals, oxygen p orbitals and the zinc d orbitals. Experimentally, the C-doped ZnO films showed ferromagnetism with a Curie temperatures higher than 400 K, and the measured magnetic moment of the samples is about 1.5–3.0 μ_B per carbon, in well agreement with the theoretical prediction. They have proposed a similar hole-mediated mechanism of the magnetic origin in this system as in Mn-doped GaAs: the holes in the O 2p states which are introduced by the substitution of C atoms at O sites can mediate with the localized spin alignment of C atoms via a long-range p–p interaction. In addition, they calculated the ZnO codoped with C and N, with both elements substituting at O sites. The results showed that the ferromagnetism could be enhanced with the codoping compared to single carbon-doped ZnO, which further confirmed the hole-mediated ferromagnetism.

This work has inspired a series of researches on the carrier-mediated interactions with p electrons instead of the conventional d or f electrons. Based on the density-functional theory and the GGA, Shen et al. [51] predicted that ZnO:N is FM in a low nitrogen concentration with the magnetic moment of 1.0 μ_B atom^{-1}. The magnetism comes from the mediation of local magnetic coupling of N atoms with the spin polarized carriers in p states nearby O atoms. They also proposed a Be-codoping mechanism to enhance the ferromagnetism of N-doped ZnO.

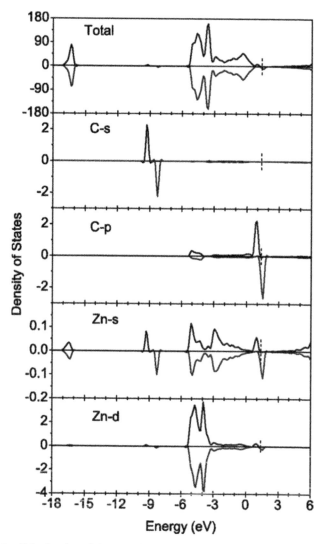

FIGURE 7.4 Calculated total (top panel) and local density of states for the carbon dopant and a neighboring Zn atom. The Fermi level is indicated by the dashed vertical line (reprinted with permission from Ref. [50]). (Reprinted with permission from Pan, H.; Yi, J. B.; Lin, J. Y.; Feng, Y. P.; Ding, J.; Van, L. H.; Yin, J. H. Room-Temperature Ferromagnetism in Carbon-Doped ZnO. *Phys. Rev. Lett.* 2007, *99*, 127201. © 2007 by the American Physical Society.)

As mentioned before, it is quite difficult to fabricate p-type doping in wide band gap semiconductor-based DMSs under normal circumstances. However, first principles calculations have shown the possibilities of various nonmagnetic p-type dopants in ZnO by substituting group-I elements (Li, Na, and K) for Zn sites.

Ferromagnetism was obtained in Li-doped ZnO by Yi et al. [52]. They combined both theoretical and experimental approach to investigate the observed ferromagnetism. Based on DFT, the ferromagnetism could be attributed to the existence of cation vacancies. According to their calculations, the V_{Zn} in ZnO prefers a spin-polarized state, and each vacancy carries a magnetic moment of 1.33 μ_B. By calculating the formation energies and magnetic properties of different kinds of defects and defects complexes, they are able to give a possible understanding of the evolution of electronic and magnetic properties of Li-doped ZnO. Interstitial Li is the major defects at low Li-doping concentration and at low oxygen partial pressure, which can introduce extra electrons to the system, leading to n-type behavior. With the increasing of Li doping, Li begins to substitute the Zn site, and Li_{Zn} can be the dominant defect under moderate PO_2. Holes can also be generated via substituting and can dominate over electrons due to the increasing of Li concentration. As a result, the Li:ZnO behaves p-type. However, neither the Li interstitial nor substitutional defects are magnetic. Therefore, no magnetism will be observed in the systems at low PO_2. Defects complex such as $(Li_I + Li_{Zn} + V_{Zn})$ begins to form with the enhancement of dopants concentration under high oxygen chemical potential. This complex can lower the formation energy of Zn vacancy (V_{Zn}) and contributes a magnetic moment of 1.1 μ_B which is originated from the vacancy. FM coupling of these magnetic moments can be mediated by holes introduced by Zn vacancies and Li substitutional doping, bringing to a FM ZnO at room temperature. X-ray diffraction (XRD) and positron annihilation spectroscopy (PAS) studies were carried out to further confirm the cation vacancy-induced ferromagnetism as well.

Similarly, Na-doped ZnO was investigated using DFT with LSDA. The calculations of total DOS of the supercell with various single defects revealed that both Zn vacancy (V_{Zn}) and O_{int} are magnetic defect that may contribute to the observed ferromagnetism (see Fig. 7.5). With the calculation of the formation energy of relevant supercell, one can find that supercell with either a V_{Zn} or oxygen interstitial is difficult to generate under the normal circumstances. This is in consistence with the fact that pristine ZnO samples are not FM. It should be noted that Na substitutional can keep stable throughout the whole range of oxygen chemical potential, and it is found to be the most stable defect at low oxygen chemical potential. Since Na_{Zn} introduces holes, the system is therefore p-type. However, at high oxygen chemical potential, more complicated defects complex will form with the increasing dopants concentration.

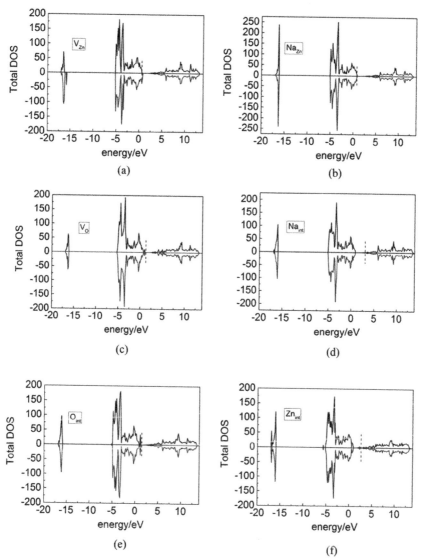

FIGURE 7.5 DOS of supercell with (a) Zn vacancy, (b) Na substitutes Zn, (c) oxygen vacancy, (d) Na institutional, (e) oxygen institutional, and (f) Zn institutional of supercell with different defects. The Fermi level is indicated by the vertical dashed lines.

A significant decrease in the formation energy of over 5.1 eV can be achieved when the (Na_{Zn} + Na_{int} + V_{Zn}) complex is formed, which is quite the same as in Li:ZnO system. This defect complex has an overall magnetic moment of 0.5 μ_B, whereas the calculations also indicate that (Na_{Zn} + Na_{int}) and (Na_{int} + V_{Zn}) complex both do not produce magnetic moment. It is also pointed out that even though (Na_{Zn} + V_{Zn}) complex enjoys rather high magnetization of about 1.49 μ_B, its contribution to RTFM could be neglected due to the whole-range high formation energy.

Since oxygen interstitial prefers a spin-polarized state, formation energies of supercell with oxygen interstitial related defects were calculated as well. The formation energies of the other complexes are relatively high in the whole range of chemical potential, further confirming that (Na_{Zn} + Na_{int} + V_{Zn}) defect complex is the origin of the ferromagnetism.

RTFM was successfully achieved in K-doped ZnO systems as well from first principles calculations based on DFT. The results show that the magnetization in these systems is originated from the O 2p hole states around Zn vacancies, which is the same as that in Li and Na-doped ZnO. Moreover, K dopants have shown unique functions on the ferromagnetism since the substitutional K can induce magnetic moments to the system by forming partial V_{Zn} via lattice distortion. Hence K-doped ZnO can be magnetic at low doping concentrations. Further investigation indicates that the defects complex (K_{Zn} + K_{int} + V_{Zn}) has a low formation energy, therefore, stabilizing Zn vacancies and inducing magnetism at higher doping concentrations. In addition, the calculations show that the K dopants prefer a large separation, suggesting uniform distribution. Experimentally, K-doped ZnO nanorods were fabricated using a hydrothermal method and RTFM was observed. 2-at% K-doped ZnO has the largest saturation magnetization, consistent with first principles calculations.

The above studies have shown that defects in ZnO may induce magnetic ordering under certain conditions. The study on the effects of native point defect in ZnO have been extensively evaluated via ab-initio studies [25,53,54] as well as experimental researches [53,55]. These defects can be used to produce and tune the magnetic ordering in ZnO-based materials, which is the so-called defect engineering, promising for the fabrication of spintronics materials.

7.5.2 TiO_2

Titanium dioxide (TiO_2) has known to be another wide band gap semiconductor, which has three natural structures that maintain the octahedral environment of Ti ions (Fig. 7.6): rutile, anatase, and brookite in the order of abundance. In principle, brookite TiO_2 tends to be unstable, while rutile and anatase phases are stable phases. Though many attempts have been made to determine the relative stability of rutile and anatase from both the experimental and theoretical points of view [56], it still remains contro-versial. Rutile is thought to be the most stable phase since the anatase phase is less thermodynamically stable and unlikely to form bulk with high crystallinity.

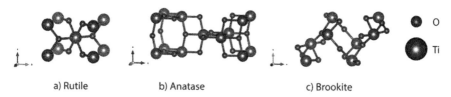

a) Rutile   b) Anatase   c) Brookite

FIGURE 7.6   Crystal structure of TiO_2 in (a) rutile, (b) anatase, and (c) brookite phase.

Both rutile and anatase TiO_2 crystallize in a tetragonal structure, with lattice constant of $a = 4.59$ Å and $c = 2.96$ Å, and $a = 3.78$ Å and $c = 9.52$ Å separately. The band-gap energy of anatase TiO_2 is slightly higher (~3.2 eV) than that of rutile phase (~3 eV) [57].

7.5.2.1 MAGNETISM IN TM-DOPED TiO_2

Ferromagnetism has been widely reported in Co-doped in both rutile and anatase TiO_2 phases [58–60]. At first, magnetic behaviors were only observed in Co-doped anatase TiO_2, but not in rutile phase.

In 2001, Matsumoto et al. [58] fabricated Co-doped anatase TiO_2 films which exhibited high-temperature (up to 400 K) ferromagnetism via a combi-natorial PLD molecular beam epitaxial (MBE) technique. Their magnetic measurements showed that each Co atom carried a magnetic moment of 0.32 μ_B when Co concentration is up to 8%. Later, Chambers et al. [59] synthe-sized Co-doped anatase TiO_2 film which had a magnetic moment of 1.26 μ_B Co^{-1} via the oxygen-plasma-assisted MBE technique. Theses researches casted doubt on that the magnetization of the Co atoms might depend on the local structure around it. However, by using reactive co-sputtering, Park et al. [60] deposited FM Co-doped rutile TiO_2 thin film. The magnetic moment per Co atom was estimated to be about 0.94 μ_B. More importantly, their investigation revealed a Curie temperature higher than 400 K for Co doping concentration of 12%.

In fact, Co-doped TiO_2 was one of the initial attempts in oxide semiconductors, which motivated a mass of TM-doped-oxide-based DMSs. Since then, intensive studies have focused on the structural, magnetic, electronic, and other properties of TiO_2 both experimentally and theoretically.

Besides Co doping, other TMs like Fe, Mn, Ni, V, and Cu [61,62,63] were added into TiO_2 host, to achieve potential RTFM. Experimental studies have reported several possible room temperature FM behaviors in these materials.

7.5.2.2   THE ORIGIN OF FERROMAGNETISM IN TM-DOPED TiO_2

Due to intrinsic complexities and fewer investigations, the microscopic mechanism of the long-range magnetic ordering in TM-doped TiO_2 is still unclear. As mentioned before, carrier-mediated ferromagnetism has been widely reported in III–V-based DMSs [1,23]. Experimental measurements of the anomalous Hall effect in co-doped TiO_2 have shown a connection between carrier density and ferromagnetism [64]. Their findings have supported that the carrier-mediated mechanism is the origin of the FM ordering of the Co magnetic moments. However, other reports have shown the formation of FM clusters at different ambient conditions. TEM pictures have demonstrated the existence of Co clusters [65]. Based on the definition of superparamagnetism, Co dopants can be single domain particles and contribute to ferromagnetism with large magnetic moments.

Fe as a dopant in TiO_2 has also led to some interesting magnetic properties. Adequate reports have suggested that TM dopants and the oxygen vacancies play an important role in high Curie temperature in TM-doped TiO_2 [61,66,67]. The effects of oxygen vacancies in these materials have been carefully examined. Suryanarayanan et al. prepared 3% Co- [67] and 5% Fe-doped [68] TiO_2 thin films in both anatase and rutile types on sapphire substrates, and they found the magnetic moments of these samples could be strongly enhanced after vacuum annealing. The magnetization of anatase samples increased to 0.36 μ_B Co^{-1}, 0.46 μ_B Fe^{-1}, and rutile samples to 0.68 μ_B Co^{-1}, 0.48 μ_B Fe^{-1}, respectively. The FM Curie temperature of these samples is above 350 K. Moreover, the magnetic moment of rutile samples can be reduced greatly as a result of reheating in air. Their experimental results indicated that the observed ferromagnetism may be due to the oxygen-related defects.

Subsequently, first principles calculations were carried out to study the effects of oxygen vacancies in TM-doped TiO_2 systems. Park et al. [66] investigated the electronic structures of Co-/Mn-/Fe-/Ni-doped anatase TiO_2 using LDSA and LDSA + U + SO band calculation. They predicted rather active magnetic behaviors of the oxygen vacancy near Co site other than near Ti site, with the spin magnetic moment of 2.53 μ_B/Co, and the spin state originates from the five neighboring oxygen anions. Weng *et al.* [69] demonstrated first principles calculations concerning the origin of the Co-doped TiO_2. Their results showed that oxygen vacancy could enhance the ferromagnetism. In contrary to the results from Park, V_O near Ti sites was predicted to influence the physical properties of the system significantly as well, while V_O prefers to generate near Co dopants with a magnetic moment of 0.9 μ_B.

Errico et al. [70] looked into the structural, electronic, and magnetic properties of TM-doped rutile TiO_2 with different doping concentrations (25% and 6.25%) using ab-initio methods. They predicted Mn-, Fe-, and Co-doped TiO_2 to be FM, and the ferromagnetism mainly comes from the p–d interaction between the impurity element and the surrounding oxygen atoms. Quite stable values of magnetic moments of these systems without oxygen are obtained with different doping concentrations (3.0 μ_B, 2.6 μ_B, and 1.0 μ_B for Mn-, Fe-, and Co-doped TiO_2, respectively). The calculated magnetic moments decrease from the order of Fe, Co to Ni, in agreement with the experimental results of Hong et al. [71]. However, according to Errico's calculations, the oxygen vacancy tends to locate in the nearest-neighboring site to the dopants if there is any, in consistent with Weng's calculation [69]. Further calculations suggest that dopants may lower the formation energies of oxygen vacancies and the dopants related defects are more likely to generate with the presence of oxygen vacancies as well. Strong interaction is observed between oxygen vacancies and impurities. In the cases of Mn, Fe, and Co-doped TiO_2, the interaction increases the local magnetic moment of the TMs, while it induces magnetic behaviors in Ni- and Cu-doped TiO_2. The oxygen vacancies are believed to be able to provide adequate carriers to produce FM ordering.

Chen et al. [72] further confirmed the existence of vacancy-enhanced ferromagnetism in Fe-doped–rutile TiO_2 system. Based on their studies, the FM coupling between two Fe atoms can be enhanced by vacancies through two ways. On the one hand, Fe ions can form a shallow impurity state in the form of Fe–V_O defects complex with the nearest-neighboring vacancy; on the other hand, the Fe dopants can capture the vacancy electrons, and therefore enhance the FM double exchange interaction. The charge density and spin density of Fe:TiO$_2$ with d_{Fe-Fe} = 3.57 Å without or with oxygen vacancy in the plane contained Fe–O–Fe or Fe–V$_O$–Fe are illustrated in Figure 7.7 [72]. Combined with the calculation of the exchange energy between Fe–Fe pairs (Fig. 7.8 [72]), the idea that the reversal of AFM coupling to FM coupling is dominated by oxygen vacancies can be further confirmed.

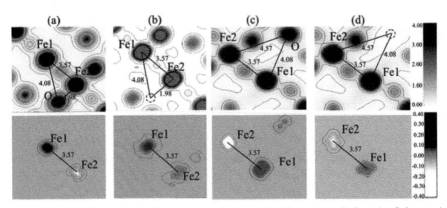

FIGURE 7.7   Charge density (top panel) and spin density (lower panel) (in unit of electron/cell volume) for 2Fe and 2Fe + V$_O$ with d_{Fe-Fe} = 3.57 Å. (a and b) Reversal of AFM ordering to FM ordering when V$_O$ is at a NN location of Fe$_2$. (c and d) The AFM ordering was retained when V$_O$ is farther away from the 2 Fe. All the charge densities are plotted from the plane which contained bond Fe–O–Fe or Fe–V$_O$–Fe (reprinted with permission from Ref. [72]). (Reprinted with permission from Chen, J.; Rulis, P.; Ouyang, L. Z.; Satpathy, S.; Ching, W. Y. Vacancy enhanced ferromagnetism in Fe-doped rutile TiO$_2$, *Phys. Rev. B* 2006, 74, 5. © 2006by the American Physical Society.)

Although pure Cu is nonmagnetic, RTFM appeared in Cu-doped TiO$_2$ thin films as reported by Duhalde et al. [61]. The magnetic moment is relatively large, about 1.5 μ_B per Cu atom. Large magnetic moment was only found when dopant Cu is close to the oxygen vacancy site from ab-initio calculations. Their calculations suggested that the magnetism may depend on the defects complex formed by Cu substitutional and oxygen vacancy, similar to the mechanism of Fe-doped TiO$_2$ rutile thin films mentioned previously. In addition, their calculation also suggests that doping with Cu can stabilize the formation of oxygen vacancies as well.

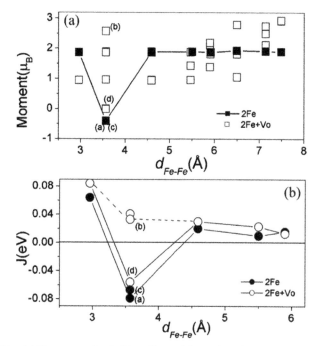

FIGURE 7.8 (a) Fe moment in the Fe + V_O model as a function of d_{Fe-Fe} (open square). The data for 2 Fe with no V_O are also shown (solid square). (b) Exchange energy J in the 2 Fe (solid circle) and 2 Fe + V_O (open circle) models. The data points marked (a)–(d) correspond to the four cases with d_{Fe-Fe} fixed at 3.57 Å shown in Figure 7.8 (reprinted with permission from Ref. [72]). (Reprinted with permission from Chen, J.; Rulis, P.; Ouyang, L. Z.; Satpathy, S.; Ching, W. Y. Vacancy enhanced ferromagnetism in Fe-doped rutile TiO_2, *Phys. Rev. B* 2006, *74*, 5. © 2006 by the American Physical Society.)

Based on the previous works, Li et al. [73] performed first principles calculations within the GGA using the PAW method to explore the possible mechanism of ferromagnetism in Cu-doped TiO_2 system. In particular, the Cu–V_O defect complex was taken into consideration. This complex would form CuO_4, in which Cu and the four most neighboring O ions attract each other after structural relaxation. By comparing the spin-density distributions of oxygen vacancies and the complex, they found the majority magnetic moment of V_O originated from the Ti ions. However, in Cu–V_O defects complex the magnetism arises from the doped Cu ions. Two possible mechanisms can elucidate the observed magnetism from the calculations: one comes from the hybridization between the Cu 3d orbitals and oxygen 2p orbitals when energy levels situate near the Fermi level and the other originates from the spin polarization of the 3d orbital of Cu ions and the 2p orbital of O ions at lower energy levels.

First principles study was also performed on V-doped anatase TiO_2 by Du et al. [74] using both GGA and GGA + U approach. Based on the calculated band structure and DOS of V-doped anatase model, it is suggested that the magnetism found in this system is quite complicated. Instead of simply depending on the direct spin-spin interaction between the V ions, a combination of short-range superexchange coupling between the nearest neighbor V ions and long-range BMP percolation induced by oxygen vacancy account for the existing ferromagnetism.

Droubay et al. [63] grew Cr-doped anatase TiO_2 films on $LaAlO_3$ (0 0 1) substrates and observed room-temperature ferromagnetism with a magnetic moment of 0.60 μ_B Cr^{-1}, regardless of the Cr concentration varying from 2% to 16%. The as-deposited films are highly insulating. They claimed that ferromagnetism came from the F-center exchange interaction, by which F-center electrons mediated with the Cr spins, resulting in FM coupling. This mechanism is quite different from the other DMSs since no itinerant carriers or superexchange is required.

7.5.2.3 MAGNETISM IN NONMAGNETIC ELEMENT-DOPED TiO2

Ferromagnetism has also been observed in nonmagnetic-element-doped TiO_2. N- and Ar-implanted rutile TiO_2 have been fabricated and exhibited ferromagnetism [75]. Stimulated by the theoretical and experimental facts that C- and N-doped ZnO may exhibit RTFM, Li et al. [76] and Yang et al. [77] predicated the possibility of RTFM in C-doped TiO_2 in rutile and anatase structure, respectively. Li's calculations indicated that in C-doped rutile system, the replacement of oxygen atoms with carbon could induce a magnetic moment of about 2.0 μ_B C^{-1}. Yang explored the possibility of FM ordering of carbon-doped anatase system via DFT calculations. It was found that the FM coupling interaction strongly depended on the distance of adjacent C atoms and C–Ti–C angle. Their results showed that each substitutional carbon could generate a magnetic moment of 2.0 μ_B when the C–C distance lies between 3 and 4 Å. A strong FM coupling occurs between the moments when two C atoms form a slightly bend C–Ti–C unit (<C–Ti–C = 160°).

First principles calculations of nitrogen-doped anatase TiO_2 was demonstrated by Tao et al. [78] to find out the possible RTFM. A total magnetic moment of 1.0 μ_B N^{-1} was obtained in their study. By plotting the spin density distribution configurations, carrier-mediated exchange mechanism was proposed to explain the ferromagnetism: the holes in O 2p states induced by nitrogen doping can couple with the N 2p spins via p–p interaction and therefore induce ferromagnetism. Subsequently, Yang et al. [79] performed density functional studies to investigate the FM and AFM states of N-doped TiO$_2$ in both anatase and rutile structures. A series of N-doped TiO$_2$ models are taken into consideration. Normally, it is believed that the substitution of N atoms in O sites occurs when two N^{3-} anions replace three O^{2-} anions and an oxygen vacancy. However, According to Yang's calculation, N dopants at O sites exist as N^{2-} ions with net spin moments of 1.0 μ_B, which is in well consistence with Li's results. The magnetic coupling interaction between two spin-polarized N^{2-} ions is explored in different groups classified by N–N distance and <N–Ti–N angle. Generally, when the two dopants are coordinated to a common Ti atom, strong FM or AFM ordering can be achieved by the spin exchange coupling between the spin moments of two N dopants. When the <N–Ti–N angle is smaller than ~99°, an AFM coupling is introduced via superexchange interaction between the p orbitals of N dopants. FM coupling is favored when the <N–Ti–N angle is greater than ~102° due to a p–p electron hopping interaction between the nitrogen and oxygen. Both Tao and Yang's studies of N-doped TiO$_2$ have shown similarities to a N-doped ZnO system [51].

Possible RTFM was also predicted in Li-doped anatase TiO$_2$ by Tao et al. [80] using first principles calculations. The found at the lithium dopants could generate holes in O 2p orbitals and resulted in magnetic moments. However, the stability of ferromagnetism depends greatly on direction and distance.

7.5.3 In$_2$O$_3$

Indium oxide (In$_2$O$_3$) is another well-known transparent oxide semiconductor with a wide direct band-gap (3.75 eV [81]). The crystalline form of In$_2$O$_3$ is a cubic structure as shown in Figure 7.9, with experimental lattice constant equal to 10.117 ± 0.001 Å [82]. It is a high symmetry structure with In^{3+} at a six-folds coordinated position and O^{2-} occupying a fourfold coordinated position. This structure is rather complicated and quite different from the other oxide-based DMS hosts. However, In$_2$O$_3$ with the cubic structure can be ideal for industrial applications since high crystallinity samples can grow easily on low-cost substrates like MgO [83]. It is reported that pristine In$_2$O$_3$ in both bulk crystal and thin film forms have shown a high level of n-type conductivity (10^{18}–10^{20} carriers cm^{-3}). Moreover, when Sn substitutes

In atoms, we can obtain the indium tin oxide (ITO), which is one of the most commonly used transparent semiconductors.

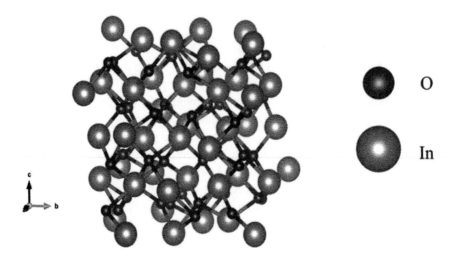

FIGURE 7.9   Crystal structure of cubic In_2O_3.

7.5.3.1   MAGNETISM IN TM-DOPED In_2O_3

According to the reported experimental results, 3d TMs (V, Cr, Mn, Fe, Co, Ni, and Cu) have been introduced into In_2O_3, and RTFM has been widely observed in these systems [11, 84–87].

Origin of the ferromagnetism in TM-doped In_2O_3 is also controversial: It may come from the charge transfer via superexchange interaction, or the mobile-electron-mediated coupling, or FM clusters. Moreover, the ferromagnetism can also be related to oxygen vacancies since strong magnetic ordering has been observed in undoped In_2O_3 in both nanocrystals and thin films form. d^0-induced ferromagnetism is therefore expected.

He et al. [86] synthesized Fe and Cu-codoped-In_2O_3 thin films and observed intrinsic RTFM in 20% Fe-doped In_2O_3. Fabrication and magnetic properties of Fe-doped In_2O_3 bulk ceramics with Cu codoping were reported by Yoo et al. [87]. The continuous shrink of the lattice constant of the samples with the increase of the doping concentrations of Fe showed the effective incorporation of Fe ions in TiO_2 host. High-temperature ferromagnetism (Curie temperature over 750 K) was observed in samples with high concentration of magnetic ions (up to 20%), and the ferromagnetism was claimed to be related to the magnetic impurities. In addition, RTFM was also achieved in 6%Ni-doped In_2O_3 by Hong et al. [83]. The mechanism of the ferromagnetism is still not clear.

TM doping has drawn great attention since the dopants can introduce interesting properties to the systems. The ionic radii of V (0.79 Å), Cr (0.8Å), Mn (0.83 Å), and Fe (0.78 Å) in 2+ oxidation state ion are close to the Shannon–Prewitt radius of In ions which is about 0.80 Å. While the ionic radii for these dopants in 3+ oxidation state is 0.67, 0.64, 0.62, and 0.59 Å, respectively, similar to the radius of Sn^{4+} in ITO. The substitutional effects of these dopants have fascinated many researchers' interests.

Gupta et al. [85] explored the substitutional effect of Ti, V, and Cr in In_2O_3 and the possible origin of the magnetism in this system from both theoretical calculations and experimental researches. Due to the complexity of cubic structure, different configurations of the doping positions were considered in their calculations. V- and Cr-doped In_2O_3 both exhibit room temperature FM behaviors, while Ti-doped In_2O_3 has a rather weak magnetization at room temperature. They predicted that the substitution of vanadium could yield ferromagnetism in In_2O_3. The structure with V–V separation of 5.06 Å (the third nearest Vanadium neighbor) is the most stable configuration. Moreover, this structure shows strong FM coupling of 151 meV. The calculated magnetic moment is about 2 μ_B V^{-1}. The p–d interaction leads to the FM coupling.

In particular, F and Cr-doped [81] In_2O_3 exhibits high Curie temperatures about 750 K and 900 K, separately. Fe-doped In_2O_3 has attracted the most attention because of the high Curie temperature (700 K), high solubility of Fe in In_2O_3 (>20%), and a homogeneous solid solution can be realized at least up to 15% Fe doping [86]. Guan et al. [88] found that pure Fe-doped In_2O_3 has an AFM ground state, but with oxygen vacancies and Cu codoping, the system can lead to a weak FM coupling from first principles calculations. They investigated the electronic structures and FM stability, and found that the FM exchange interaction could be mediated by a hybridization of Fe 3d states, Cu 3d states and O 2p states in a form of Fe1–O1–Cu–O2–Fe2 coupling chain. The vacancies and the codoping behavior can therefore stabilize the ferromagnetism.

Raebiger et al. [89] investigated the electronic structure, doping, and magnetism in 3d TM-doped In_2O_3 and ZnO using a band-structure-corrected theory with nonlocal external potentials and on-site Coulomb energy corrections, which can guarantee the accuracy of the occupation of the 3d-induced levels. They found that the incompletely 3d levels which are resonant inside the conduction band could cause the long-range FM interactions.

7.5.3.2   MAGNETISM IN NONMAGNETIC ELEMENT-DOPED In_2O_3

Similar to other oxide-based DMSs, the magnetism in TM-doped In_2O_3 can be related to the precipitates or secondary phases, which is undesirable for technical applications. Dopants without partially filled d or f ion may be introduced into In_2O_3 to achieve the intrinsic ferromagnetism to avoid the ferromagnetism from clusters or secondary phases.

Guan et al. [90] predicted steady RTFM in N-doped In_2O_3 using first principles calculations. Based on their calculations of electronic structures and magnetic properties, the nitrogen dopants were found to be responsible to the ferromagnetism. The magnetic coupling is mediated by the holes which are generated by dopants via p–p interactions in short N–N separations. Their calculation also suggested that nitrogen dopants may induce a total magnetic moment of 1.0 μ_B per N, which is mainly localized on the doped N atoms.

Experimentally, Koida et al. [91] found that when the carrier density is about 10^{20} cm^{-3}, the H-doped In_2O_3 could have a high mobility of 100 cm^2 V^{-1}·s^{-1}. Later, Limpijumnong et al. [92] conducted ab-initio study of the role of hydrogen impurities. Three different oxidation states of hydrogen were considered together with several different locations of the dopants. Their calculations indicated high stability of both interstitial and substitutional hydrogen. Moreover, the H dopants in interstitial sites acted as shallow donors in In_2O_3, as well as the substitutional H at oxygen sites. Their calculations agree well with the experimental observations of Koida [91].

Nonconventional magnetism in a pristine or alkali-doped In_2O_3 is observed from theoretical calculations based on DFT [93]. The magnetism in undoped In_2O_3 relies on the indium vacancies instead of the oxygen vacancies. All the alkali (V_{In}, Li, Na, and K)-doped In_2O_3 can induce strong FM moments which are localized on the first shell of O atoms around alkali substitutional sites. Similar interaction chain like Cu-codoped system [88] is required to produce carrier for mediating the long range FM coupling. Moreover, In_2O_3 with Mg and Ca dopants were predicted to be magnetic as well.

7.5.2 SnO₂

Tin dioxide (SnO_2) is always in a rutile form with tin atoms at octahedral coordinated positions and oxygen atoms at trigonal planar coordinated positions [94] (as depicted in Fig. 7.10). The lattice constant of SnO_2 is $a = b = 4.373$ Å, $c = 3.185$ Å from experiments [95]. SnO_2 is a typical n-type semiconductor with a wide band gap of about 3.7 eV. It is a multifunctional oxide which exhibits many interesting properties, such as transparency, conductivity, and ferromagnetism at low dimensional structures [96].

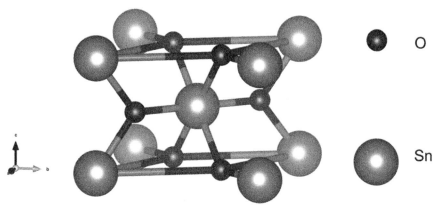

○ O

● Sn

FIGURE 7.10   Crystal structure of rutile SnO_2.

7.5.3.3   MAGNETISM IN TM-DOPED SnO₂

A number of TM-doped SnO_2 thin films have been heatedly discussed for their potential DMS property at room temperature.

Ogale and Choudhary et al. [97] used pulsed laser deposition to fabricate Co-doped SnO_2 films and they reported RTFM with $T_C = 650$ K and observed a giant magnetic moment of about $(7.5 \pm 5)\mu_B$ per Co atom. The magnetic moment is believed to be originated from both cobalt orbital and surrounding atoms.

Hays and Punnoose et al. [98] in 2005 observed FM behavior in SnO_2 samples with low doping concentration of Co (less than 1%). However, their XRD and photoelectron spectroscopy results showed that the magnetization would decrease when increasing the doping concentration. On the contrary, Liu et al. [99] observed a high magnetic moment of 2.37 μ_B Co^{-1} in samples doped with ~6.3% Co in 2007, whereas the moment falls rapidly into merely 0.1 μ_B per Co atom when the concentration of Co is increased to 7%.

Coey et al. [100] also fabricated Fe-doped SnO_2 films and found the materials are FM (1.8 μ_B Fe^{-1}) with a Curie temperature of 610 K. Fitzgerald et al. [101] reported RTFM in Mn-, Fe-, and Co-doped SnO_2 and the magnetic moment is 0.11 or 0.95 μ_B per Mn or Fe atoms, with the Curie temperature of 340 and 360 K, respectively.

Ni-doped SnO_2 thin films are usually fabricated on (0 0 1) $LaAlO_3$, (0 0 1) $SrTiO_3$, and R-cut Al_2O_3 substrates using pulsed laser deposition technique from a ceramic target. All these films have shown RTFM under the same oxygen pressure as during deposition [102]. Samples grown on $LaAlO_3$ substrates have a large magnetic moment of about 2 μ_B Ni^{-1} and $T_C = 400$ K. This large saturated magnetization is believed to be intrinsic since Ni metal has a moment of only 0.6 μ_B. Due to the arguments in experimental results, first principles calculations were carried out to shed light on the mechanism of ferromagnetism in TM-doped SnO_2 [103–105].

Based on DFT, Wang et al. [103] investigated the electronic and magnetic properties of both Co- and Fe-doped SnO_2 and found FM behaviors in these systems. The long-range magnetic interaction between Co ions can be reduced by decreasing the Co–Co distance, while in Fe-doped SnO_2 system, the exchange interaction can oscillate with the Fe–Fe distance. Their results also indicated that increasing the dopants concentration could not enhance the Curie temperature.

Later, Wang et al. performed first principles calculations in SnO_2 doped with TMs (V, Mn, Fe, and Co). They observed a strong influence of oxygen vacancy on the magnetic properties in Fe- and Co-doped materials, and the defect sites can easily attract the Fe, and Co ions. Zhang et al. [105] applied the LSDA plus Hubbard U scheme to explore the origin of the ferromagnetism in Co-doped SnO_2. As expected, they found that oxygen vacancies are critical in determining the magnetic order. With the presence of oxygen vacancies, a strong carrier-mediated ferromagnetism interaction is induced as a result of the charge transfer between the vacancy and dopant. Consequently, the materials display a long-range FM ordering which is governed by intrinsic defects.

Mn-doped SnO_2 is theoretically predicted to be paramagnetic [106] and experimentally it indeed exhibits a large magnetoresistance and interesting paramagnetic behaviors at low temperature [107]. This paramagnetism is believed to be due to the shortage of FM interactions between Mn ions. However, Fitzgerald and Venkatesan et al. [101] reported a ferromagnetism with $T_C = 650$ K made by the same material. Some researchers claimed that the intrinsic defects mediated the ferromagnetism [108]. The spectrum of X-ray absorption near-edge spectroscopy research carried out by Liu et al. [109] indicates that Mn ions substituted Sn sites are in oxidation states. The magnetic–hysteresis curves showed that the saturated magnetization of the film decreased fell along with the decrease in carrier concentration. This result reveals that the origin of Mn-doped SnO_2 film is related to carrier-mediated model. Therefore, the carrier concentration can affect the ferromagnetism [109].

More recently, Espinosa et al. [110] carried out the investigations of the origin of the magnetism in undoped and Mn-doped SnO_2 thin films using both experimental techniques and computational methods. Combined their observations with calculations, they found that the Sn vacancies are responsible for the measured magnetic moments, while oxygen vacancies did not contribute to the magnetization and conductivity of the materials.

DFT with Hubbard-like term (DFT + U) was introduced to Cr-doped SnO_2 system with different dopant concentrations. The p–d interaction between the Cr ions and O ions was predicted to be associated with the magnetic behavior [111].

7.5.4 MAGNETISM IN NONMAGNETIC ELEMENT-DOPED SnO_2

The reported large magnetic moments in SnO_2 systems doping with low-concentration dopants have casted doubt on the origin of the ferromagnetism. It is more likely to be originated from the electronic or lattice defects other than secondary phases or magnetic clusters [112]. Since there is no 3d electron in SnO_2, the RKKY interaction should not play a key role in determining the ferromagnetism.

In 2008, theoretical studies by Rahman et al. [113] predicted that cation intrinsic vacancy in perfect SnO_2 could lead to RTFM. Their calculations showed that SnO_2 with perfect bulk structures did not have magnetism, while Sn vacancies can induce a large magnetic moment of about 4.00 μ_B.

Xiao et al. [114] predicted that N-doped SnO_2 preferred a spin-polarized state with a magnetic moment of about 1.0 μ_B per nitrogen atom. The substitution behavior of N atoms at oxygen sites can result in p–p long-range interaction between the N 2p bands and the nearby O 2p bands, and therefore contribute to the magnetic coupling. Moreover, their calculations also indicated that the magnetic coupling between nitrogen dopants is AFM. Rahman et al. [115] investigated the magnetism of carbon-doped SnO_2 via first principles calculations. They predicted that C could induce moments of about 2 μ_B at (0 0 1) SnO_2 surfaces, while no magnetism could be observed when the carbon atom is located at subsurface oxygen sites. The p orbitals of carbon atom could interact with the surface and subsurface oxygen atoms therefore resulted in the magnetism. Later, Hong et al. [116] reported on the structural and magnetic properties of C-doped SnO_2 and gave direct experimental feedback to the prediction. They found that the ferromagnetism in SnO_2 could be enhanced by doping with carbon, and the dopants introduced defect-induced magnetism (about 3.91 μ_B) to the system. Their results of thickness dependences on magnetism also suggested that C defects should exist in deeper layers instead of only at/near the surface as described in the previous prediction.

7.5.4 HfO₂

Hafnium oxide (HfO_2), also known as hafnia, was introduced by Intel in 2007 as a replacement for silicon oxide as a gate insulator in field-effect transistors. It has a wide band gap of 5.3–5.7 eV [117]. HfO_2 is isomorphic to ZrO_2 and they share the same high-pressure phase transition sequences. HfO_2 has a monoclinic structure with the $P21/c$ space group [118], with the increase of the pressure, it then becomes orthorhombic-I (space group $Pbca$) [119], orthorhombic- II (space group $Pnma$). The orthorhombic structure is quite stable under 1800 K [120]. HfO_2 is cubic with space group $Fm3m$ at high temperature (above 2700 K), which transforms to the tetragonal form with space group $P42/n$ at about 2570 K [121]. Typical structures for HfO_2 are listed in Figure 7.11.. Among which, the monoclinic structure attracts the most interest in research.

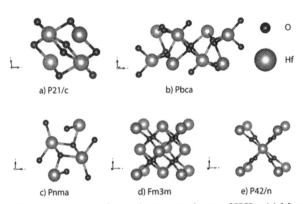

a) P21/c   b) Pbca   O   Hf

c) Pnma   d) Fm3m   e) P42/n

FIGURE 7.11 Crystal structures of some important phases of HfO_2: (a) Monoclinic $P21/c$, (b) orthorhombic $Pbca$, (c) orthorhombic $Pnma$, (d) cubic $Fm3m$, and (e) tetragonal $P42/n$.

7.5.4.1 MAGNETISM IN UNDOPED HfO₂

Unprecedented FM order was discovered in pure HfO_2-thin films deposited on R-cut sapphire substrate by Venkatesan et al. [122]. The reported magnetic moment was about 0.1 μ_B per formula unit. No previous reports have shown magnetism in oxides which is composed of ions with closed-shell configurations. This unique phenomenon is named d^0 magnetism since the Hf ion in HfO_2 has an oxidation state of +4, suggesting that Hf atoms have an empty d shell and the magnetism is not induced by the partially filled d orbitals. After this report, d^0 magnetism was subsequently discovered in ZnO, CaO, and other oxides. Intrinsic defects were proposed to be responsible for the ferromagnetism because of the nonmagnetic nature of HfO_2. For instance, the partially filled d orbitals in Hf can coordinate with the intrinsic defects like oxygen vacancies, and therefore contribute to the observed magnetism.

First principles band structure calculations carried out by Pammaraju et al. [123] showed that the observed ferromagnetism in HfO_2 is mainly originated from the intrinsic point defects via a short-range magnetic interaction between the Hf vacancies and the p orbitals of oxygen. Strong FM coupling with a magnetic moment of about 3.5 μ_B was found in the isolated cation vacancies on the first and second nearest-neighbor sites of HfO_2. However, later experimental results by Joey et al. [124] indicated that hafnium vacancies are much less than oxygen vacancies due to the high-charge state though they observed abundant hafnium vacancies at grain boundaries.

Zheng et al. [125] then conducted a series of first principles studies on the native point defects in hafnia and zirconia and illustrated the relationship between the formation energy of various defects and the chemical potential as a function of Fermi level. Similar behaviors of the defects were found in both HfO_2 and ZrO_2, the oxygen vacancies, oxygen interstitials, Hf/Zr vacancies, and Hf/Zr interstitials could form under different chemical potentials and Fermi levels.

Chen et al. [126] demonstrated first principles simulations on the surface of HfO_2, and obtained d^0 ferromagnetism in the O rich nonstoichiometric (1 1 1) surface of HfO_2. They attributed the ferromagnetism to the 2p bands of oxygen atoms since they have large spin exchange energy. Ab-initio investigations of low-index surfaces of the HfO_2 with tetragonal and cubic structures are presented by Beltrán et al. [127]. Their calculations suggested that only fully O-terminated surfaces are half-metallic. However, the calculated magnetic moments are only 0.13 μ_B Å^{-2}, which are much lower than the other reports. The origin of the ferromagnetism is proposed to be a combination of 2p bands of oxygen ions and the Coulomb repulsion between the holes in partially filled p bands and Hund's rule coupling.

The d^0 induced ferromagnetism is also expected in surfaces of other simple oxides, such as ZnO, In_2O_3, Al_2O_3, SnO_2, and TiO_2 [122,128].

7.5.4.2   MAGNETISM IN DOPED HfO$_2$

Theses surprising results have led to extensive research interests in clarifying the possible contribution of oxygen vacancies in tuning the ferromagnetism in HfO_2 system. Hong et al. [129] deposited Fe-doped monoclinic HfO_2 thin films with Fe concentrations of 1% and 5% via PLD techniques. RTFM was reported in these systems as well as the undoped system grown under the same conditions. They found the samples with different doping concentrations have different magnetic moments, and the oxygen pressure and heat treatments have strong impact on the magnetization. Their observations revealed the effects of intrinsic defects like oxygen vacancies are the main source for the magnetism. Subsequently, Zhang et al. [130] investigated the effects of oxygen vacancies in HfO_2 or ZrO_2 doping with metallic ion (Al, Ti, or La) via first principles calculations and found that the formation energy of oxygen vacancy can be decreased after doping.

Later, Weng et al. [131] performed a series of first principles calculations to investigate the effect of cation vacancies and the substitution of the nonmagnetic elements (K, Sr, and Al) in Hf sites on the magnetic properties of HfO_2, thus understand the relationship between the magnetism in HfO_2 and the number of doped holes. Results indicated that oxygen atoms near the dopant sites could coordinate with the surrounding Hf atoms forming a planar trigonal-like structure. Holes were introduced into the p orbitals of these oxygen atoms, and resulted in a spin-split. They found ferromagnetism could be achieved by introducing holes into systems with narrow p bands near Fermi level. This research has shown great potential in finding new FM materials in nonmagnetic d^0 systems.

7.6   ISSUES AND OUTLOOK

7.6.1   ISSUES IN FIRST PRINCIPLES CALCULATIONS IN OXIDE-BASED SEMICONDUCTORS

First principles prediction is regarded as a powerful tool to develop the structures and properties of magnetic oxides. Moreover, with the development of computational theories, high accuracy of simulations can be achieved in investigating the electronic structures of materials and the microscopic mechanisms. However, there are two issues remained to be solved in most calculations based on DFT: one comes from the exchange–correlation functionals that applied for approximation; the other is due to the limitation of the supercell size of the model [132].

7.6.1.1 BAND-GAP ERRORS

Electronic band structure is one of the most fundamental physical properties of a material. It has always been an important target in first principles calculations to have an accurate description of the band structures. Most of the first principles studies based on the Kohn–Sham density-functional theory are carried out by conventional local spin–density approximations, where the occupation of the 3d-induced levels is incorrect due to spurious charge spilling into the misrepresented host conduction band and have only considered magnetism and carrier doping separately. While the semilocal (gradient expansion) spin-density approximations (GGA) cannot provide accurate relationship between the derivative discontinuities and the occupation number in particles as well. These deficiencies in these approximations will lead to incorrect descriptions of the band gap. The band gap is always underestimated with these conventional functionals. And this underestimation is greatly related to the properties of the systems, especially the chemical bonds [133]. However, it is definitely detrimental to the defect-related calculations. For instance, the positions of the defect transition level maybe inaccurate, and the calculated formation energies of the systems can be problematic as well. The artificial self-interaction and the deficiency of discontinuity concerning different particle numbers can be main contributors of the band gap error, as well as the error in interpreting the Kohn–Sham gap as the true band gap. However, the error still exists even when the Kohn–Sham Hamiltonian is used with the exact exchange–correlation potential [134].

Therefore, efforts have been devoted to improve the description of band gaps within the framework of Kohn–Sham DFT or its generalized formalisms. DFT with Hubbard-like term (DFT + U) [135] is introduced to systems to provide better description of magnetic moments, internal degrees of freedom and electronic band structure features. In this method, two groups of the electrons are considered: localized states and the delocalized states. The Hubbard-like U term in the Hamiltonian respect to strong Coulomb repulsion is used in localized states, while a standard orbital-independent one-electron potential is applied to describe the delocalized states. The determination of the value of U is rather important since specific examples have suggested that we may obtain unphysically large values of U when fitting the band gap [136]. The DFT + U method is physically intuitive [137], but it can only relate to the nonlocality in the exchange potential. Hence, it cannot provide the understanding satisfactorily.

GW method, in which the exchange–correlation self-energy $\sum_{xc}$ is simply a product of G and the screened Coulomb interaction (W), is currently the most accurate first-principles approach to describe electronic band properties of extended systems. Compared to LDA/GGA, the computational efforts required for GW calculations are much heavier, so that its applications have been limited to relatively small systems.

On the other hand, the Hybrid functionals like non-local Hartree–Fock scheme exchange can remedy the errors caused by derivative discontinuity. Typical hybrid functionals includes the HSE06, PBE0, HF, B3LYP, etc. They can pay respect to the occupation number which pulls the conduction band maximum down and describe the positions in conduction band correctly.

Many new exchange–correlation energy functionals have been developed in recent years. By choosing the appropriate exchange–correlation potential, first principles calculations can provide better descriptions for the electronic band structures, though there is still something to be improved. It is foreseeable that first principles calculations will become an important tool for understanding band structures together with the experimental observations.

7.6.1.2   SUPERCELL SIZE LIMITATION

In general, a periodically repeated supercell is adopted to simulate the structure of the material and its properties. However, spurious interactions may emerge due to the periodic supercell boundary, especially when defects are introduced into the systems. In addition, the ionic relaxations are artificially introduced into the supercell, to which the errors in the geometry optimization and relaxation and elastic energies can be accounted.

Average potential correction is the simplest way to eliminate the periodic monopole-background potential contribution. Multiple expansion scaling laws have been introduced for electrostatic corrections. Freysoldt et al. [138] proposed a three-step ansatz including the L^{-3} term in the expansion, which can correct the size limitations without any empirical values. This method has been proven to be easily and well performed in many systems during DFT calculations.

7.6.2 SUMMARY AND OUTLOOK

The presence of the magnetism in nonmagnetic oxides has opened the gate for developing new materials, such as DMS, for next generation of devices. In order to implement technical applications of these new materials, a well understanding of origin of the magnetism is indispensable. First-principles calculations based on DFT have been widely applied in solving the existing questions and exploring the potential possibilities of oxide-based semiconductors. With the unceasing development of computational theory, and the great enhancement of computer power, first principles calculations will become more and more important for the research of spintronics materials like DMSs. Combining the experimental works with the computational studies, magnetic and electronic properties of oxide-based materials can be understood systematically. In addition, first principles calculations can provide a possible guidance for searching other promising materials.

KEYWORDS

- semiconductor devices
- field-effect transistor
- diluted magnetic semiconductor
- first principles calculations
- intrinsic ferromagnetism

REFERENCES

1. Munekata, H.; Ohno, H.; Von Molnar, S.; Segmüller, A.; Chang, L.; Esaki, L. *Phys. Rev. Lett.* **1989,** *63,* 1849.
2. Dietl, T.; Ohno, H.; Matsukura, F.; Cibert, J.; Ferrand, D. *Science* **2000,** *287,* 1019.
3. Ohno, H. *Science* **1998,** *281,* 951.
4. Hehre, W. J. *Acc. Chem. Res.* **1976,** *9,* 399.

5. Hohenberg, P.; Kohn, W. *Phys. Rev.* **1964**, *136*, B864.

6. Kohn, W.; Sham, L. J.; *Phys. Rev.* **1965**, *140*, A1133.

7. Fermi, E. *Rend. Accad. Naz. Lincei* **1927**, *6*, 32; Thomas, L. H. The Calculation of Atomic Fields. In: Presented at the *Mathematical Proceedings of the Cambridge Philosophical Society*, 1927.

8. Levy, M. *Proc. Nat. Acad. Sci.* **1979**, *76*, 6062.

9. Lundqvist, S.; March, N. H. *Theory of the Inhomogeneous Electron Gas*; Plenum Press: New York, 1983; Perdew, J. P.; Parr, R. G.; Levy, M.; Balduz, Jr., J. L. *Phys. Rev. Lett.* **1982**, *49*, 1691.

10. Dirac, P. A. Note on Exchange Phenomena in the Thomas Atom. In: Presented at the *Mathematical Proceedings of the Cambridge Philosophical Society*, 1930.

11. Ceperley, D.; Alder, B. *J. Phys. Colloq.* **1980**, *41*, C7.

12. Perdew, J. P.; Yue, W. *Phys. Rev. B* **1986**, *33*, 8800.

13. Perdew, J. P.; Ziesche, P.; Eschrig, H. *Electronic Structure of Solids' 91*, vol 11, Akademie Verlag: Berlin, 1991.

14. Filippi, C.; Umrigar, C.; Taut, M. *J. Chem. Phys.* **1994**, *100*, 1290.

15. Perdew, J. P.; Burke, K.; Ernzerhof, M. *Phys. Rev. Lett.* **1996**, *77*, 3865.

16. Kresse, G.; Furthmüller, J. *Comput. Mater. Sci.* **1996**, *6*, 15; Kresse, G.; Furthmüller, J. *Phys. Rev. B* **1996**, *54*, 11169; Troullier, N.; Martins, P. E. *Phys. Rev. B* **1991**, *43*, 1993.

17. Hamann, D.; Schlüter, M.; Chiang, C. *Phys. Rev. Lett.* **1979**, *43*, 1494.

18. Vanderbilt, D. *Phys. Rev. B* **1990**, *41*, 7892.

19. Blöchl, P. E.; Jepsen, O.; Andersen, O. K. *Phys. Rev. B* **1994**, *49*, 16223.

20. Katayama-Yoshida, H. Fukushima, T. Dederichs, P. H. Sato, K. Toyoda, M. Kizaki, H. Dinh, V. A. *J. Korean Phys. Soc.* **2008**, *53*, 1.

21. Dietl, T.; Ohno, H.; Matsukura, F. *Phys. Rev. B* **2001**, *63*, 195205.

22. Durst, A. C.; Bhatt, R.; Wolff, P. *Phys. Rev. B* **2002**, *65*, 235205; Angelescu, D.; Bhatt, R. *Phys. Rev. B* **2002**, *65*, 075211; Bednarski, H.; Spałek, J. *Acta Phys. Pol., A* **2011**, *120*, 967.

23. Calderon, M. J.; Sarma, S. D. *Ann. Phys.* **2007**, *322*, 2618.

24. Madelung, O. *Semiconductors: Group IV Elements and III–V Compounds*, vol 17; Springer-Verlag: Berlin 1982.

25. Kohan, A.; Ceder, G.; Morgan, D.; Van de Walle, C. G. *Phys. Rev. B* **2000**, *61*, 15019.

26. Zhang, S.; Wei, S.-H.; Zunger, A. *Phys. Rev. B* **2001**, *63*, 075205; Oba, F.; Nishitani, S. R.; Isotani, S.; Adachi, H.; Tanaka, I. *J. Appl. Phys.* **2001**, *90*, 824; Lee, E.-C.; Kim, Y.-S.; Jin, Y.-G.; Chang, K. *Phys. Rev. B* **2001**, *64*, 085120.

27. Look, D. C.; Hemsky, J. W.; Sizelove, J. *Phys. Rev. Lett.* **1999**, *82*, 2552.

28. Fukumura, T.; Jin, Z.; Ohtomo, A.; Koinuma, H.; Kawasaki, M. *Appl. Phys. Lett.* **1999**, *75*, 3366.

29. Fukumura, T.; Jin, Z.; Kawasaki, M.; Shono, T.; Hasegawa, T.; Koshihara, S.; Koinuma, H. *Appl. Phys. Lett.* **2001**, *78*.

30. Jin, Z.; Fukumura, T.; Kawasaki, M.; Ando, K.; Saito, H.; Sekiguchi, T.; Yoo, Y.; Murakami, M.; Matsumoto, Y.; Hasegawa, T. *Appl. Phys. Lett.* **2001**, *78*, 3824.

31. Jin, Z.; Murakami, M.; Fukumura, T.; Matsumoto, Y.; Ohtomo, A.; Kawasaki, M.; Koinuma, H. *J. Cryst. Growth* **2000**, *214*, 55.

32. Herng, T. S.; Wong, M. F.; Qi, D.; Yi, J.; Kumar, A.; Huang, A.; F. Kartawidjaja, C.; Smadici, S.; Abbamonte, P.; Sánchez-Hanke, C. *Adv. Mater.* **2011**, *23*, 1635.

33. Sharma, P.; Gupta, A.; Rao, K.; Owens, F. J.; Sharma, R.; Ahuja, R.; Guillen, J. O.; Johansson, B.; Gehring, G. *Nat. Mater.* **2003**, *2*, 673; Buchholz, D.; Chang, R. P.; Song, J.; Ketterson, J. *Appl. Phys. Lett.* **2005**, *87*.

34. Park, J. H.; Kim, M. G.; Jang, H. M.; Ryu, S.; Kim, Y. M. *Appl. Phys. Lett.* **2004**, *84*, 1338.

35. Lee, H.-J.; Jeong, S.-Y.; Cho, C. R.; Park, C. H. *Appl. Phys. Lett.* **2002**, *81*, 4020.

36. Park, M. S.; Min, B. *Phys. Rev. B* **2003**, *68*, 224436.

37. Ueda, K.; Tabata, H.; Kawai, T. *Appl. Phys. Lett.* **2001**, *79*, 988.

38. Sato, K.; Katayama-Yoshida, H. *Jpn. J. Appl. Phys.* **2000**, *39*, L555.

39. Shiba, H. *Prog. Theor. Phys.* **1971**, *46*, 77.

40. Spaldin, N. A. *Phys. Rev. B* **2004**, *69*, 125201.

41. Mofor, A. C.; El-Shaer, A.; Bakin, A.; Waag, A.; Ahlers, H.; Siegner, U.; Sievers, S.; Albrecht, M.; Schoch, W.; Izyumskaya, N. *Appl. Phys. Lett.* **2005**, *87*, 062501.

42. Ferrand, D.; Cibert, J.; Wasiela, A.; Bourgognon, C.; Tatarenko, S.; Fishman, G.; Andrearczyk, T.; Jaroszyński, J.; Koleśnik, S.; Dietl, T. *Phys. Rev. B* **2001**, *63*, 085201.

43. Chien, C.-H.; Chiou, S. H.; Guo, G.; Yao, Y.-D. *J. Magn. Magn. Mater.* **2004**, *282*, 275.

44. Feng, X.; *J. Phys.: Condens. Matter* **2004**, *16*, 4251.

45. Ye, L.-H.; Freeman, A.; Delley, B. *Phys. Rev. B* **2006**, *73*, 033203.

46. Hou, D.-L.; Ye, X.-J.; Meng, H.-J.; Zhou, H.-J.; Li, X.-L.; Zhen, C.-M.; Tang, G.-D. *Appl. Phys. Lett.* **2007**, *90*, 142502.

47. Ando, K.; Saito, H.; Jin, Z.; Fukumura, T.; Kawasaki, M.; Matsumoto, Y.; Koinuma, H. *J. Appl. Phys.* **2001**, *89*.

48. Han, S.; Song, J.; Yang, C.-H.; Park, S.; Park, J.-H.; Jeong, Y.; Rhie, K. *Appl. Phys. Lett.* **2002**, *81*, 4212.

49. Shim, J. H.; Hwang, T.; Lee, S.; Park, J. H.; Han, S.-J.; Jeong, Y. *Appl. Phys. Lett.* **2005**, *86*, 082503.

50. Pan, H.; Yi, J. B.; Lin, J. Y.; Feng, Y. P.; Ding, J.; Van, L. H.; Yin, J. H. *Phys. Rev. Lett.* **2007**, *99*, 127201.

51. Shen, L.; Wu, R.; Pan, H.; Peng, G.; Yang, M.; Sha, Z.; Feng, Y. *Phys. Rev. B* **2008**, *78*, 073306.

52. Yi, J. B.; Lim, C. C.; Xing, G. Z.; Fan, H. M.; Van, L. H.; Huang, S. L.; Yang, K. S.; Huang, X. L.; Qin, X. B.; Wang, B. Y.; Wu, T.; Wang, L.; Zhang, H. T.; Gao, X. Y.; Liu, T.; Wee, A. T. S.; Feng, Y. P.; Ding, J. *Phys. Rev. Lett.* **2010**, *104*, 137201.

53. Xing, G.; Lu, Y.; Tian, Y.; Yi, J.; Lim, C.; Li, Y.; Li, G.; Wang, D.; Yao, B.; Ding, J. *AIP Adv.* **2011**, *1*, 022.

54. Janotti, A.; Van de Walle, C. G. *Appl. Phys. Lett.* **2005**, *87*, 122102.

55. Khalid, M.; Ziese, M.; Setzer, A.; Esquinazi, P.; Lorenz, M.; Hochmuth, H.; Grundmann, M.; Spemann, D.; Butz, T.; Brauer, G. *Phys. Rev. B* **2009**, *80*, 035331.

56. Muscat, J.; Swamy, V.; Harrison, N. M. *Phys. Rev. B* **2002**, *65*, 224112; Calatayud, M.; Mori-Sánchez, P.; Beltrán, A.; Pendás, A. M.; Francisco, E.; Andrés, J.; Recio, J. *Phys. Rev. B* **2001**, *64*, 184113; Ranade, M.; Navrotsky, A.; Zhang, H.; Banfield, J.; Elder, S.; Zaban, A.; Borse, P.; Kulkarni, S.; Doran, G.; Whitfield, H. *Proc. Natl. Acad. Sci.* **2002**, *99*, 6476; Navrotsky, A.; Kleppa, O. *J. Am. Ceram. Soc.* **1967**, *50*, 626.

57. Burdett, J. K.; Hughbanks, T.; Miller, G. J.; Richardson, Jr., J. W.; Smith, J. V. *J. Am. Chem. Soc.* **1987**, *109*, 3639.

58. Matsumoto, Y.; Murakami, M.; Shono, T.; Hasegawa, T.; Fukumura, T.; Kawasaki, M.; Ahmet, P.; Chikyow, T.; Koshihara, S.-Y.; Koinuma, H. *Science* **2001**, *291*, 854.

59. Chambers, S. A.; Thevuthasan, S.; Farrow, R. F. C.; Marks, R. F.; Thiele, J. U.; Folks, L.; Samant, M. G.; Kellock, A. J.; Ruzycki, N.; Ederer, D. L.; Diebold, U. *Appl. Phys. Lett.* **2001**, *79*, 3467.

60. Park, W. K.; Ortega-Hertogs, R. J.; Moodera, J. S.; Punnoose, A.; Seehra, M. *J. Appl. Phys.* **2002**, *91*, 8093.

61. Duhalde, S.; Vignolo, M.; Golmar, F.; Chiliotte, C.; Torres, C. R.; Errico, L.; Cabrera, A.; Renteria, M.; Sánchez, F.; Weissmann, M. *Phys. Rev. B* **2005**, *72*, 161313.

62. Hong, N. H. *J. Magn. Magn. Mater.* **2006**, *303*, 338; Kim, K. J.; Park, Y. R.; Park, J. Y.; *J. Korean Phys. Soc.* **2006**, *48*, 1422.

63. Droubay, T.; Heald, S. M.; Shutthanandan, V.; Thevuthasan, S.; Chambers, S. A.; Oster-walder, J. *J. Appl. Phys.* **2005**, *97*, 3.

64. Toyosaki, H.; Fukumura, T.; Yamada, Y.; Nakajima, K.; Chikyow, T.; Hasegawa, T.; Koinuma, H.; Kawasaki, M. *Nat. Mater.* **2004**, *3*, 221; Higgins, J.; Shinde, S.; Ogale, S.; Venkatesan, T.; Greene, R. *Phys. Rev. B* **2004**, *69*, 073201.

65. Shinde, S. R.; Ogale, S. B.; Higgins, J. S.; Zheng, H.; Millis, A. J.; Kulkarni, V. N.; Ramesh, R.; Greene, R. L.; Venkatesan, T. *Phys. Rev. Lett.* **2004**, *92*, 4.

66. Park, M. S.; Kwon, S.; Min, B. *Phys. Rev. B* **2002**, *65*, 161201.

67. Suryanarayanan, R.; Naik, V.; Kharel, P.; Talagala, P.; Naik, R. *Solid State Commun.* **2005**, *133*, 439.

68. Suryanarayanan, R.; Naik, V.; Kharel, P.; Talagala, P.; Naik, R. *J. Phys.: Condens. Matter* **2005**, *17*, 755.

69. Weng, H.; Yang, X.; Dong, J.; Mizuseki, H.; Kawasaki, M.; Kawazoe, Y. *Phys. Rev. B* **2004**, *69*, 125219.

70. Errico, L.; Rentería, M.; Weissmann, M. *Phys. Rev. B* **2005**, *72*, 184425.

71. Hong, N. H.; Sakai, J.; Prellier, W.; Hassini, A.; Ruyter, A.; Gervais, F. *Phys. Rev. B* **2004**, *70*, 195204.

72. Chen, J.; Rulis, P.; Ouyang, L. Z.; Satpathy, S.; Ching, W. Y. *Phys. Rev. B* **2006**, *74*, 5.

73. Li, Q.; Wang, B.; Woo, C.; Wang, H.; Zhu, Z.; Wang, R. *Europhys. Lett.* **2008**, *81*, 17004.

74. Du, X. S.; Li, Q. X.; Su, H. B.; Yang, J. L. *Phys. Rev. B* **2006**, *74*, 4.

75. Cruz, M.; da Silva, R.; Franco, N.; Godinho, M. *J. Phys.: Condens. Matter* **2009**, *21*, 206002.

76. Li, Q. K.; Wang, B.; Zheng, Y.; Wang, Q.; Wang, H. Phys. Status Solidi Rapid Res. Lett. **2007**, *1*, 217.

77. Yang, K.; Dai, Y.; Huang, B.; Whangbo, M.-H. *Appl. Phys. Lett.* **2008**, *93*, 132507.

78. Tao, J.; Guan, L.; Pan, J.; Huan, C.; Wang, L.; Kuo, J.; Zhang, Z.; Chai, J. Wang, S. *Appl. Phys. Lett.* **2009**, *95*, 062505.

79. Yang, K.; Dai, Y.; Huang, B.; Whangbo, M.-H. *Chem. Phys. Lett.* **2009**, *481*, 99.

80. Tao, J.; Guan, L.; Pan, J.; Huan, C.; Wang, L.; Kuo, J. *Phys. Lett. A* **2010**, *374*, 4451.

81. Philip, J.; Punnoose, A.; Kim, B.; Reddy, K.; Layne, S.; Holmes, J.; Satpati, B.; Leclair, P.; Santos, T.; Moodera, J. *Nat. Mater.* **2006**, *5*, 298.

82. Marezio, M. *Acta Crystallogr.* **1966**, *20*, 723.

83. Nguyen, H. H.; Sakai, J.; Ngo, T. H.; Brizé, V. *Appl. Phys. Lett.* **2005**, *87*, 102505.

84. Hong, N. H.; Sakai, J.; Huong, N. T.; Ruyter, A.; Brize, V. *J. Phys.: Condens. Matter* **2006**, *18*, 6897.

85. Gupta, A.; Cao, H. T.; Parekh, K.; Rao, K. V.; Raju, A. R.; Waghmare, U. V. *J. Appl. Phys.* **2007**, *101*, 09N513.

86. He, J.; Xu, S.; Yoo, Y. K.; Xue, Q.; Lee, H.-C.; Cheng, S.; Xiang, X.-D.; Dionne, G. F.; Takeuchi, I. *Appl. Phys. Lett.* **2005,** *86,* 052503.

87. Yoo, Y. K.; Xue, Q.; Lee, H.-C.; Cheng, S.; Xiang, X.-D.; Dionne, G. F.; Xu, S.; He, J.; Chu, Y. S.; Preite, S. *Appl. Phys. Lett.* **2005,** *86,* 042506.

88. Guan, L. X.; Tao, J. G.; Xiao, Z. R.; Zhao, B. C.; Fan, X. F.; Huan, C. H. A.; Kuo, J. L.; Wang, L. *Phys. Rev. B* **2009,** *79,* 184412.

89. Raebiger, H.; Lany, S.; Zunger, A. *Phys. Rev. B* **2009,** *79,* 165202.

90. Guan, L. X.; Tao, J. G.; Huan, C. H. A.; Kuo, J. L.; Wang, L. *Appl. Phys. Lett.* **2009,** *95,* 012509.

91. Koida, T.; Fujiwara, H.; Kondo, M. *Jpn. J. Appl. Phys.* **2007,** *46,* L685.

92. Limpijumnong, S.; Reunchan, P.; Janotti, A.; Van de Walle, C. G. *Phys. Rev. B* **2009,** *80,* 193202.

93. Guan, L.; Tao, J.; Huan, C.; Kuo, J.; Wang, L. *J. Appl. Phys.* **2010,** *108,* 093911.

94. Greenwood, N. N.; Earnshaw, A. *Chemistry of the Elements*, Elsevier, 1997.

95. Baur, W. H. *Acta Crystallogr.* **1956,** *9,* 515.

96. Kılıç, Ç.; Zunger, A. *Phys. Rev. Lett.* **2002,** *88,* 095501; Presley, R.; Munsee, C.; Park, C.; Hong, D.; Wager, J.; Keszler, D. *J. Phys. D: Appl. Phys.* **2004,** *37,* 2810.

97. Ogale, S.; Choudhary, R.; Buban, J.; Lofland, S.; Shinde, S.; Kale, S.; Kulkarni, V.; Higgins, J.; Lanci, C.; Simpson, J. *Phys. Rev. Lett.* **2003,** *91,* 077205_1.

98. Hays, J.; Punnoose, A.; Baldner, R.; Engelhard, M. H.; Peloquin, J.; Reddy, K. *Phys. Rev. B* **2005,** *72,* 075203.

99. Liu, X.; Sun, Y.; Yu, R. *J. Appl. Phys.* **2007,** *101,* 123907.

100. Coey, J.; Douvalis, A.; Fitzgerald, C.; Venkatesan, M. *Appl. Phys. Lett.* **2004,** *84,* 1332.

101. Fitzgerald, C.; Venkatesan, M.; Douvalis, A.; Huber, S.; Coey, J.; Bakas, T. *J. Appl. Phys.* **2004,** *95,* 7390.

102. Hong, N. H.; Ruyter, A.; Prellier, W.; Sakai, J.; Huong, N. T. *J. Phys.: Condens. Matter* **2005,** *17,* 6533.

103. Wang, X.; Zeng, Z.; Zheng, X.; Lin, H. *J. Appl. Phys.* **2007,** *101,* 09H104.

104. Wang, X.; Dai, Z.; Zeng, Z. *J. Phys.: Condens. Matter* **2008,** *20,* 045214.

105. Zhang, C.-W.; Yan, S.-S. *J. Appl. Phys.* **2009,** *106,* 063709.

106. Wang, X. L.; Dai, Z. X.; Zeng, Z. *J. Phys.: Condens. Matter* **2008,** *20,* 045214.

107. Kimura, H.; Fukumura, T.; Koinuma, H.; Kawasaki, M. *Phys. E: Low Dimens. Syst. Nanostruct.* **2001,** *10,* 265; Kimura, H.; Fukumura, T.; Kawasaki, M.; Inaba, K.; Hasegawa, T.; Koinuma, H. *Appl. Phys. Lett.* **2002,** *80,* 94.

108. Khare, N.; Kappers, M. J.; Wei, M.; Blamire, M. G.; MacManus-Driscoll, J. L. *Adv. Mater.* **2006,** *18,* 1449.

109. Liu, S. J.; Liu, C. Y.; Juang, J. Y.; Fang, H. W. *J. Appl. Phys.* **2009,** *105.*

110. Espinosa, A.; Sánchez, N.; Sánchez-Marcos, J.; de Andrés, A.; Muñoz, M. C. *J. Phys. Chem. C* **2011,** *115,* 24054.

111. Stashans, A.; Puchaicela, P.; Rivera, R. *J. Mater. Sci.* **2014,** *49,* 2904.

112. Fitzgerald, C. B.; Venkatesan, M.; Dorneles, L. S.; Gunning, R.; Stamenov, P.; Coey, J. M. D.; Stampe, P. A.; Kennedy, R. J.; Moreira, E. C.; Sias, U. S. *Phys. Rev. B* **2006,** *74,* 115307.

113. Rahman, G.; García-Suárez, V. M.; Hong, S. C. *Phys. Rev. B* **2008,** *78,* 184404.

114. Xiao, W.-Z.; Wang, L.-L.; Xu, L.; Wan, Q.; Zou, B. *Solid State Commun.* **2009,** *149,* 1304.

115. Rahman, G.; García-Suárez, V. M. *Appl. Phys. Lett.* **2010,** *96,* 052508.

116. Hong, N. H.; Song, J.-H.; Raghavender, A.; Asaeda, T.; Kurisu, M. *Appl. Phys. Lett.* **2011**, *99*, 052505.
117. Bersch, E.; Rangan, S.; Bartynski, R. A.; Garfunkel, E.; Vescovo, E. *Phys. Rev. B* **2008**, *78*, 085114.
118. Hann, R. E.; Suitch, P. R.; Pentecost, J. L. *J. Am. Ceram. Soc.* **1985**, *68*, C.
119. Adams, D. M.; Leonard, S.; Russell, D. R.; Cernik, R. J. *J. Phys. Chem. Solids* **1991**, *52*, 1181.
120. Ohtaka, O.; Fukui, H.; Kunisada, T.; Fujisawa, T.; Funakoshi, K.; Utsumi, W.; Irifune, T.; Kuroda, K.; Kikegawa, T. *J. Am. Ceram. Soc.* **2001**, *84*, 1369.
121. Leger, J. M.; Atouf, A.; Tomaszewski, P. E.; Pereira, A. S. *Phys. Rev. B* **1993**, *48*, 93.
122. Venkatesan, M.; Fitzgerald, C.; Coey, J. *Nature* **2004**, *430*, 630.
123. Pemmaraju, C. D.; Sanvito, S. *Phys. Rev. Lett.* **2005**, *94*, 217205.
124. Coey, J. M. D.; Venkatesan, M.; Stamenov, P. Fitzgerald, C.; Dorneles, L. *Phys. Rev. B* **2005**, *72*, 024450.
125. Zheng, J. Ceder, G.; Maxisch, T.; Chim, W.; Choi, W. *Phys. Rev. B* **2007**, *75*, 104112.
126. Chen, G.; Zhang, Q.; Gong, X.; Yunoki, S. d_0 Ferromagnetic Surface in HfO_2. *J. Phys. Conf. Ser.* **2012**, *400*, 032088.
127. Beltrán, J. I.; Muñoz, M. C.; Hafner, J. *New J. Phys* **2008**, *10*, 063031.
128. Coey, J. *Solid State Sci.* **2005**, *7*, 660; Zhang, S.; Ogale, S. B.; Yu, W.; Gao, X.; Liu, T.; Ghosh, S.; Das, G. P.; Wee, A. T.; Greene, R. L.; Venkatesan, T. *Adv. Mater.* **2009**, *21*, 2282; Sundaresan, A.; Rao, C. *Nano Today* **2009**, *4*, 96; Peng, H.; Li, J.; Li, S.-S.; Xia, J.-B. *Phys. Rev. B* **2009**, *79*, 092411.
129. Hong, N. H.; Poirot, N.; Sakai, J. *Appl. Phys. Lett.* **2006**, *89*.
130. Haowei, Z.; Bin, G.; Shimeng, Y.; Lin, L.; Lang, Z.; Bing, S.; Lifeng, L.; Xiaoyan, L.; Jing, L.; Ruqi, H.; Jinfeng, K. Effects of Ionic Doping on the Behaviors of Oxygen Vacancies in HfO_2 and ZrO_2: A First Principles Study. In: Presented at *Simulation of Semiconductor Processes and Devices, 2009. SISPAD '09. International Conference on*, 9–11 Sept. 2009, 2009.
131. Weng, H.; Dong, J. *Phys. Rev. B* **2006**, *73*, 132410.
132. Nieminen, R. M.; *Modell. Simul. Mater. Sci. Eng.* **2009**, *17*, 084001.
133. Aryasetiawan, F.; Gunnarsson, O. *Rep. Prog. Phys.* **1998**, *61*, 237.
134. Godby, R.; Schlüter, M.; Sham, L. *Phys. Rev. B* **1988**, *37*, 10159.
135. Liechtenstein, A.; Anisimov, V.; Zaanen, J. *Phys. Rev. B* **1995**, *52*, R5467.
136. Lany, S.; Zunger, A. *Phys. Rev. B* **2008**, *78*, 235104.
137. Baraff, G.; Schlüter, M. *Phys. Rev. Lett.* **1985**, *55*, 1327.
138. Freysoldt, C.; Neugebauer, J.; Van de Walle, C. G. *Phys. Rev. Lett.* **2009**, *102*, 016402.

INDEX